Progress in Colloid & Polymer Science

Vol. 95 (1994)

PROGRESS IN COLLOID & POLYMER SCIENCE

Editors: F. Kremer (Leipzig) and G. Lagaly (Kiel)

Volume 95 (1994)

# Surfactants and Colloids in the Environment

Guest Editors:

M. J. Schwuger
and F.-H. Haegel (Jülich)

SPRINGER-VERLAG BERLIN
HEIDELBERG GMBH

Die Deutsche Bibliothek — CIP-Einheitsaufnahme

**Surfactants and colloids in the environment**
/ guest ed.: M. J. Schwuger and F.-H. Haegel. —

(Progress in colloid & polymer science; Vol. 95)
ISBN 978-3-662-15674-2    ISBN 978-3-7985-1668-7 (eBook)
DOI 10.1007/978-3-7985-1668-7
NE: Schwuger, Milan J. [Hrsg.]; GT

ISBN 978-3-662-15674-2
ISSN 0340-255 X





Chemistry editor: Dr. Maria Magdalene Nabbe; English editor: James C. Willis; Production: Holger Frey, Bärbel Flauaus.

Type-Setting:
Graphische Textverarbeitung,
Hans Vilhard,
64753 Brombachtal

# Preface

From September 28 through October 1, 1993, the German Colloid Society (Kolloid-Gesellschaft) held its biannual General Meeting at the Research Center of Jülich (KFA). About 200 participants from Germany, other European countries, and overseas attended this meeting. Forty-three lectures and more than 50 posters on the main topic "Surfactants and Colloids in the Environment" and on different fields of colloid chemistry were presented. The meeting was supported by several companies and the Research Center of Jülich. Thanks to donation and the receipts from an industrial exhibition, we were able to sponsor the participation of several colleagues from Eastern Europe and to invite some protagonists of colloid chemistry from Germany and abroad.

The meeting was inaugurated by the president of the Kolloid-Gesellschaft, Prof. Milan Schwuger, Jülich, who welcomed the participants on behalf of the society. Greetings were expressed by Prof. Ernst Pöppel, member of the board of directors of the Research Center and Dr. Edda Müller, responsible assistant director of the Federal Ministry for Environment, Nature Conservation and Reactor Safety. Both speakers emphasized the importance of colloid chemistry in environmental research from the scientific and political points of view.

The opening ceremony ended with the first highlight of the conference. Three prizes were awarded to outstanding colloid scientists by the Kolloid-Gesellschaft. The Wolfgang Ostwald Prize was conferred on Prof. Wolfgang Helfrich from the "Freie Universität Berlin" for his excellent studies on vesicles, membranes, and liquid crystals. This award is made for extraordinary lifetime achievements in colloid science. Professor Dušan Vučelić, Belgrade, was honored for his contributions to zeolite synthesis and characterization with the Steinkopff Prize which is sponsored by the publisher of this journal and marks achievements in environmental colloid chemistry. The Zsigmondy Grant for young scientists was conferred on Dr. Erwin Klumpp, Jülich, for his studies on the interaction of soil colloids, pollutants and surfactants.

The scientific program started with the lectures of the two senior laureates. Wolfgang Helfrich presented his latest theoretical studies on "Pathways of Vesiculation". Dušan Vučelić spoke on "Physico-chemical Aspects of Zeolite-A Synthesis and Application for Environmentally Safe Detergents". Brief biographies of these two scientists and their contributions also open this volume.

Contributions on the physico-chemical behavior of colloids and surfaces in the environment and their use in environmental technology demonstrate the high standard and usefulness of colloid science in this field. Latest results on the interaction of heavy metals and organics with soil minerals, transport properties of pollutants, and the relevance of surfactants as effective modifiers for natural interfaces were presented. Also, the use of colloids and surfactants in environmental technologies for soil remediation and waste water treatment was emphasized. This volume comprises a choice of papers dealing with these subjects.

The studies presented in the open section were widespread and, in most cases, fundamental. Many of the lectures on general colloid chemistry, however, made clear that the main subject of this meeting was not a specific one isolated from other fields of colloid chemistry, but that general colloid and environmental science can inspire each other.

Latest results on the influence of charges on coagulation, adsorption and ion exchange which were presented during the conference will certainly influence the understanding of the behavior of colloids and surfaces in the environment and in remediation techniques.

Improvement of surfactant analysis will enhance the possibilities for the observation of surfactants in the environment, e.g., enable early detection of possible accumulation or differentiation of the components of complex mixtures. Novel surfactant systems with interesting properties and applications were presented which will be environmentally important in the future or can be used favorably in remediation techniques.

Many contributions dealt with alkylpolyglucosides, a class of surfactants which is considered to be increasingly important for applications due to their low toxicity, good degradability, and other special properties. Another field of interest was microemulsions. They may play an increasing role as reaction and extraction media without hazardous organic solvents.

Besides the scientific program, personal relationships between participants could be initiated and deepened during two evening receptions, one held in the Research Center and the other in the "Ludwig Forum" in Aachen. The spirit of inspiration we felt while in this museum of modern art meshed perfectly with the diligence demonstrated in so many scientific contributions. Both elements will be needed for genuine solutions of environmental problems.

Before we invite you to delve into this volume on "Surfactants and Colloids in the Environment", we want to acknowledge all who contributed to the success of this meeting. We want to express special thanks to the members of the scientific committee, Prof. G. Peschel, Dr. W. von Rybinski, Dr. S. Storp, Prof. H. Versmold, and Prof. D. Woermann, who did a good job in planning a blanced program.

We are also grateful to the companies BASF, C3-Analysentechnik, Partikel-Analytik-Meßgeräte GmbH, and Perkin Elmer for supporting us by gift or advertisements in the program booklet, and to Collotec, Lauda, Krüss, Malvern, Mütek, and Südchemie for their participation in the industrial exhibition. The publishers John Wiley & Sons and Steinkopff-Verlag also contributed financially to the meeting.

Further, we acknowledge the help of Mrs. R. Mengels and Dr. B. Krahl-Urban from the Conference Service of the Research Center for their support in organizing the meeting. And, although space does not allow to name everyone, we thank all the helping hands without which such a conference cannot succeed.

Franz-Hubert Haegel (Jülich)

# Contents

Progr Colloid & Polym Sci (1994) 95:1—2
© Steinkopff Verlag 1994

# Opening address by the president
# of the Kolloid-Gesellschaft
# Prof. M. J. Schwuger

I would like to welcome you to the 36th General Meeting of the Kolloid-Gesellschaft, in Jülich, and wish you a pleasant stay and very successful scientific discussions. I particularly welcome Dr. Edda Müller, representing the Minister for the Environment, Nature Conservation and Nuclear Safety, and Prof. Dr. Ernst Pöppel, representing the Board of Directors of the Research Centre Jülich.

This year's colloid conference is unique in many respects. It is the first such conference following Germany's reunification. The 1991 General Meeting in Mainz had a special character as a joint meeting with ECIS and, therefore, cannot be regarded as representative.

Interface and colloid chemistry was already at an advanced stage of development in the eastern part of Germany before World War II. This is illustrated by the schools of Leipzig, Dresden, and Berlin which enjoyed worldwide reputation. This reputation was also maintained during the division of Germany, thus strong, synergistic impact may be expected for the future of colloid and surface chemistry in Germany. The significance of interface and colloid chemistry in the new Federal States has been taken into account by the appointment of Hans Sonntag (Berlin) and Hans-Jörg Jacobasch (Dresden) to the Board of the Kolloid-Gesellschaft. We are particularly pleased about the foundation of the Max Planck Institute for Interface and Colloid Research, which is being established near Potsdam.

Overcoming the division of Germany went hand in hand with overcoming the division of science in East and West. Before World War II, the conferences of the Kolloid-Gesellschaft were always also an opportunity for scientists from East and West to meet each other. I am particularly pleased to welcome scientists from practically all East European countries in addition to our western friends who have regularly presented their papers on these occasions. I hope that the conferences of the Kolloid-Gesellschaft will, in the future continue to be events where scientists from all over the world can meet each other. The fresh start this year is greatly appreciated; we are in a position to welcome nearly 300 representatives from 17 nations. It has been a beneficial tradition of the Kolloid-Gesellschaft to coopt members of other, especially East European associations onto the Board. Following this tradition, we elected Imre Dékány (University of Szeged), current chairman of the Hungarian Colloid Chemistry Association, to our Board to succeed the late Erwin Wolfram.

We hope that the present situation will give a lasting, strong impetus to the development of interface and colloid chemistry in Germany and Europe.

A further particularity of this meeting is its topic: "Surfactants and

Colloids in the Environment." Our special field of research has decisively influenced progess in chemistry and technology in the 19th and 20th centuries. The current industrial society would be inconceivable without this special discipline since surfactants and colloids are of great importance in connection with both everyday products and the latest technologies. Let me just mention detergents, cosmetics, and recording tapes, which have become indispensable in everyday life. The great technological processes of raw material and energy production, of plastics and lacquer fabrication would also be impossible without surfactants and colloids. Their enormous scientific significance is reflected in the production figures. Approximately 5.5 million tons of surfactants were produced last year in the statistically better accessible West European countries, the USA and Japan. Their significance is due to their particular surface-active properties.

Surfactants can reach the environment via different paths, either directly, e.g., in connection with pesticides, or indirectly adsorbed in sewage sludge applied as fertilizer in agriculture. Surfactants are also encountered in rivers and lakes. Apart from synthetic surfactants, nature also produces surface-active substances by biodegradation, and even the human organism daily produces approx. 20 g of surfactants in the form of bile acids.

Various interactions lead to mobilization, remobilization, and immobilization processes for organic and/or inorganic pollutants in nature. These physicochemical processes are vital for keeping our waters and drinking water clean. The soils and sediments are sinks for various chemicals whose penetration into waters must be prevented. Since the extremely hot summer in 1959, when hills of foam primarily consisting of surfactant-protein complexes were formed on our rivers and weirs, attention has been focused on the biodegradation of surfactants. This rather narrow approach must be regarded as too simplifying. It should just be mentioned that the adsorption kinetics or organic pollutants is many times accelerated in the presence of surfactants, and adsorption processes which take one to several weeks can take place within minutes. Such qualitative and quantitative differences clearly show that these processes can take place much more rapidly than would be possible by biodegradation. We therefore consider it our task to combine, for the first time, all the knowledge available in the field of physicochemical interactions in the environment and to indicate future perspectives for the necessary physicochemical investigations in the environment based on the results of our conference.

I wish all participants a fruitful exchange of ideas and thank you for your interest in this highly topical subject.

Progr Colloid & Polym Sci (1994) 95:3—4
© Steinkopff Verlag 1994

# Wolfgang Ostwald Prize 1993 awarded to Wolfgang Helfrich

Professor Dr. Wolfgang Helfrich, Berlin, received the Wolfgang Ostwald Prize of the Kolloid-Gesellschaft for his fundamental work on complex fluids and colloids, in particular for his contributions to the cooperative behavior of liquid crystals and membrane systems.

W. Helfrich is a real pioneer in these interdisciplinary research fields. On the one hand, he made novel experimental discoveries, on the other hand, he developed basic theoretical concepts which turned out to be crucial for research in this area.

W. Helfrich has an extraordinary intuition which often led him to unusual insights. In many cases, his colleagues first had some difficulties to follow his line of thought. Thus, it took more than 10 years until his basic work on the physics of membranes was recognized in the physics community. This has completely changed, however, during the last couple of years, and many researchers have now picked up his ideas and extended them in new directions.

## Curriculum vitae

Wolfgang Helfrich was born on March 25, 1932, in Munich. He studied physics at the universities of Munich, Tübingen, and Göttingen. Then he joined the group of Nikolaus Riehl, the "Father of the fluorescent strip lamp" at the University of Munich, where he received his doctorate in 1961 for his experimental work on "Space-charge-limited currents in organic crystals". In 1967, he completed his "Habilitation" in experimental physics.

From 1967 until 1970, he worked at the RCA Laboratories in Princeton where he was a member of an interdisciplinary group of physicists, chemists, and engineers studying the electro-optical properties of liquid crystals. From 1970 to 1973, he was at Hoffmann-La Roche in Basel, where he invented, together with Martin Schadt, a novel liquid crystal display. For this invention, he received the Hewlett-Packard Europhysics Award in 1976. Since 1973, W. Helfrich has worked and taught at the Freie Universität Berlin. His scientific achievements were also honored with the Hennessy Prize for innovation ("science pour l'art") in 1993.

## Liquid crystals

Wolfgang Helfrich has discovered and explained several basic effects in the field of liquid crystals. An overview of these phenomena, which carry his name, is given in the monograph "The physics of liquid crystals" of Pierre-Gilles de Gennes, winner of the Nobel Prize of Physics in 1991.

A very important result of W. Helfrich in this field was his explanation of the so-called "dynamic scattering mode". This state arises from an instability which appears in a nematic liquid crystalline layer between two electrodes. As soon as the voltage exceeds a threshold value, convection rolls are formed in the layer. This instability was discovered experimentally at the end of the 1960s. The underlying mechanism, however, remained mysterious until W. Helfrich gave a relatively simple explanation, in which the electrical conductivity caused by impurities played an important role. These phenomena of electro-convection have been rediscovered recently in the context of hydrodynamic pattern formation.

Further work on the electro-optical properties of liquid crystals led to the above-mentioned invention of a novel liquid crystal display. This display also consists of a thin layer of a nematic liquid crystal which is situated between two electrodes. However, the molecules are now anchored at the two opposite electrodes in such way that their preferred direction is twisted within the layer. This distortion of the nematic phase can be changed by the applied voltage leading to a change of the optical properties. Almost all liquid crystal displays used today in watches and other devices are based on this simple and rather robust construction principle.

## Membrane systems

The new field of membrane physics was co-founded and substantially formed by Wolfgang Helfrich. It deals with the structure and dynamics of membranes and vesicles in aqueous solution.

The shape of lipid vesicles often deviates from a sphere and, thus, cannot be determined by interfacial tension. Already 20 years ago, W. Helfrich developed a "fluid shell theory" in which this shape is controlled by the curvature and, thus, by the bending rigidity of the membrane. Meanwhile, this approach has led to a very fruitful interaction of experiment and theory and, thus, to a quantitative understanding of the vesicle shape.

Fluid membranes are very flexible and already curve as a result of thermal fluctuations. This flickering of the membrane can be observed directly in the optical microscope. In a stack or bunch of membranes, these thermally excited bending modes lead to an entropically induced repulsion between the membranes as first predicted by W. Helfrich. This force has been confirmed by scattering experiments on lamellar phases in oil water mixtures.

Lipid membranes in a stack are bound together by attractive van der Waals forces. A systematic theory for the interplay of attractive forces and entropic repulsion leads to the theoretical prediction of a critical adhesion or unbinding transition. Such a transition was experimentally observed by Wolfgang Helfrich and Michael Mutz for stacks of sugar lipid membranes.

W. Helfrich also realized that lateral tension applied to a membrane at first smoothes the bending undulations and pulls area out of the membrane without affecting its molecular packing. His quantitative predictions for this effect were confirmed by experiments on vesicles. In addition, a lateral tension should strongly reduce the entropically induced repulsion of the membranes. Therefore, the adhesion of membranes can be induced by such a tension. W. Helfrich used this concept of tension-induced adhesion in order to explain the experimentally observed behavior of membrane bunches. However, he then found some discrepancies between theory and experiment which led him to the idea that lipid membranes can possess a "hidden" area reservoir. Some preliminary experimental evidence suggests that this reservoir is provided by a superstructure of the membrane on suboptical length scales.

W. Helfrich has made many more contributions to membrane physics which I cannot describe here in detail because of the restricted space. A few examples are: experimental and theoretical investigations on the behavior of cylindrical vesicles which are similar to semi-flexible polymers; theoretical contributions to the influence of the Gaussian curvature on the membrane shape; the prediction that the bending rigidity is decreased by fluctuations, an effect which was later rediscovered in the string theories of high energy physics; the experimental observation of sponge-like phases of lipid membranes; and theoretical studies on the influence of the electric field on the bending rigidity.

Even though membrane science is still a rapidly developing research field which provides many intriguing and open problems, it is already clear today that further progress in this area will be based, in an essential way, on the concepts developed by Wolfgang Helfrich.

R. Lipowsky
Research Center Jülich and MPI Potsdam

Progr Colloid & Polym Sci (1994) 95:5—6
© Steinkopff Verlag 1994

# Steinkopff-Prize 1993
# recipient Dušan Vučelić

Professor Dr. Dušan Vučelić was born July 17, 1938 in Belgrade, where he completed a Master of Science degree at the Faculty of Physical Chemistry of Belgrade University in 1964. In 1970, he received his doctorate in Physical Chemistry, his thesis being "NMR Relaxation of Molecules at Interphases," a field with only a few publications at that time. Subsequently, he became assistant professor of solid state physical chemistry at Belgrade University. Since 1981, Dušan Vučelić has held a full professorship in Physical Chemistry of Solid State and Molecular Biophysics. He spent most of his scientific life at Belgrade University and at the Research Institute of General and Physical Chemistry, of which he was the principal founder, with the exception of three short periods: in 1976, he worked at the Inorganic Department of Oxford University, Great Britain; in 1980, at the Magnetic Resonance Laboratory of Stanford University, USA, and in 1991 as a visiting professor at the School of Physics, Lomonosov University, Russia.

His main achievements are in the field of solid interfaces and molecular biophysics as documented in more than 100 papers in leading international journals. He followed and confirmed the almost forgotten van der Waals hypothesis about sorbed molecules being a separate thermodynamic phase rather than independently bound molecules. Between 1969 and 1975, he determined many physico-chemical paramters for sorbate-zeolite systems, as well enthalpies, entropies, and heat capacities as properties of dynamics and relaxation. The phase transitions of sorbed interphases have been found to occur at temperatures below those for the corresponding transitions in the bulk state. The discovery of this phenomenon, especially for water, is of great importance, not only for physical chemistry, but also for molecular biology. During this research on phase transitions, he discovered the existence of a variety of cation energy levels within the zeolite framework. The cation sublattice was shown to depend strongly on the competition between sorbed molecules and cations. This has led to the hypothesis that competition "excites" cations to higher energy levels, thus making them more mobile. Dušan Vučelić was subsequently able to elegantly confirm this by demonstrating increases in conductivity at elevated temperatures. By this means, water complexes were found within zeolite cages long before their direct observation by neutron diffraction.

NMR and non-isothermal methods have always been Prof. Vučelić's main tools for research on complex systems. His intensive use of both these techniques has led to deeper understanding of the extremely complex kerogen molecules. It has been shown that the almost established hypothesis of the origins of the high catalytic activity within amorphous phases of oil shales and kerogens is not universally valid. In reality, two systems exist, with high and low catalytical activity, and free radicals playing the main role in the latter. In the early 1980s, Prof. Vučelić became interested in biophysics. Interphases provided the bridge to this field because, starting from the membrane, the whole cell may be regarded as an interphase. This approach rapidly led to some outstanding results. First, a combined bio-thermo-tropic effect was demonstrated for folic and fatty acids. Second, the well known physiological phenomenon that long chain fatty acids adsorb through the wall of the digestive tract, in contrast to the short chain ones which pass directly through the venal port, was explained on the basis of thermodynamics, e.g., only long chain fatty acids form micelles. Finally, the expectation of perturbation of water transport through membranes by the stronger hydrogen bond of $D_2O$ was confirmed and demonstrated for calcium channels.

A stronger tendency towards applied research has always been a part of Prof. Vučelić's character. Hence, in the past few years he has tried to apply pure biophysical methods to medical research. This has resulted in a new method for the diagnosis of skin diseases based on the propagation of acoustic waves through complex skin interphases and to a new hypothesis about the cause of Balkan nephropathy.

Laboratory research and its application alone was not sufficient to fulfill Vučelić's practical spirit, and so he turned early towards industry. Many hundreds of his devices, based on solid state and sorption chemistry, have been incorporated into different production lines world wide. His main achievement in industry has been the construction of zeolite plants all over the world, based on this know-how.

As a consequence, his technical designs and chemical processes must logically be cited here. In Birač (former Yugoslavia), a large complex of five plants was finished by 1990/91. There, zeolite-A was produced in huge reactors and the plant became the world's largest producer (220 000 ton/year) as well as the quality leader (with Degussa). The competition has, up to now, found it impossible to emulate the production process, based on cheap but highly impure Bayerish liquid. A similar approach has been used in Italy, where the process is based on the meta-stable cation distribution discovered during Vučelić's fundamental research, thus giving zeolite with a high exchange rate but without increasing the amount of expensive caustic soda. As a result of environmental demands (energy and material savings), a concentrated and compact product is the future goal. For this type of product the understanding of the surface properties of solids are of key importance, and it is for this reason that Prof. Vučelić sees his fundamental and applied research as being crucial. Many new products, with a variety of surface properties, have already been produced in his laboratories and tested on the pilot plant scale. It is to be hoped and expected that Vučelić will produce this new generation of modified zeolites and compound similar to zeolites on an industrial scale.

Of all his posts and awards, Prof. Vučelić's favorite is his position as President of the IUPAB Commission for "Radiation and Environemental Biophysics," because he views it as "the ultimate duty of scientists to look into the far future and to protect humanity."

Consequently the Steinkopff-Prize, dedicated to achievements in solving environmental and industrial problems via colloid and interfacial methods, was presented in 1993 to Prof. Dušan Vučelić.

M. J. Schwuger (Research Center Jülich)

Progr Colloid & Polym Sci (1994) 95:7—13
© Steinkopff Verlag 1994

W. Helfrich

# Pathways of vesiculation

Wolfgang Helfrich
Fachbereich Physik,
Freie Universität Berlin,
Arnimallee 14,
14195 Berlin, FRG

**Abstract** Possible ways of producing bilayer vesicles are reviewed from the theoretical point of view. The emphasis is on vesiculation driven by a negative bending energy of spherical vesicles, called spontaneous vesiculation. A new model is presented to explain the existence of onions, i.e. concentric spheres of equidistant membranes.

**Key words** vesicles — onion-like membrane structures — bending elasticity

## Introduction

Vesicles are utilized today to encapsulate drugs and other agents for protection and targeted delivery. They are made of closed fluid bilayers, typically a single one, but sometimes several or many. Lipid vesicles are often called liposomes, especially when multilamellar. A book on liposomes written by D. D. Lasic and geared to their medical applications has just been published [1].

In recent years there has been much progress in the physics of vesicles, both experimentally and theoretically. It is no longer necessary to tear up bilayers by sonication or other means to produce fragments which close to form small vesicles. Methods have been found to make the spherically curved state of suitable bilayers energetically more favorable than the flat state. This permits spontaneous vecisulation if the energy barriers associated with cutting and resealing the bilayer are thermally surmountable.

Another method of transforming extended bilayers into vesicles is simple dilution. A positive bending energy of the spherical vesicle is compensated, in a large enough volume of water, by the ncgative free energy associated with its entropies of translation and size distribution. Budding, well-known from cell biology, is another, dynamic method which could be envisaged to produce vesicles. It requires a spontaneous curvature of the bilayer, which can occur only if the two monolayers composing the bilayer, including the adjacent solutions, are out of thermodynamic equilibrium with each other.

In the following we consider vesiculation, in particular spontaneous vesiculation, and we concentrate on theoretical aspects. Apart from referring to existing theory (a fair review would require more time and space), we will propose a new theoretical model for "onions". Simons and Cates [2], who introduced this graphic name for multilamellar vesicles, were the first to deal with onions in thermodynamic equilibrium. Unlike these authors, we will assume the bending energies of the spheres to be negative.

## Bending energies

In order to express the condition for spontaneous vesiculation in terms of curvature elastic moduli, we start from the usual formula for the bending energy per unit area of fluid bilayer

$$g = \frac{1}{2} \kappa J^2 + \bar{\kappa} K .$$  (1)

Here, $J = c_1 + c_2$ and $K = c_1 c_2$ are the total and Gaussian curvatures, respectively, with $c_1$ and $c_2$ being the principal curvatures, $\kappa$ is the bending rigidity and $\bar{\kappa}$

the elastic modulus of Gaussian curvature. Equation (1) holds for the symmetric bilayer, the two sides of which are by definition equal or, more precisely, in thermodynamic equilibrium with each other. According to (1), the bending energy of a spherical vesicle is

$$E_{sph} = 4\,\pi\,(2\,\kappa + \bar{\kappa})\,, \tag{2}$$

while all other closed shapes have larger energies. Vesiculation will be called spontaneous when $E_{sph}$ is negative, i.e., $2\,\kappa + \bar{\kappa} < 0$. Since $\kappa > 0$, without known exception, a sufficiently negative $\bar{\kappa}$ is required for spontaneous vesiculation.

Whenever the condition of spontaneous vesiculation is satisfied, the bilayer may be expected to split up into smaller and smaller vesicles. Obviously, Eq. (1), being quadratic in the principal curvatures, does not tell us where this process stops. The fact that there is a limit can be taken into account, in a natural but approximative way, by adding to (1) a quartic term [3]. For the surface of a sphere of radius $r$, we may write

$$g = (2\,\kappa + \bar{\kappa})\,\frac{1}{r^2} + \frac{\kappa'}{r^4} \tag{3}$$

and

$$E_{sph} = 4\,\pi\left(2\,\kappa + \bar{\kappa} + \frac{\kappa'}{r^2}\right)\,, \tag{4}$$

where $\kappa'$ is positive. Assuming a population of spheres of equal size, we have the total bending energy

$$\frac{A}{r^2}\left(2\,\kappa + \bar{\kappa} + \frac{\kappa'}{r^2}\right) \tag{5}$$

with $A$ being the total bilayer area. The radius $r^*$ minimizing this energy obeys

$$(r^*)^2 = -\frac{2\,\kappa'}{2\,\kappa + \bar{\kappa}}\,. \tag{6}$$

In the ease of a negative $\kappa'$, a positive energy term of order $1/r^6$ could stabilize the size of the vesicles.

The bending rigidity $\kappa$ has been measured for biological model membranes, i.e. lipid bilayers, and for the bilayers of standard surfactants which in microemulsions form monolayers between hydrocarbon and water. Lipid molecules usually possess two hydrocarbon chains between 14 and 18 carbon atoms long and their bending rigidity is typically $(0.5-1)\,\cdot\,10^{-19}$ J at room temperature, i.e. at $kT = 4\,\cdot\,10^{-21}$ J [4]. The bilayers of standard surfactants, having a single, relatively short hydrocarbon chain, are more flexible with bending rigidities near or a few times $kT$ [5]. Another way of lowering the bending rigidity is to mix surfactants of different spontaneous curvature [6]. As a rule, the bending rigidity is determined from the mean square amplitudes of individual thermal undulation modes or collectively from the storage of area by all undulation modes as a function of lateral tension. The higher flexibility of the bilayers of standard surfactants agrees with the finding that they are more easily broken and resealed than lipid bilayers.

The modulus of Gaussian curvature has no effect on thermal undulations as its integral over a closed surface depends only on the genus of the latter. It has been determined indirectly from the shape of pierced unilamellar lipid vesicles [7] and, for monolayers, from phase equilibria between an interface and one or two bulk phases [8]. There are no measurements of the fourth order modulus $\kappa'$.

## Spontaneous vesiculation

On the sole basis of bending energies, one expects a multiply self-connected bilayer (e.g., a cubic phase) for $0 < \bar{\kappa}$, a planar multilayer system for $-2\,\kappa < \bar{\kappa} < 0$, and vesicles for $\bar{\kappa} < -2\,\kappa$ [8]. Since $\kappa$ is generally positive, it is mainly the sign and magnitude of $\bar{\kappa}$ that determine the topology of a bilayer. The modulus of Gaussian curvature, although hard to measure, has a simple physical meaning. It is the second moment of the stress profile $s\,(z)$ of the flat bilayer [10],

$$\bar{\kappa} = \int z^2\,s\,(z)\,dz\,, \tag{7}$$

$z$ being a coordinate normal to the layer. This relationship is valid if the zero and first moments vanish, which are lateral tension and the (negative) product of spontaneous curvature and bending rigidity, respectively.

As the stress profile of the flat symmetric bilayer is mirror symmetric with respect to the bilayer mid-plane, it is sufficient to integrate (with $z = 0$ at the mid-plane) over the upper monolayer. Its stress profile may be divided into two or three parts. First, the region of the hydrocarbon chains which, because of their parallel alignment, produce negative stress (push). Second, the hydrocarbon/water interface with a $\delta$-function-like positive stress (pull). Third, the region of the polar heads of the amphiphilic molecules and beyond. The last region may be negligible if there is little repulsion between the polar heads because they are electrically neutral and of a smaller cross-section than the hydrocarbon chains. However, it can be predominant in (7) if the heads are thick and, particularly, if they are charged and facing an aqueous medium of large Debye length. These are the situations in which spontaneous vesiculation has, in fact, been observed.

The monolayers have a spontaneous curvature $J_0$ and a surface of inextension at a distance $\pm z_0$ from the bilayer mid-plane. Using these quantities, one can transform (7) into [11]

$$\bar{\kappa} = 2\,(\bar{\kappa}_{\mathrm{m}} - 2\,z_0\,\kappa_{\mathrm{m}}\,J_0)\,, \tag{8}$$

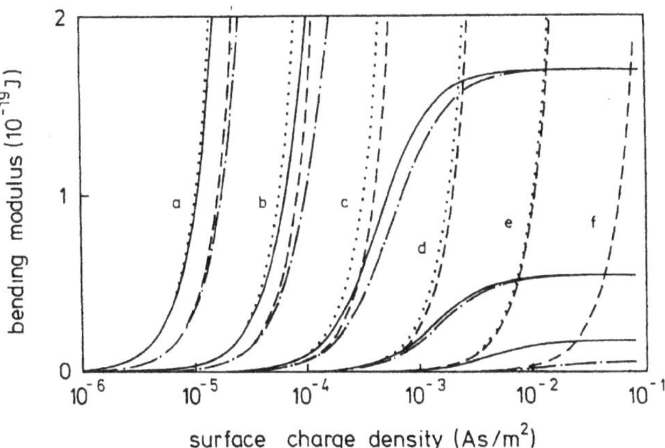

**Fig. 1** Electrical part of the bending rigidity versus surface charge for various Debye lenths: a) $\lambda_D = 1$ µm, b) $\lambda_D = 300$ nm, c) $\lambda_D = 100$ nm, d) $\lambda_D = 30$ nm, e) $\lambda_D = 10$ nm, f) $\lambda_D = 3$ nm. The results of Debye-Hückel theory are shown by dotted and dashed lines for zero and maximum electrical coupling, respectively, between the two sides of the curved bilayer. The results of Poisson-Boltzmann theory are indicated by solid and by dashed-dotted lines for zero and maximum electrical coupling, respectively. The calculations hold for $kT = 4 \cdot 10^{-21}$ J, $\varepsilon_{\text{bilayer}} = 2\,\varepsilon_0$, $\varepsilon_{\text{water}} = 80\,\varepsilon_0$. Both the surface charge and the surface of inextension were taken to coincide with the monolayer hydrocarbon/water interface. (From ref. [17].)

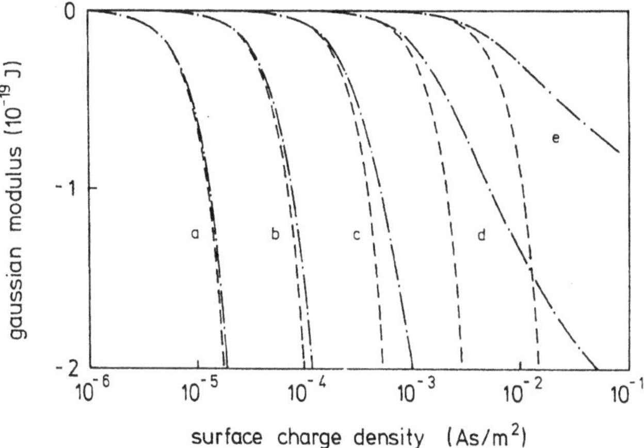

**Fig. 2** Electrical part of the modulus of Gaussian curvature versus surface charge density. The dashed and dashed-dotted lines represent the results of Debye-Hückel and Poisson-Boltzmann theory, respectively. The Debye lengths and other parameters are the same as in Fig. 1. The bilayer thickness was taken to be 4 nm. The results apply if the compensating mechanical tension resides in the interface (and has no effect on the "mechanical" part of the bending rigidity). (From ref. [17].)

where the monolayer moduli are marked by the subscript $m$ and $\kappa_m = (1/2)\,\kappa$. The relationship indicates that spontaneous vesiculation can be brought about by a sufficiently positive monolayer spontaneous curvature. Monolayer curvature is counted positive when it is convex toward the water.

The contributions of electrostatic double layers to the bending moduli are extremely variable. Moreover, they can be calculated, with certain reservations to be mentioned at once. The results obtained in Debye-Hückel [12—14] and Poisson-Boltzmann approximation [15—17] are plotted in Figs. 1 and 2 versus surface charge density for various Debye lengths. They were computed on the assumption that the (uniform) surface charge and the surface of inextension coincide with the hydrocarbon/water interface. Both elastic moduli are modified by surface charge and the increase of $\kappa$ is always larger than the decrease of $\bar{\kappa}$ in Debeye-Hückel theory. Fortunately, at high but attainable surface charge densities and fairly small Debye lengths, Poisson-Boltzmann theory predicts the electrical part of $-\bar{\kappa}$ to be several times that of $\kappa$. Specifically, the negative electrical contribution to $\bar{\kappa}$ can be so large that the condition for spontaneous vesiculation, $2\,\kappa + \bar{\kappa} < 0$, is satisfied even for lipid bilayers. Inspection shows, in addition, that $\bar{\kappa}$ is likely to be larger than given in Fig. 2. This is because the surface of inextension of the monolayer is probably not at the interface but somewhere inside the layer. A precise theory is complicated because the negative lateral tension produced by the electric double layer has to be compensated by a positive "mechanical" tension. We know neither the stress profile of this compensating tension nor its effect on the "mechanical" part of bending elasticity. It is commonly implied (but not stated) that the compensating stress resides in the surface of inextension and has no effect on mechanical elasticity.

Spontaneous vesiculation has been known for some time, but the reason for it was not clear. One of the early methods of detecting small vesicles was NMR [18]. Electron microscopy was also used early on [19]. More recent investigations determine phase diagrams and characterize the phases by means of electron microscopy. Perhaps the first of these is a study by Kaler et al. [20] who examined mixtures of cetyl trimethylammoniumtosylate and sodium dodecylbenzenesulfonate. When dissociated in water, the salts are oppositely charged surfactants. There are two vesicular phases at dilutions of less than 5 wt % of surfactant with miscibility gaps towards lamellar phases at lower dilutions. No vesicular phase was found near the equimolar line of charge neutrality. We think that the vesiculation was spontaneous and induced by surface charge. However, its absence at charge neutrality might also be due to the formation of an insoluble 1 : 1 complex of the two surfactants [20].

Subsequently, Hoffmann, Meyer and coworkers identified spontaneous vesicles in a ternary system containing dodecyldimethylaminoxide as surfactant and hexanol as cosurfactant [21]. Increasing the cosurfactant concentration at fixed surfactant content, they found a continuous range of lamellar phases. Small unilamellar vesicles were followed by onions, a mixture of onions and extended bilayers, and finally extended bilayers only. The phases occurred in a triangle extending into the water corner far. Across miscibility gaps, they were accompanied by a phase of cylindrical micelles and a disordered multiply self-connected bilayer (sponge or $L_3$ phase) at lower and higher cosurfactant concentrations, respectively. Although surfactant and cosurfactant are electrically neutral, there seems to be a strong steric repulsion between the surfactant polar heads in the flat layer before their spacing is increased by intercalated cosurfactant molecules.

## Other pathways of vesiculation

There are several ways of generating usually spherical vesicles even when the bending energy of the symmetric bilayer is positive. The most common method in biochemical studies has been sonication since the advent of suitable sonicators. The bilayers are torn up by the sound waves and their patches close again because of the high edge energies of most bilayers. The size of the vesicles thus obtained decreases with the strength and duration of sonication, the limiting vesicle radius being about 10 nm. Lipid vesicles of this size are usually highly stressed, i.e., the bending energies per unit area are very large, and this makes them susceptible to mutual fusion and aggregation (the latter being, perhaps, due to semifusion, i.e. fusion of the outer monolayers).

Another way of enforcing vesiculation is to squeeze a (lipid) dispersion through an extruder with a pore size needed to generate vesicles of the desired radius [22].

A rather gentle method to induce vesiculation despite a positive bending energy of the symmetric bilayer is to destroy the symmetry, at least temporarily, by producing a spontaneous curvature. This will give rise to the formation of buds, i.e., spheres connected to the rest of the bilayer by a narrow constriction. If the buds break off they become vesicles. A special kind of budding is that of smaller vesicles from a larger one. The formation of buds up to the fully constricted state has recently been studied in great detail on the theoretical side [23—25]. In experiments with lipid bilayer vesicles, the spontaneous curvature needed to induce budding was obtained by changing the temperature [26]. Another means of producing spontaneous curvature would be an exchange of the outer aqueous medium. A difference in aqueous media has been invoked to explain the vesiculation from multilayer systems of suitable lipid bilayers that followed a change of pH [18]. However, it was also pointed out that spontaneous curvature is not necessary if the vesiculation is spontaneous. It is difficult to imagine that the mechanism of budding transforms a whole multilayer system into vesicles.

A positive vesicular bending energy does not rule out the existence of vesicles in thermodynamic equilibrium.

Any planar multilayer system may be expected to split up into vesicles at sufficient dilution because of the entropies of translation and size distribution. In practice, the sponge phase ($L_3$ phase) was found to intervene in ternary or quasi ternary surfactant/water systems before the vesicular phase was reached upon further dilution at a bilayer concentration of about 1 wt % [27, 28]. This is surprising because at lower dilutions it is only the bilayer composition that controls which of the phases exists. A possible reason for the preference of the sponge near the vesicular phase may be an easier accommodation of interspersed vesicles (especially the larger ones) in the sponge than in the planar phase. (Also, the (local) coexistence of the sponge with small vesicles will be associated with a partial demixing of surfactant and cosurfactant. Both effects could help to explain why the $L_3$ phase bends around the tip of the planar phase in the water corner.) Two theories have been proposed for the phase transition from sponge to vesicles. One is based on the Ising model and allows a second-order phase transition [29, 30]. The other is in terms of a first-order phase transition from the sponge to the vesicular phase, the bending energy of the vesicles being significantly larger than $kT$ at the transition [31].

## Properties of vesicles

The membranes of lipid bilayer vesicles are usually not in thermodynamic equilibrium. The exchange of lipids between the membranes of different vesicles is known to be very slow, with relaxation times up to hours or days. Rupture of lipid bilayers usually requires high lateral tension, and thermally excited pores seem to be very rare. Therefore, the number of vesicles is constant or changes very slowly. Equilibration including changes of topology appears to be much faster with the bilayers of ordinary surfactants, but the various relaxation times remain to be measured.

Even when the number of vesicles cannot change, their sizes will obey a size distribution function, provided there is a free exchange of surfactant molecules within and between the bilayers. In dealing with size distributions, we assume for simplicity that the bilayer consists of a single species of amphiphilic molecules. The distribution function is easy to calculate if the vesicles are unilamellar and spaced wide enough to behave like an ideal gas. It is of a universal shape if, in addition, the vesicles are on average so large that the bending energy of the spheres is not affected by the fourth-order term in (3). The size distribution $w(N)$ then takes the form [3]

$$w(N) \sim Ne^{+\lambda N/kT} , \tag{9}$$

Progr Colloid Polym Sci (1994) 95:7—13
© Steinkopff Verlag 1994



where $N$ is the number of amphiphilic molecules forming the vesicle membrane and $\lambda(<0)$ is the chemical potential. The factor $N$ in (9) stems from a dependence of the effective bending rigidity $\kappa$ on the number $N$ of molecules in the bilayer [3, 32]

$$\kappa = \kappa_0 - \frac{kT}{8\pi} \ln \frac{N}{2}. \tag{10}$$

Here, $\kappa_0$ is the bare bending rigidity which would be valid in the (hypothetical) absence of thermal undulations. Other authors derived a logarithmic correction of $\kappa$ [33] which is three times that in (10) and an analogous correction of $\bar{\kappa}$ [34] which we did not find. Adopting their numbers would yield in (9) $N^{3/4}$ instead of $N$ for the factor before the exponential function. If the vesicles are small on average, the fourth order term in (3) may have to be taken into account with a positive sign [3]. This sharpens the size distribution by suppressing the smallest vesicle sizes. The chemical potential $\lambda$ in (9) takes the value that conserves the total number of surfactant molecules.

If the number of vesicles is free to change, the same types of size distributions derive from minimizing the total free energy under the constraint of a fixed total number of molecules [35]. The total free energy is the sum of the free energies of the different vesicle sizes. Each free energy contains the bending energy (2) or (4) corrected according to (10) and a contribution due to translational entropy. The latter may be expressed by

$$-kT \ln \frac{V}{v_s \, n_N}, \tag{11}$$

where $n_N$ is the number of vesicles of size $N$, $v_s$ the volume of a surfactant molecule and $V$ the total volume of the surfactant/water system.

We remark that the total energy need not be negative for the vesicular phase to be stable since the competing phases, the planar multilayer system or the sponge phase can have positive free energies. If van der Waals and electrostatic forces are negligible, as should be the case for dilute systems with a small enough Debye length, there is only undulatory interaction in the planar multilayer system. Its energy per unit area is

$$g_{und} = 0.2 \, \frac{(kT)^2}{\kappa \, z^2}, $$

where $z$ is the mean membrane spacing. Logarithmic corrections of this formula and the more complicated case of the sponge phase are treated elsewhere [31].

Even stiff membranes ($\kappa > kT$) will form vesicles if the vesicular bending energy (2) is negative. Let us start from the planar multilayer system to discuss this situation. If $r^*$ as given by (6) is on the order of the mean spacing $z$, loosely packed unilamellar vesicles of this radius

can take up all the available bilayer area. However, close packed onions seem a better solution whenever $r^* \gg z$. In order to estimate the size of these onions, we now make the following three assumptions. First, the mean stacking period $p$ in the onions is uniform and the same as in the original planar multilayer system. Second, all onions are of equal size and the deviations of the outer skins from sphericity in a close packed array are ignored. Third, any entropies of translation and size distribution of the individual onion skins are neglected. (One reason for this neglect is the fact that in a multilayer system the local fluctuations of membrane density are part of the undulations and cannot be treated separately.) With these simplifications, the total bending energy of the system may be written as

$$\frac{V}{\frac{4\pi}{3}(mp)^3} \left[ 4\pi(2\kappa + \bar{\kappa})m + 4\pi\frac{\kappa'}{p^2} \right.$$
$$\left. \cdot \left(1 + \frac{1}{4} + \frac{1}{9} \cdots + \frac{1}{m^2}\right)\right], \tag{12}$$

where $m$ is the number of skins in each onion. Restricting ourselves to $m \gg 1$, we replace the finite sum by

$$\sum_{m=1}^{\infty} \frac{1}{m^2} = \frac{\pi^2}{6}. \tag{13}$$

Minimizing the total energy (13) by varying $m$ then leads to

$$m = -\frac{\pi^2}{4} \frac{\kappa'}{2\kappa + \bar{\kappa}} \frac{1}{p^2} = \frac{\pi^2}{8} \frac{(r^*)^2}{p^2}, \tag{14}$$

where use is made of (6). The result suggests that the onion size diverges as the negative energy $2\kappa + \bar{\kappa}$ approaches zero.

Multilamellar vesicles with many equally spaced shells have, in fact, been seen in electron microscopy [21]. The equal spacing points to the effect of undulatory interaction. (The smoothness of the membranes seen in the pictures is probably an artefact resulting from membrane area contraction during the rapid cooling [36].) A strong repulsion between membranes is needed to avoid "empty shells" in the center of the multilamellar vesicle where the positive contribution of the quartic term to the vesicular bending energy is largest. A primitive criterion for the filling of the $l$-th shell is provided by the inequality

$$\frac{2\kappa + \bar{\kappa}}{l^2 p^2} + \frac{\kappa'}{l^4 p^4} < 0.2 \, \frac{(kT)^2}{\kappa \, z^2}, \tag{15}$$

which displays the bending energy (3) and undulatory interaction energy (10) per unit area on the left and right, respectively. When the first term is omitted, which seems reasonable for the smallest shells, (14) becomes

$$l^4 > 5 \left(\frac{\kappa}{kT}\right)^2 \frac{\kappa'}{\kappa} \frac{1}{p^2} \tag{16}$$

for dilute systems with practically $z = p$. Estimating $(\kappa'/\kappa)^{1/2}$ to be on the order of 10 nm, we find it possible to satisfy the criterion for the innermost shell ($l = 1$), e.g. with $\kappa \approx kT$ and $p \gg 10$ nm. Note, however, that for onions with 30 or more skins the negative bending energy per skin will be less than $kT$ for $\kappa$, $|\bar{\kappa}|$ on the order of $kT$.

It is easy to see that the size of the onions decreases dramatically if the innermost shell is not filled, which amounts to dropping the first term in the series (12). To generalize the theory, one has to take $p$ to be variable and express the total bending energy in terms of the radii of the outermost and innermost onion skins. We refrain from this exercise in view of many other uncertainties. For instance, the nonspherical shapes of the outermost skins, because of their additional bending energies, should cause the phase transition from the onion phase to the planar multilayer system to be discontinuous.

The first theory to explain the existence of onions was developed by Simons and Cates [2]. These authors considered onions with slightly positive bending energies. They omitted nonharmonic terms but took into account the logarithmic dependence of the bending energy on vesicle size and the entropies of translation and size distribution, focusing on the case of two to five skins. Our theory may be regarded as complementary to theirs.

## How to encapsulate

If spontaneous vesiculation is to be used for encapsulation, some problems may arise from the dynamics of the shape transformation. Let us consider planar bilayers which have been charged (e.g. by a new pH) or changed in composition (e.g. by withdrawing cosurfactant) so that the vesicular bending energy becomes negative. The formation of vesicles can still be hampered by energy barriers. For instance, creating a spherical bud connected by a narrow constriction to the rest of the bilayer requires the energy $8\pi\kappa$ which is positive although $4\pi(2\kappa + \bar{\kappa})$ is negative.

Fortunately, a strongly negative modulus of Gaussian curvature favoring vesiculation is likely, because of (8), to be associated with a large positive monolayer spontaneous curvature that lowers the energy of bilayer edges. The resulting "brittleness" of the bilayer has been revealed by electron micrographs of onions. In some systems, the vesicle membranes were perforated or looked like skeletons [21]. A brittleness of those bilayers may also be anticipated on experimental grounds since the vesicular phase is accompanied, on the side of lower cosurfactant content, by a phase of cylindrical micelles. Generally, a

bilayer edge should be similar in structure and energy to one half of a cylindrical micelle. The brittleness of the bilayer should lower the energy barriers of vesiculation, permitting it to proceed, e.g., via the formation of bilayer patches which recombine and close to form vesicles. (Patches of perforated bilayer as intermediate, unstable structures seem natural if the vesicles originate from cylindrical micelles.)

Vesicles made of brittle bilayers will be leaky. In order to keep encapsulated material inside, they must be sealed. This can be accomplished by changing the conditions again so that the planar state is preferred over vesicles. The change may have to be rapid in order to prevent the formation of large onions or flat bilayer.

## Conclusion

We have listed possible pathways of vesiculation and considered in some detail spontaneous vesiculation as defined by a negative bending energy of the bilayer sphere, i.e., $4\pi(2\kappa + \bar{\kappa}) < 0$. To meet this condition, a generally positive bending rigidity, $\kappa$, has to be overwhelmed by a negative modulus of Gaussian curvature, $\bar{\kappa}$. Recalling that $\bar{\kappa}$ equals the second moment of the stress profile, we discussed surface charge and, more generally, monolayer spontaneous curvature as means to achieve a strongly negative $\bar{\kappa}$.

The possibility of tuning $\bar{\kappa}$ is not the only advantage offered by mixed bilayers. Another is a decrease of the bending rigidity which arises when the surfactant molecules differ by their monolayer spontaneous curvatures [6, 37]. It reflects an adjustment of the local surfactant concentrations to the local monolayer curvature or, so to speak, of spontaneous to actual monolayer curvature.

In addition, the possibility of demixing favors phase separation. The coexistence of micelles and bilayers in two separate phases has recently been explained in terms of the demixing associated with differences in monolayers curvature [38]. A significant demixing in the same phase between small vesicles and extended bilayer, planar or sponge, could upon up new ways of controlling spontaneous vesiculation.

The present treatment of "spontaneous" onions contains several simplifications, among them the concept of fixed equidistant "shells" and the neglect of deviations of the onion skins from the spherical shape. There are many questions which remain to be addressed. For the purpose of encapsulation, it could be interesting to find surfactant bilayers forming onions with a hollow interior, i.e. several empty shells.

**Acknowledgement** I am grateful to M. M. Kozlov for discussions.

## References

1. Lasic DD (1993) Liposomes: from Physics to Applications. Elsevier, Amsterdam
2. Simons BD, Cates ME (1992) J Phys II France 2:1439
3. Helfrich W (1986) J Physique 47:321
4. See, e.g., Evans E, Rawicz W (1990) Phys Rev Lett 64:2094
5. See, e.g., Roux D, Nallet F, Freyssingeas E, Porte G, Bassereau P, Skouri M, Marignan J (1992) Europhys Lett 17:575
6. See, e.g., Kozlov MM, Helfrich W (1992) Langmuir 8:2792
7. Lorenzen S, Servuss RM, Helfrich W (1986) Biophys J 50:565
8. See, e.g., Strey R, Colloid and Polymer Sci, in press
9. Helfrich W, Harbich W (1987) in: Meunier J, Langevin D, Boccara N (eds) Physics of Amphiphilic Layers. Springer Proc Physics 21:58
10. Helfrich W (1981) in: Balian R, Kléman M, Poirier JP (eds) Physics of Defects. Les Houches Session XXXV, 1980, North-Holland Publishing Company 713
11. Petrov AG, Bivas I (1985) Prog Surf Sci 16:389
12. Winterhalter M, Helfrich W (1988) J Phys Chem 92:6865
13. Kiometzis M, Kleinert H (1989) Phys Lett A 140:520
14. Duplantier B (1990) Physica A 168:179
15. Lekkerkerker HNW (1989) Physica A 159:319
16. Mitchell DJ, Ninham BW (1989) Langmuir 5:1121
17. Winterhalter M, Helfrich W (1992) J Phys Chem 96:327
18. Hauser H (1989) Proc Natl Acad Sci USA 86:5351; and references cited therein
19. Talmon Y, Evans DF, Ninham BW (1983) Science 221:1047
20. Kaler EW, Murthy AK, Rodriguez BE, Zasadzinski JAN (1989) Science 245:1371
21. Munkert U, Hoffmann H, Thunig C, Meyer HW, Richter W, Prog Coll Polymer Sci, in press
22. See, e.g., MacDonald RC, MacDonald RI, Menco BPM, Takeshita K, Subbaro NK, Hu L (1991) Biochim Biophys Acta 1061:297
23. Seifert U, Berndl K, Lipowsky R (1991) Phys Rev A 44:1182
24. Miao L, Fourcade B, Rao M, Wortis M, Zia RKP (1991) Phys Rev A 43:6843
25. Wiese W, Harbich W, Helfrich W (1992) J Phys Cond Matter 4:1647
26. Käs J, Sackmann E (1991) Biophys J 60:825
27. Porte G, Marignan J, Bassereau B, May R (1988) J Physique 49:511
28. Gazeau D, Bellocq AM, Roux D, Zemb T (1989) Europhys Lett 9:447
29. Cates ME, Roux D, Andelman D, Milner S, Safran S (1988) Europhys Lett 5:733; Erratum ibid 7:94
30. Roux D, Coulon C, Cates ME (1992) J Phys Chem 96:4174
31. Helfrich W (1994) J Phys Condens Matter 6:A79
32. Helfrich W, J Physique 48:285
33. Peliti L, Leibler S (1985) Phys Rev Lett 54:960
34. Kleinert H (1986) Phys Lett 116A:57
35. Hervé P, Roux D, Bellocq AM, Nallet F, Gulik-Krzywicki T (1993) J Phys II France 3:1255
36. Klösgen B, Helfrich W (1993) Eur Biophys J 22:329
37. Safran SA, Pincus P, Andelman D, MacKintosh F (1991) Phys Rev A 43:1071
38. Andelman D, Kozlow MM, Helfrich W (1994) Europhys Lett 25:231

Progr Colloid & Polym Sci (1994) 95:14—38
© Steinkopff Verlag 1994

D. Vučelić

# Physico-chemical aspects of zeolite-A synthesis and application for environmental safe detergents

D. Vučelić
Belgrade University,
Faculty of Physical Chemistry
and Institute of General
and Physical Chemistry,
P.O. Box 550,
Studentski trg 16,
Belgrade, Yugoslavia

**Abstract** Synthesis of zeolites is characterized by a long lag crystallization period. However, this period is not dead; many precrystallization processes are active, leading to nucleation, followed by rapid crystallization. The most lag-active species are complexes of aluminum and, to a lesser extent, silicon and sodium. If these complexes are exposed to non-equilibrium conditions or outside stress, modified zeolite-A can be obtained. This zeolite has a few properties important for surfactants: significantly higher rate of $Ca^{2+}$ exchange, lattice windows large enough for $Mg^{2+}$ binding (21 mg $Mg^{2+}$ per g zeolite) and increased sorption capacity for nonionics and co-polymers. — In connection with trends (compact detergents) the outside zeolite surface area plays an important role. Only under non-equilibrium conditions can zeolite synthesis lead from low (1.3—2.6 $m^2/g$) to high surface area (9—12 $m^2/g$). Modification can also be made, respective catalytic or inhibitory properties for bleaching substances like perborate or peracids. — Appearance of huge quantities of zeolite in the environment (precostal and sludge fertilized field) might have negative side-effects. However, results with six microorganisms (*Staphylococcus aureus* — gram positive, *Escherichia coli* — gram negative, *Bacillus subtilis*, *Thrichophyton mentagrophytes*, *Candida albians* and *Aspergillus fumigatus*) show significant bactericidal effects. The effect is highest for the dangerous *Thrichophyton mentagrophytes* fungi; already 0.1% of zeolite suspension at 20°C is sufficient. In this respect, and due to high affinity for heavy metals, zeolite suspension at 20°C is sufficient. In this respect, and due to the high affinity for heavy metals, zeolite behaves as an environmental cleaner. Results with different algal species are not conclusive within the limited laboratory period of examination that has been conducted.

**Key words** zeolite-A — synthesis — impure raw material — particle size — ion exchange capacity

## Introduction

Although less than 20 years have passed since the first patents [1, 2] of zeolite-A as inorganic builders, its usage in laundry detergents reached 1.1 million tons in 1993, and is still growing [3].

Zeolite-A synthesis in the laboratory was known long before its usage in detergents [4]. Being in crystal form, it seems that physical properties should be stable and unchangeable. Based on that, in the beginning, involvement of zeolite as insoluble material in washing powder demanded significant changes of detergent formulations. As formulations became more sophisticated and new trends appeared (compact detergent) new

Progr Colloid Polym Sci (1994) 95:14—38
© Steinkopff Verlag 1994

physico-chemical characteristics of zeolite itself were needed. Fortunately, zeolite-A properties can be varied to some degree, not only during synthesis, but also in the processing (filtration and drying), so these requirements have been achieved. This is especially valid for two important factors: rate of exchange and crystal-surface.

Many papers and an extensive review [5] of zeolite cleaning action and zeolite-based detergents have been published. Surprisingly, in spite of millions of tons of production, except for craftsmans patents, very few papers and no review of synthesis from this point of view have been published so far.

There are two, basically different, syntheses: solute-hydrothermal and precrystallization of solid kaolin (or other natural aluminosilicates). The latter, due to difficulties in obtaining high quality, has not been used in Europe (and only partially in the USA) and is not discussed in this paper. Readers are referred to papers [6—10].

## Mechanisms and kinetics of hydrothermal zeolite-A crystallization

From the chemical point of view the reaction is very simple; mixing silicate- and aluminate-solution in excess of water leads to small zeolite crystals of fixed stoichiometry. If the silicate-solution is water glass with a ratio $SiO_2/Na_2O$ of 2 (very common in detergent zeolite production) and aluminate-solution obtained from $Al(OH)_3$ dissolution in $NaOH$ (also usual) with a ratio $Na_2O/Al_2O_3$ of 2, the chemical reaction can be presented in a simple way:

$$Na_2O \cdot 2\ SiO_2\ (aq.) + Al_2O_3 \cdot 2\ Na_2O\ (aq.)$$

$$\xrightarrow{H_2O} Na_2O \cdot Al_2O_3 \cdot 2\ SiO_{2(cryst.)} + 2\ Na_2O\ (aq.) \quad (1)$$

Equation (1) is very useful for many purposes:

a) analytical determination of all reactants is presented in oxide form;
b) stability of aluminate solutions depends on $Na_2O/Al_2O_3$ ratio;
c) crystallization field is determined within the $Na_2O/Al_2O_3/SiO_2/H_2O$ ratio;
d) some properties such as crystal size, exchange rate, and induction period are sensitive to some ratio.

Factors a, b, and c are important for production, and for the application properties. Species from Eq. (1), of course, do not exist in solution. For the crystallization mechanism, reaction with real species should be presented. In an ideal case, when waterglass consists of monomeric species only, the reaction is given by:

$$NaH_3SiO_4\ (aq.) + [Al(OH)_4]^-Na^+\ (aq.)$$

$$\xrightarrow{H_2O} Na[AlO_2 \cdot SiO_2] \cdot 3\ H_2O_{(cryst.)} + Na^+ + OH^-. \quad (2)$$

Although the chemical reaction given by Eq. (2) seems quite simple, the underlying mechanism is very complex. Immediately after mixing tetrahedral silicate monomers:

$$\begin{array}{c} OH \\ | \\ NaO-Si-OH \\ | \\ OH \end{array}$$ and tetrahedral aluminate:

$$\begin{array}{c} OH \\ | \\ HO-Al-OH \qquad Na^+, \\ | \\ OH \end{array}$$

an aluminosilicate gel is precipitated. Crystallization begins after a long "induction time," from 1 to 20 h, depending on the experimental conditions (temperature, $Al_2O_3$, $H_2O$ and $SiO_2$ concentrations).

There are two basic hypotheses about a possible mechanism:

### Gel hypothesis

Flaningen and Breck [11], Flaningen [12], McNicol et al. [13]. Nucleation "precursors" are slowly formed during gel ripening (induction period). Once formed through reorganization of silica- and alumina-tetrahedral units, they initiate the nucleation process followed by rapid crystallization and crystal growth embedded in the amorphous gel matrix. The possibility of crystallization directly from the gel without liquid phase [14] gives strong support to this hypothesis.

### Liquid phase hypothesis

First proposed by Barrer [15] and Kerr [16, 17], and later by many others [18], nucleation is the result of polymerization of aluminate-silicate-tetrahedral and possibly more complex ions in the liquid phase. When the "current" of nuclei exceeds a critical size, fast crystal growth starts, supplied continuously with silicate and aluminate units from gel dissolution. Aluminosilicate species found in solution by Raman and $^{29}Si$-NMR spectroscopy [19—21] and the possibility of obtaining zeolite directly from liquid phase, omitting gel stage [22], favor this concept.

At present, there is an inclination towards liquid phase crystallization, although, no doubt, both mechanisms appear simultaneously or consecutively, or one of them dominates depending on crystallization conditions.

Industrial processes always occur with overlapping mechanisms and a significant role of gel/liquid interface. The typical example is an early production process. Waterglass and aluminate-solution are simultaneously mixed at $90\,°C$ with ratios of:

$(2.8—3.1)Na_2O : Al_2O_3 : 2\ SiO_2 \cdot (70—100)\ H_2O$. Under these conditions the following results were obtained:

a) Amorphous gel is surprisingly stoichiometric: $SiO_2/Al_2O_3 = 2.3$ which corresponds to the fusion of 7 silicate with 3 aluminate tetrahedra. $Na_2O/Al_2O_3$ ratio

Scheme 1:

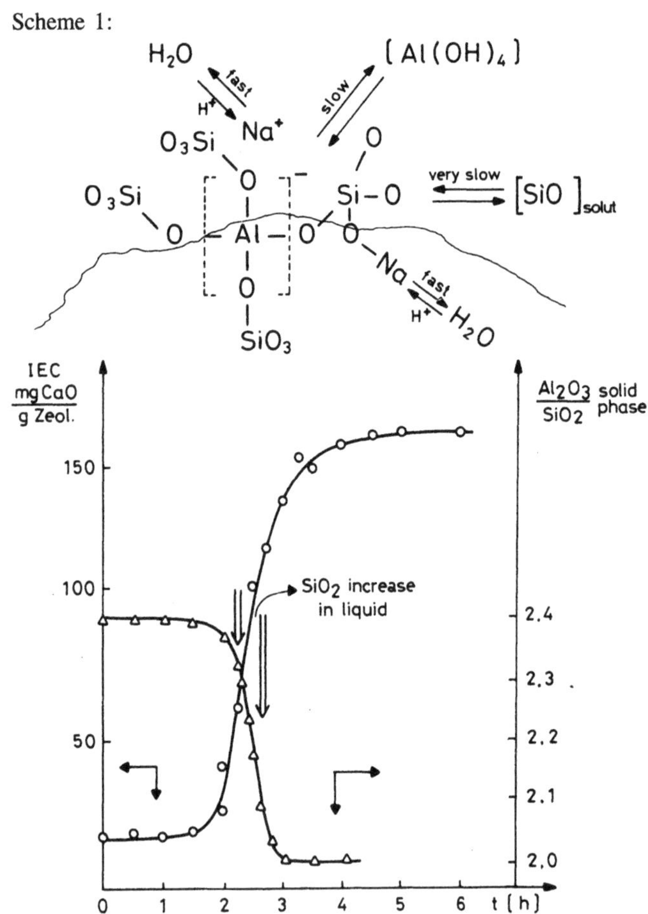

Fig. 1 Changes in $Al_2O_3/SiO_2$ ratio in solid-phase and ion exchange capacities (IEC) during zeolite-A synthesis. (Conditions: $Na_2O/Al_2O_3/SiO_2/H_2O$, 3.6/1/2/108 at 80 °C simultaneous mixing)

Scheme 2:

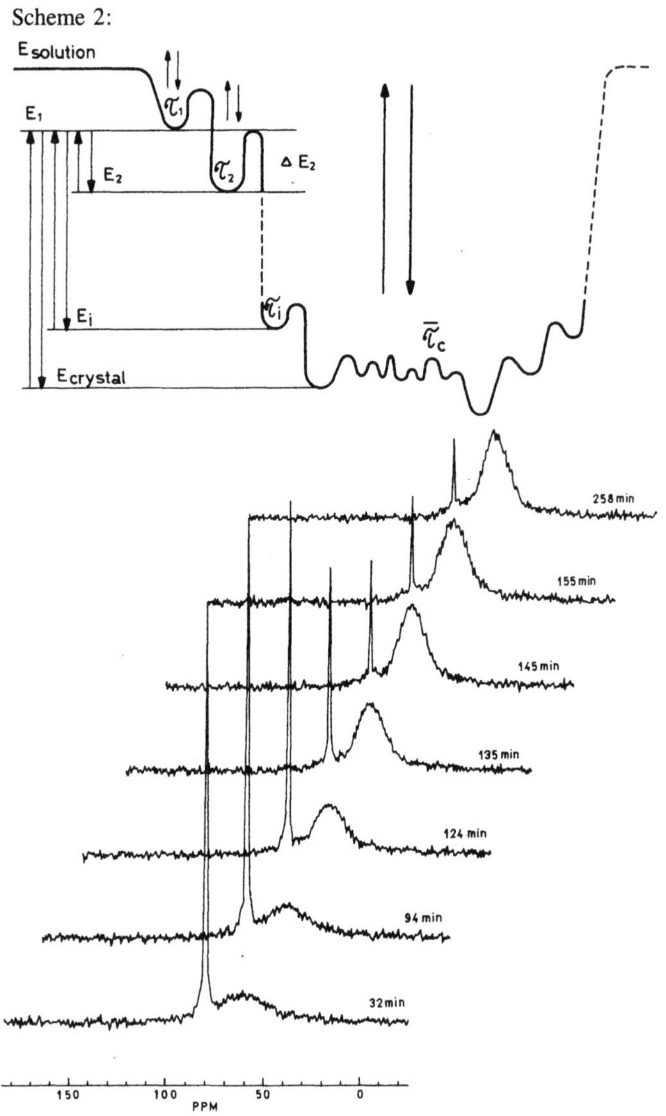

Fig. 2 $^{27}$Al-NMR spectra in the course of synthesis. (NMR-Bruker MSL 400 MHz. Synthetic conditions as in Fig. 1)

is always close to 1 and sodium can be temporarily released with water addition followed afterwards by fast auto adjustment to the previous value. Aluminum ions can be also washed out, but not so easily as sodium.

b) Around 15% of $Al_2O_3$ and 2% of $SiO_2$ are left in bulk liquid. These concentrations are stable dring the induction period.

c) The induction period lasts for about 2.5 h, followed by fast crystallization ($\sim$ 40 min). Before the crystals are large enough for the appearance of an XRD pattern there is an increase of ion exchange capacity (IEC), so this can be used as a sensitive indicator (Fig. 1).

d) Beginning of crystallization is marked by fast decrease of aluminum concentration in solution (decrease in $Al_2O_3/SiO_2$ ratio) and increase of IEC, Fig. 1 (also Figs. 2 and 3). When the concentration of Al-ions in solution drops or IEC increases to 60—70% of their final values there is always a small increase of silicate concentration in solution (2—5% of total Si present), marked in Fig. 1 with two arrows.

Under these conditions the concentration of Si in solution is too small for observation by NMR, so only $^{27}$Al-NMR can give as much information about solution as solid phase. The results are shown in Fig. 2. In solution, only one sharp line belonging to monomeric tetrahedral aluminium exists (the same is true for $^{29}$Si not shown). The intensity of this aluminum is decreasing, following the patterns in Fig. 2. The very broad peak (59 ppm) belongs to amorphous tetrahedral Al-from the gel. In the course of the action, this peak became narrower and shifted to 58.3 ppm, which is typical for tetrahedral zeolite aluminium. During the whole process (except maybe immediately after Si increase in solution) alumino-silicate species in solution were not found. It confirms that the process is going within gel phase and if solution crystallization exists, it amounts to only 2—5% of total yield.

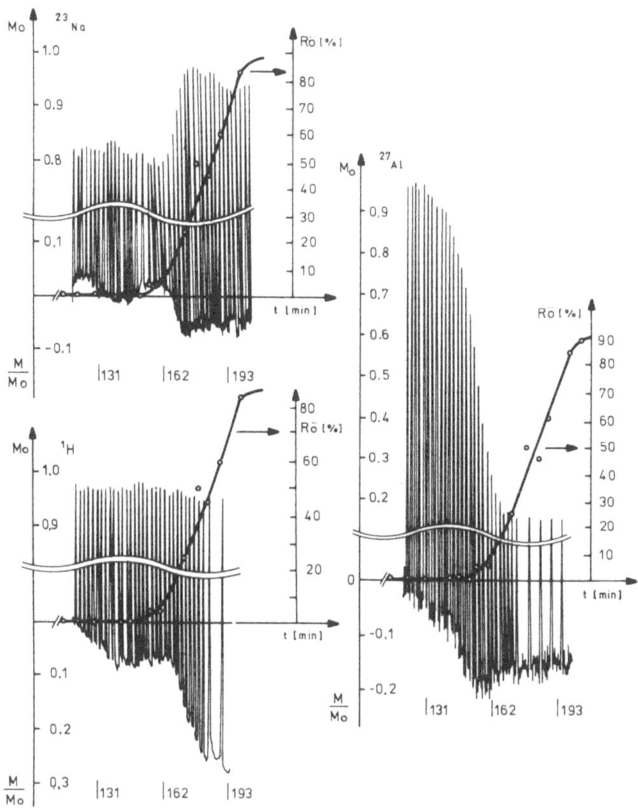

**Fig. 3** $T_1$-relaxation time and concentration of different nuclei in the course of the zeolite-A synthesis. $^1$H-$^1$H-NMR; $^{23}$Na-$^{23}$Na-NMR; $^{27}$Al-$^{27}$Al-NMR (synthesis conditions as in Fig. 1). The method is based on fast conversion from $T_1$ measurements to measurements of intensity of NMR signal $M$. At the zero time signal is dynamically cancelled so, above, abscissa intensities, and below, changes in relaxation $T_1$ are obtained

Using a special pulse sequence of $T_1$-relaxation time it is possible to monitor in milliseconds changes in the immediate surroundings of all nuclei in solution except Si (low sensitivity). Results are shown in Fig. 3. Three main conclusion are obvious:

a) There are changes in $T_1$ for all three species, during gel ripening. This is possible only if species in solution are in equilibrium with the gel during its reorganization processes. Equilibrium is fastest with large gel surface area, where, due to lowest free energy and flexibility of end chains, first crystallization nuclei occur.

b) Crystallization affects both $T_1$ and concentration of corresponding species in solution. The decrease in Al-ions corresponds to the chemical analysis. The increase in sodium concentration is an apparent one, because chemical analysis showed no changes in sodium concentration. The apparent increase is the contribution of sodium ions from the zeolite cavities. In gel, sodium is strongly fixed in proximity of [Al-

tetrahedra]$^-$ without contribution to $M_0$. In zeolite cavities sodium ions became mobile, especially in the presence of water [23, 24], thus contributing to $M_0$.

c) Behavior of the proton is very peculiar. Initially, there are changes in the proton $T_1$. Just before or when the crystallization starts, the proton is inactive. During rapid crystallization protons are again involved significantly. A very acceptable explanation is as follows. During gel ripening (reorganization, growth) many hydrogen bonds are broken and formed. This is a real induction period during which nucleation precursors are formed. In the nucleation period many nuclei are formed without involvement of hydrogen bonds. Finally, when crystals of significant size start to grow, water entrapped in cavities starts to form a few ordered complexes known as zeolite water by formation of many hydrogen bonds with lattice oxygen and coordination with sodium [25].

### Surface hypothesis

Based on the presented results, the major steps in the mechanism given by simple chemical Eq. (2) can be followed:

1) *Gel "building blocks"*:
Mixing waterglass and aluminate solutions under the above conditions in many experiments always led to an average $SiO_2/Al_2O_3$ ratio $\sim 2.33$. This demonstrates the existence of gel "building blocks" with the same average ratio, but randomly ordered (amorphous), of size 100—300 Å [12]. Many ring structures must exist because crystallization with $Li^+$ and $Na^+$ takes place, but not with larger $K^+$. The surface part of these "building blocks" is represented in Scheme 1.
The structure is porous because fast exchange between protons from $H_2O$ and all sodium in the gel exists. Under the above conditions, most free Si-O are protonated conversely, aluminate tetrahedra in the gel are almost fully deprotonated and compensated by $Na^+$ ($Na_2O/Al_2O_3$ ratio being 0.8—1). Links between "building blocks" consist of weak (hydrogen bond of Si-OH) and strong ($SiO_2$) bridges. The latter are very probably responsible for the higher level of silica in the gel ($SiO_2/Al_2O_3 \approx 2.30$). The former are the markers for the building block growth. (There are numerous proofs, for example when $Na_2O/H_2O$ ratio is increased, there is deprotonation of Si-O-H groups slowing building block growth, which leads to smaller zeolite particles at the end). Under different conditions, the first step, of course, will lead to different "building blocks" (some of them very peculiar).

2) *"Precursor" step*:
This is the main event in the basic theories of first-order phase transitions, not only in crystallization of zeolite-A. Like the "computer simulation method," in

the gel it is also an trial-and-error probe as detailed below. In solution, due to the high mobility of ions, many structures with different potential and free energies are formed within very short time (nano or pico second scale).

The interaction of species leads to a potential energy representing an ordered "precursor", the future crystal structure. The probability of finding hundreds of ions in the right positions in "one shot" is negligible. Any structure needs time to reorganize. However, high surface energy (disorder force) tends to cause dissolution, without giving time for complex reorganization and growth. Lifetime of a complex is related to the ratio between $\Delta E_{i(\text{prec.})}/\Delta E_{i(\text{surf.})}$, Eq. (3) [26].

$$\tau_i \sim A \exp\left[\frac{\Delta E_{i(\text{prec.})}}{\Delta E_{i(\text{surf.})}}\right] \tag{3}$$

$\Delta E_{i(\text{prec.})} = E_{\text{sol}} - E_i = \Delta E$ (interaction or potential energy of level "$i$" in precursor)

$E_{\text{sol.}}$ = (energy of species in solution);

$\Delta E_{i(\text{surf.})}$ = surface energy of corresponding complex.

$A$ = frequency factor.

This equation and the simplified model, schematically shown below, give a general idea about mechanism of "precursor" formation.

In a single "time shot," a few species are linked together. Interaction energy depends on their number and the correctness of their positions. The more of them, and the more of them in good positions, the greater is the drop in potential energy, the longer is the life time of the complex. Free surface energy decrease with increasing number of complex members leads to additional prolongation of lifetime. In this manner, $\tau$ behaves as a self-trial-and-error probe. Short $\tau$, "big error," leads to fast dissolution. The longer $\tau$, the longer is the supply of energy and species from solution, so the complex might eventually grow to reach a critical size, after which the whole process becomes spontaneous.

However, long $\tau$ does not exist in solution, because of extremely low probability of achieving a high number of constituents and a high number of them in good positions by the random process. In this model, "precursor" represents a complex with species in the right positions with long enough time $\tau$ to get necessary energy and material (from surroundings) to reach the critical size for crystal nuclei.

Even though the number of "shots" is higher than $10^9$—$10^{12}$ per second the induction period for such a complicated structure as zeolite-A would be too long

in solution. On the other hand, due to the slow mobility of species within gel, "shots" do not exist; only slow reorientation within building blocks is possible. There is no problem with free surface energy, because the energy of chemical bonds within "blocks" surpasses the free surface energy, but the induction period would be too long or imperfect zeolites crystals should grow.

Surprisingly except by Sand [27], the gel surface was not considered as a phase for precursors to be born and grow to crystallization nuclei, although it optimizes both concepts. Among the amorphous structure of "building blocks" some parts must be correct because they correspond to maximum interaction. These unfinished but correct parts on the surface are the sites with the highest probability for precursors to be formed. There are constant single "shots" of ions from solution. Because the problem of free surface energy dissolutions does not exist (similar to that for bulk gel), correct species at the correct positions (highest bonding energy) slowly but steadily build precursors. The competition of species, with highest binding energies, for the same site represent a self-sufficient trial and error selection.

Structures of gel are much more spongy than compact, so many precursor sites exist. The equilibrium of $SiO_4^-$ (aq.) $\leftrightarrows$ $SiO_4^-$ (gel) is shifted towards gel much more than $[Al(OH)_4]^-$ (aq.) $\leftrightarrows$ $AlO_4^-$ (gel), indicating stronger Si-O than Al-O bonds.

Consequently, surfaces of the "building blocks" consist of mainly the Si-O protonated end groups ($Na_2O/Al_2O_3$ in the gel is always between 0.9—1. It confirms domination of SiO end groups mainly responsible for the gel $SiO_2/Al_2O_3$ ratio 2.33). Shots of aluminate and sodium ions from solution complete "precursor" formation. Very probably, there is fusing in space of many neighboring Si-O end groups, so that there is not only formation of one precursor, but its growth at the same time as well. The results from Fig. 3, as well as the results from chemical analysis, are in full agreement with this mechanism: small, steady decrease of Al in solution, no changes in sodium, and nor or slight decrease of Si-ions (chemical analysis). In these two steps, crystal engineering, like genetic engineering, can be done. Decreasing protonation on the end Si-O groups results in smaller crystals and, consequently, higher rates of exchange can be obtained. Additionally, the energy field profile, at the bottom of the potential well shown above, would bring significant changes in position and binding energy of sodium. This will affect all three key parameters: IEC, rate of exchange, and selectivity of the final zeolite-A crystals.

3) *Nucleation step:*

The growing precursors cause gel shrinking and breaking of many bonds. If the size exceeds the critical radius spontaneous nucleation begins. Aluminate-ions from solution intensively support the nucleation process and the beginning of crystal growth. Both gel- and solution-silicate participate in the nucleation process. Only surface silicate between two nucleation sites from the gel is released into solution.

In this step little or no crystal engineering can be done.

4) *Crystal growth step:*

Crystals from nuclei of the size ~1 µm, with constant supply of ions from solution, grow up to 5—6 µm. (Larger crystals are thermodynamically unstable). During this step crystal surface and size can be modified.

*Kinetics:*

Althouth the mechanism is very complex, kinetic curves, usually S-shaped, can be given by a simple equation [28]:

$$Z = 1 - e^{-kt^n} \; ; \qquad (4)$$

$Z$ = ratio between mass in the gel of time $t$ ($Z_t$) and mass of the final crystals $Z_{fi}$, $k$ and $n$ are constants. Zhdanov [28] proposed a model with nuclei formed in the gel from solution precursor. Crystal growth is a pure solution process. For the crystallization of zeolite-A (conditions close to this paper) calculations based on best fitted experimental data, $n = 4$ and $k = 11.5 \cdot 10^{-9}$ h$^{-4}$, were obtained.

---

## The basic factors governing synthesis and physico-chemical properties of zeolite-A

As it was described above the mechanism of zeolite crystallization is a very complex one based on several steps each depending on both thermodynamic and kinetic factors. It is not a big surprise that many of them govern zeolite-a synthesis. The main physicochemical factors are listed below:

Primary effects

1) Composition of reactants. $Na_2O/Al_2O_3/SiO_2$ ratio;
2) Influence of water. $H_2O/Na_2O$ ratio;
3) Temperature.

Secondary effects

4) The nature of the reactants and their pre-treatment;
5) Order of mixing;
6) Gel pre-treatment (low temperature of gel aging);
7) Reaction catalysts/inhibitors;
8) Seeding.

1) Composition of reactants

The crystallization field of zeolite, Fig. 4, has been reexamined and changed significantly [18] since Milton's patents [4]. Compositonally, the ranges: $SiO_2/Al_2O_3$ 2.06—6.8 and $H_2O/Na_2O$ 30—200 are widely accepted at present [18].

However, Kostinko was the first, in his outstanding paper [29], to recognize crystalline zeolite-A as a metastable state and, consequently, bring "lifetime" as a necessary parameter. Figure 5 shows real experiments and in Fig. 6 a scheme of the general time behavior is presented. After 1 h, Fig. 5a, a mixture of amorphous, zeolite-A and X exists. After 3 h Fig. 5b, amorphous, zeolite-A and zeolite-A + hydrosodalite (HS) coexist, depending on the $Na_2O/SiO_2$ and $SiO_2/Al_2O_3$ ratios. Zeolite-A, depending on $Na_2O/Al_2O_3/SiO_2/H_2O$ ratio, can transform in time into pure HS or X-zeolite, or mixtures of the two. Transformation times decrease with temperature increase. Transformations of this kind, e.g., zeolite-A → HS can be achieved easily in less than 3 h. (Under special conditions hydrosodalite can be less stable than zeolite-A. In this process, in the course of the reaction, HS is formed first and, subsequently, the synthesis of zeolite-A is achieved via crystallization of HS. This reverse transformation is very useful when low purity raw material is involved. The production of more than 200 000 t of zeolite ZIB-1 in Birač between 1986—1991 was based on this reverse transformation. Rather than using the crystallization field, it is better to regard the influence of constituents in terms of the acceleration/deceleration of crystal transformations.

*$Na_2O/SiO_2$ influence:*

The influence is the same as for the $H_2O/Na_2O$ ratio (below), but less pronounced, especially when masked by a strong $H_2O/Na_2O$ effect or high temperature. It is not strange because the $Na_{aq}^+ \leftrightarrows HO—Si—gel$ equilibrium is affected in both cases.

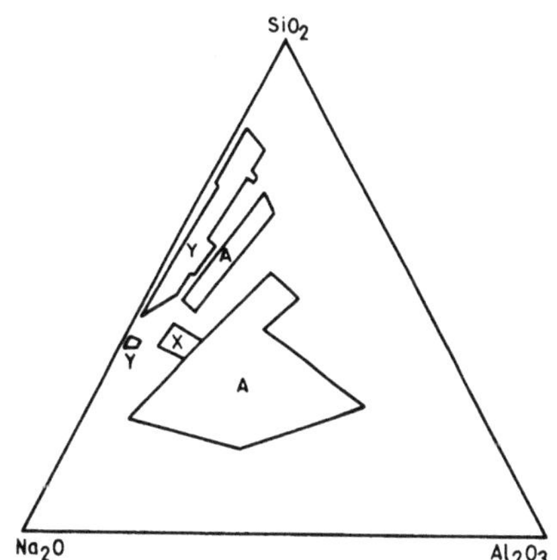


**Fig. 4** Patented batch compositions for the synthesis of zeolites-A, X, and Y. Coordinates expressed as mole percent [4]

Many experiments confirmed the strong influence of $Na_2O/SiO_2$ ratio on PSD (particle size distribution) and APS (average particle size). For $SiO_2/Al_2O_3 = 2$ (waterglass in alkaline solution, $T = 90\,°C$), a drop of $Na_2O/SiO_2$ from 2.0 to 1.2 is accompanied by APS increase from 2.5 to 4.7 μm. The effect is depressed at higher $SiO_2/Al_2O_3$ ratios, but in all cases it is significant, strongly supporting the role of $Na_2O$ as an inhibitor of the growth of SiO-groups in gel precursors and during nucleation formation.

*$SiO_2/Al_2O_3$:*
The increase of $SiO_2$ leads to an almost linear increase in reaction time, Fig. 7. The results are to be expected because an enriched ratio of $SiO_2/Al_2O_3$ in the gel results in larger building units and, consequently, more time for precursor ordering (longer induction time), and more silicate bonds to break (longer nucleation time). The increase of $Na_2O/SiO_2$ ratio, as a stronger effect, cancels the influence of $SiO_2/Al_2O_3$ starting from 2.4, Fig. 7a. An increase in $Na_2O$ increases the number of Si-ONa end groups and so smaller building blocks are formed, cancelling the effect of the $SiO_2$ surplus in the gel. $H_2O/Na_2O$ does not change the influence of the $SiO_2/Al_2O_3$ ratio as much as the $Na_2O/SiO_2$ ratio, indicating that the overall influence of water is less an equilibrium but more due to concentration increase of all constituents. The influence of $SiO_2/Al_2O_3$ on PSD and APS is the opposite to the $Na_2O/SiO_2$. However, increase in the $SiO_2/Al_2O_3$ ratio over 2.2 leads to an opposite sharp decrease in APS. Under these conditions, due to the very slow kinetics of $SiO_{4(sol.)}^- \leftrightarrows (SiO_4)_{(gel)}^-$, a very low amout of $[Al(OH)_4]^-$ is left in solution and slow precursor growth depends on slow gel reorganization, so many small nuclei appear with a release of excess $SiO_4^-$ into solution.

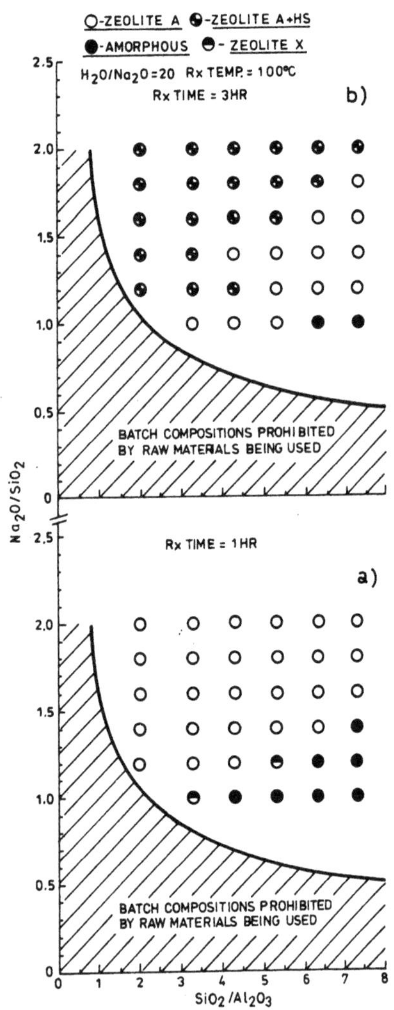



**Fig. 5** Crystallization of zeolites HS, A and X as a function of time a) after 1 h; b) after 3 h [29]

**Fig. 6** Effect of reaction time on zeolite phase transformation. Keys to symbols indicate species detected [29]

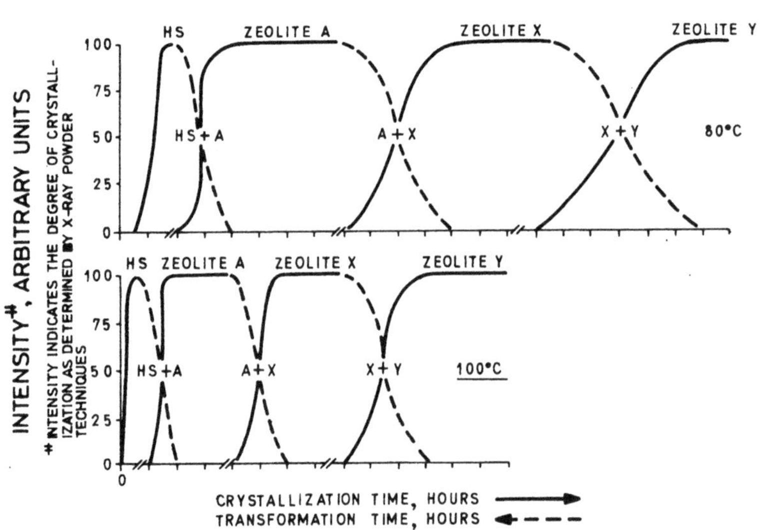

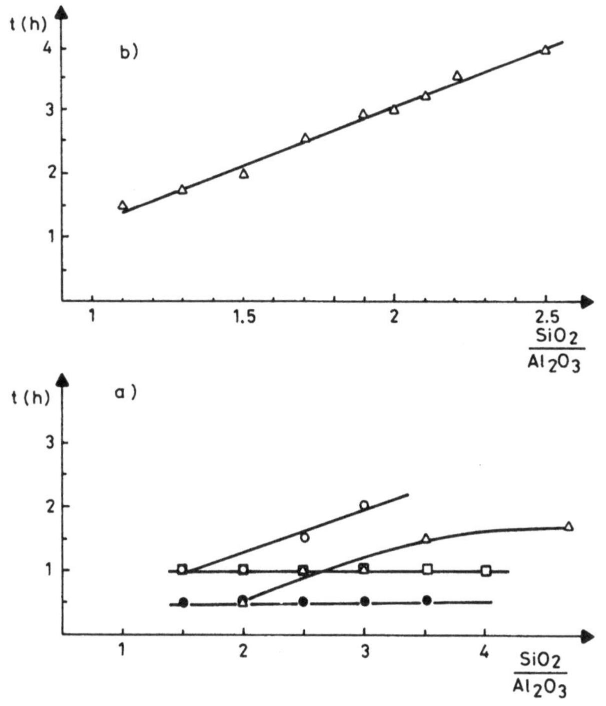

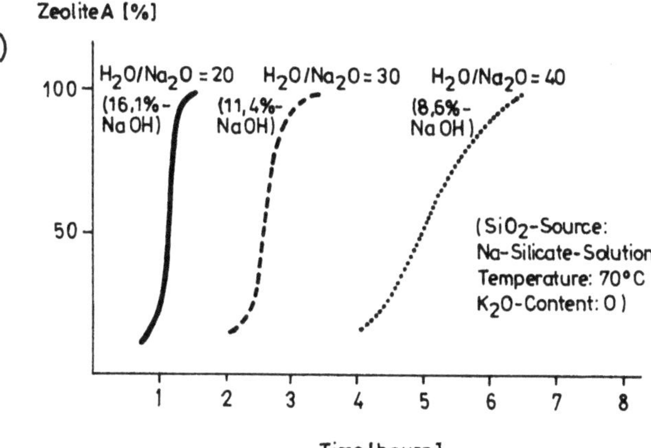



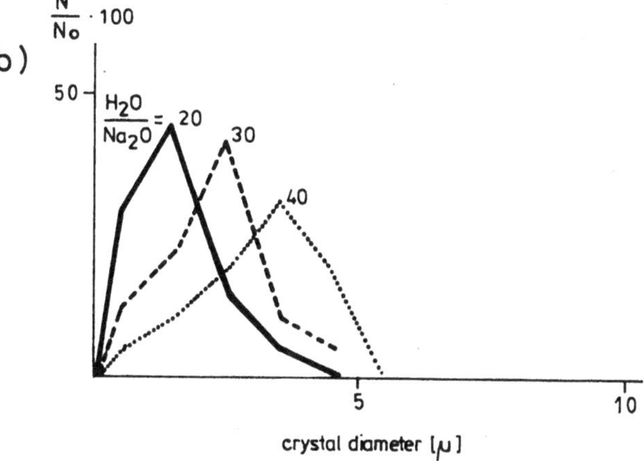



**Fig. 7** Dependence of reaction time on $SiO_2\,Al_2O_3$ ratio.
a) Relations: $\circ$ — $Na_2O/SiO_2 = 1.2$, $\square$ — $Na_2O/SiO_2 = 2.4$ at $100\,°C$ and $H_2O/Na_2O = 20$: $\times$ — $Na_2O/SiO_2 = 1.2$, $\bullet$ — $Na_2O/SiO_2 = 2.4$ at $100\,°C$ and $H_2O/Na_2O = 30$
b) $Na_2O/SiO_2 = 1.2$; $H_2O/Na_2O = 20$ at $80\,°C$

**Fig. 8** Influence of $H_2O/Na_2O$ ratio on zeolite-A crystallization rate (a) and on crystal size distribution (b)

## 2) $H_2O/Na_2O$ influence

This ratio has greatest effect on reaction time, Fig. 8a [30]. A decrease in ratio by twofold results in an almost fivefold time decrease.

The results are in good agreement with the above model. The equilibrium between $Na_2O$(sol.) $\leftrightarrows$ HO—Si—gel in which, when water decrease shifts towards deprotonation of silicate tetrahedra, automatically producing smaller building units with many Na-O-Si end groups in the gel. This results in fewer groups to order, better developed gel surface (shorter induction time); more small-sized precursors will lead to more nuclei (shortening crystallization time). Both effects are well illustrated by Fig. 8a. The effects are enhanced at higher temperature due to the increase of deprotonation with temperature.

As well as reaction time, this ratio affects, in the same way, the particle size distribution (PSD) and its mediana, Fig. 8b. The results are based on the same mechanism as for the reaction time and represent additional experimental evidence for the above hypothesis.

## 3) Temperature

This is a pure thermodynamic effect, influencing all the steps involved in the reaction time. An increase in temperature leads to a decrease in reaction time, more pronounced for the induction period than for the

crystal growth rate. This corresponds to lower activation energies of formation of precursors and nuclei, 50 kJ/mol as compared to 76 kJ/mol for the crystal growth rate [27], Fig. 9a [30].

Being a purely thermodynamic factor, temperature should not affect PSD or mean size of crystals. The results of Fig. 9b confirm the latter, but PSD became narrower on the low size side, indicating involvement of kinetic factors at low temperature.

## 4) Nature of reactants

It has been known for a long time that the source of silicate component may significantly change the reaction time and the particle size, Fig. 10 [30].

Depending on its origin, $^{29}Si$-NMR revealed the existence of different silicate anions in solutions, [31—33], Fig. 11 [33].

Silicate solutions consist of many forms: $Q^0$-monomer (A); $Q^1$-dimer (B) or other end groups; $Q_3^2$-cyclic trimer (C); $Q^2$-middle groups (D) of chain or cyclic silicate anions; $Q_6^3$-prismatic hexamers (E). The intensities ($\sim$ concentration)) and linewidths (rate of

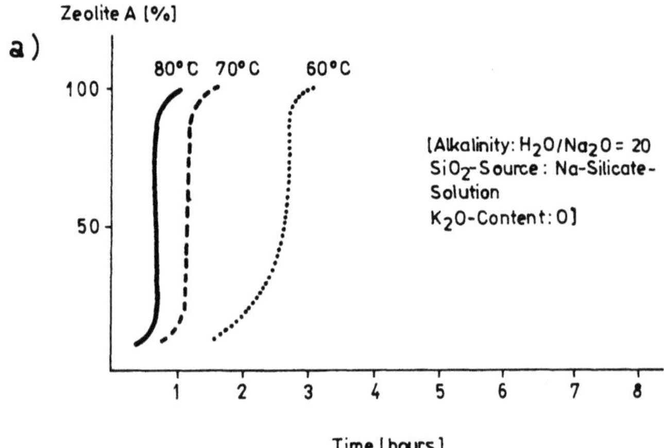


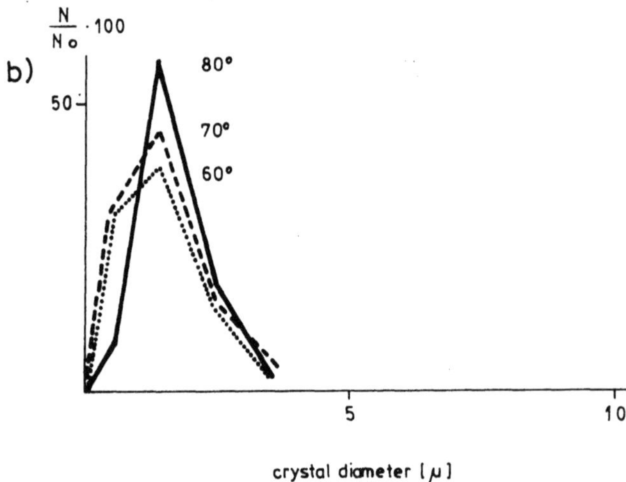


**Fig. 9** Influence of temperature on zeolite-A crystallization (a) and on crystal size distribution (b) [30]

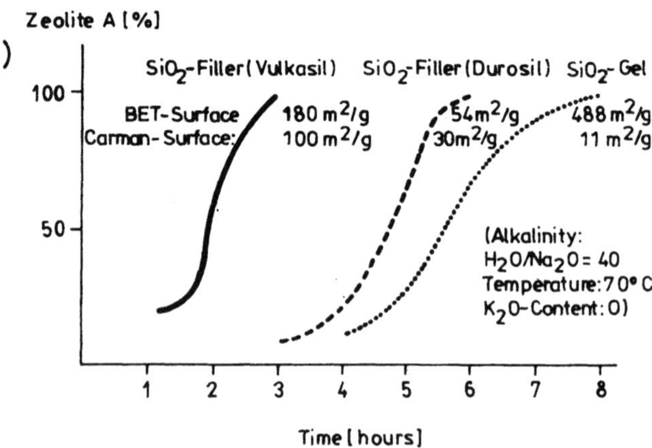


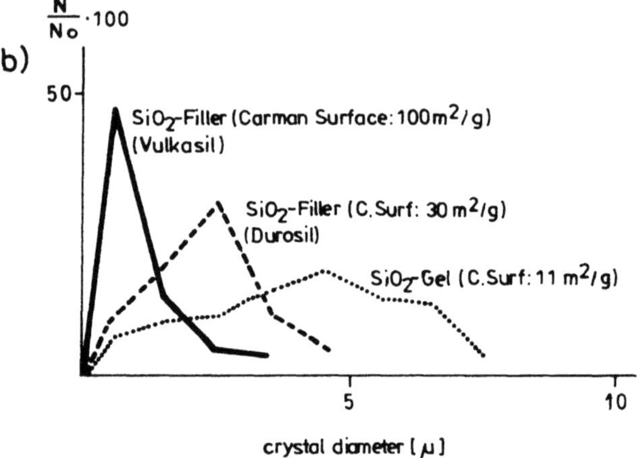


**Fig. 10** Influence of $SiO_2$ source on zeolite-A crystallization rate (a) and on crystal size distribution (b) [30]

exchange between species) of these forms are quite different for silicate solutions from different origins. (There is strong deprotonation with temperature increase, enabling polymerization). The incorporation of trimers or cyclic anions in gel will lead to many wrong structures in gel building blocks, consequently changing the reaction time and especially PSD. Similar results can be obtained with identical silicate solutions, but different $Na_2O/SiO_2$ ratios, Fig. 12 [34]. Pre-treatment, heating or cooling over some extended time (which is always the case in production) can seriously change the equilibrium of silicate species. One non-practical example, but one showing drastic changes, is shown in Fig. 13 [35].

On the other hand, the waterglass used in practical production is shown in Fig. 14. This waterglass consists mainly of polymeric units: many end groups $Q'$, cyclic trimers $Q_3^2$, with the dominant species being cyclic tetramers $Q_4^2$ and $Q_4Q_3$. Monomers represent only 1—2% of all units. For this kind of waterglass

the induction period is much longer in comparison with waterglass in which monomers, dimers, and single chain units are dominant.

Conversely to the complex behavior of silicate, aluminate-solution is very simple. it consists of monomeric units only and, so far, all attempts to prove the existence of dimers have failed.

5) Mixing order

This is an important parameter because different starting gels are obtained. Gels formed by adding silicate solution into aluminate and vice versa due to inhomogeneity, need more time in the induction period than gels formed by simultaneous mixing, Fig. 15. The former are more sensitive to the different kind of impurities and should be avoided or used depending on the situation.

Gels, also, are of different consistencies, which is important for equipment design.

6) Gel pre-treatment

Aging of gel at low temperature shortens reaction time and decreases PSD and APS [28]. The mechanism is compatible with the surface gel model. At low temperature, due to low reactivity, builiding blocks are

Progr Colloid Polym Sci (1994) 95:14—38
© Steinkopff Verlag 1994



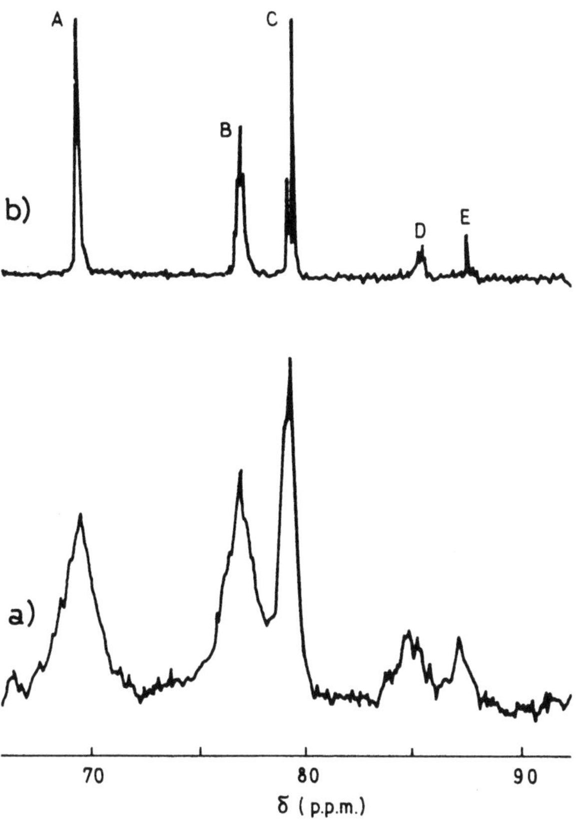

Fig. 11  $^{29}$Si-NMR spectra of sodium silicate solution Na:Si = 2 at 20°C, obtained from different sources. a) From waterglass solution Na:Si = 0.6; b) From amorphous silica [33]

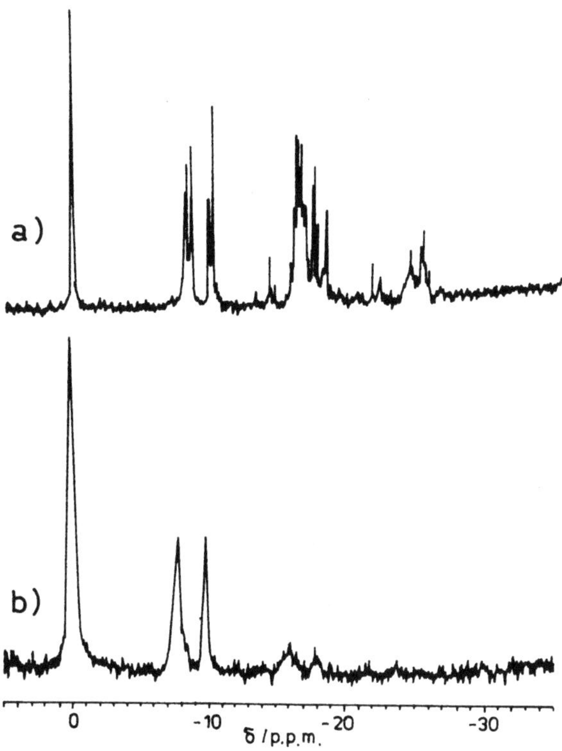

Fig. 12  $^{29}$Si-NMR spectra of sodium silicate solution (a) $SiO_2/Na_2O$ = 0.33, b) $SiO_2/Na_2O$ = 1.0 [34]

smaller and the growth of precursor to a critical size is slower, so many building blocks have time to reorient and become "precursory." In that way, more nuclei are formed and, consequently, smaller crytals are obtained. However, although aging temperature causes significant effects, it is too slow to be of practical use.

### 7) Reaction catalysts inhibitors

For a long time the influence of cations (inorganic or organic) has been known to be of significant importance for zeolite crystallization. To a much lesser extent, it is valid for anions, too. The main consequences are on: 1) type of zeolites; 2) reaction time; 3) crystal properties: PSD, APS, morphology, habitus and crystal surface characteristics. Except when cations are "templates," the basic understanding of this influence is vague. Salting-out effects and chaotropic properties of anions and cations have been proposed [36]. The lyotropic orders similar to Hofmaster's series were first reported as support for these effects for ZSM-5 zeolite [37].

NMR TECHNIQUES AND APPLICATIONS

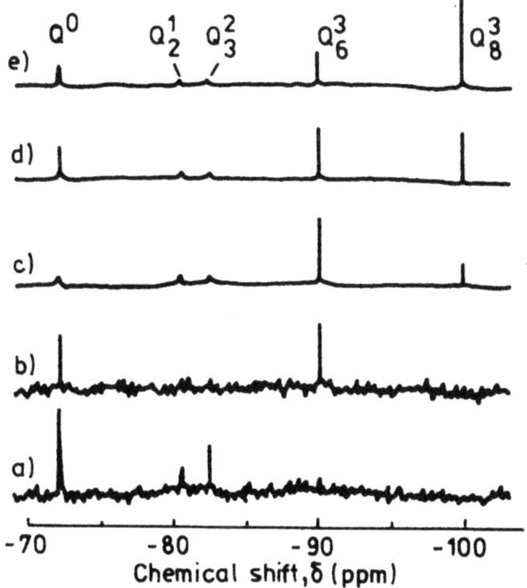

Fig. 13  $^{29}$Si-Spectra of tetramethylammonium silicate as a function of time after heating to 100°C for (a) 5 min, (b) 2 h, (c) 4.5 h, (d) 4.75 h, (e) 45 h [35]

Only a few papers exist about zeolite-A synthesis, although it was found very early that $K^+$ ions can seriously inhibit crystallization and PSD, Fig. 16 [30]. Anion effects ("active anion") have been published [38]. The importance of different silicate anions was

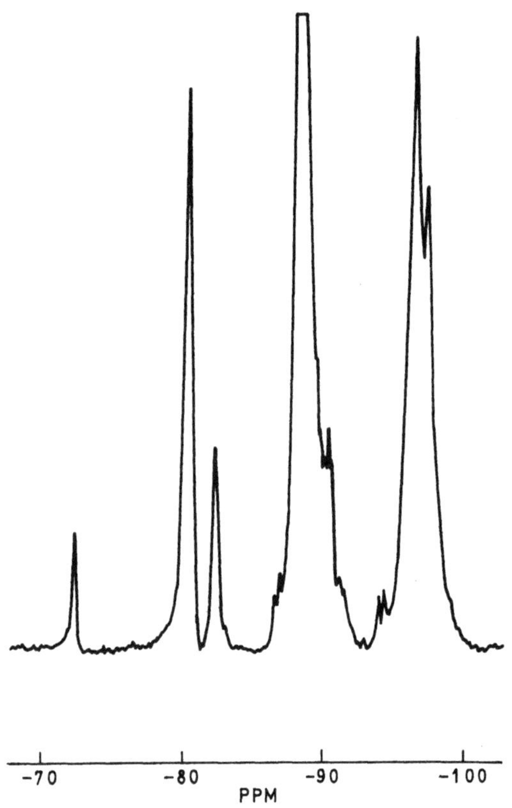

**Fig. 14** $^{29}$Si-NMR spectrum of plant waterglass $SiO_2/Na_2O = 2$

discussed above. It has been stressed (wrongly) that for high zeolite quality only monomeric or dimeric silicate anions are needed [39]. The ready availability of relatively high purity materials (waterglass and aluminum hydrate) was the main reason for the lack of interest in this important area. However, half a million tons of zeolite-A have been produced in alumina plant (Birač; know-how: Institute of General and Physical Chemistry, Belgrade) from green liquor (Bayerish liquor) as a source of the aluminate component. This process, unique at present, becomes important due to very low production cost. The process is strongly affected and directed by different anions, cations and complex organic molecules, the concentration of which can be as high as 20% or more. Depending on the constituents, the process can go completely through the liquid phase (large crystals) instead of through the gel. Many new rules are valid and some of the old ones are suppressed; for example, the influence of $Na_2O/SiO_2$. In this process, even, where the same high IEC and the same PSD are achieved, XRDs are always different, indicating different cation distributions within the zeolite cages, Fig. 17. A complex mechanism controls this kind of process, depending on the nature and quantity of foreign species.

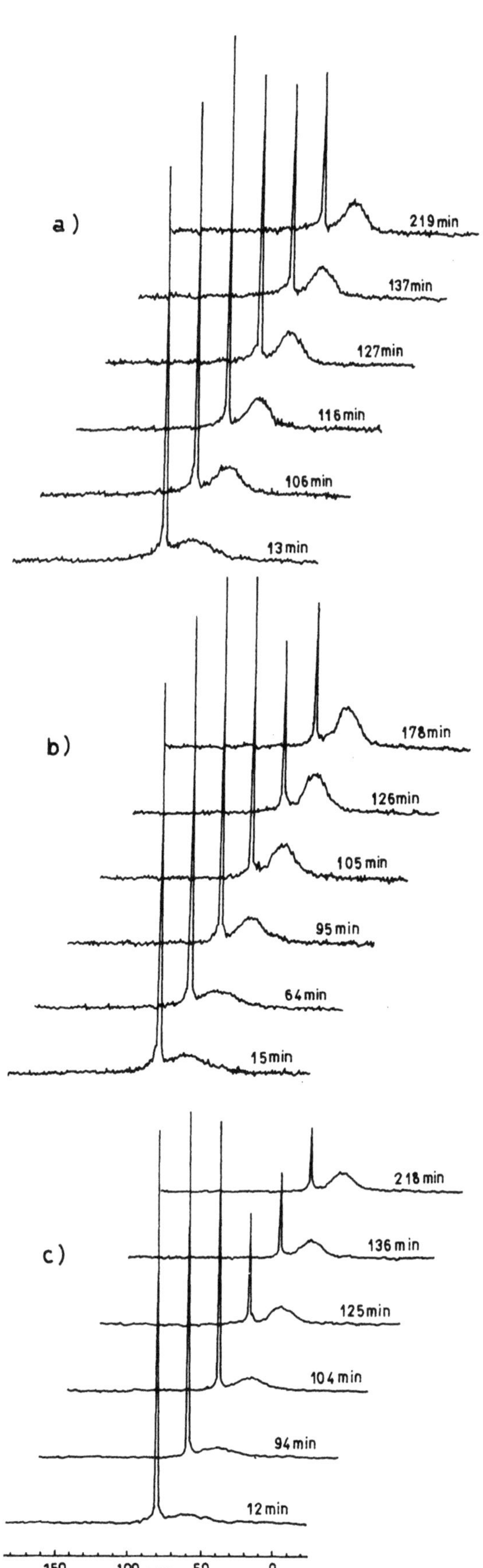

25

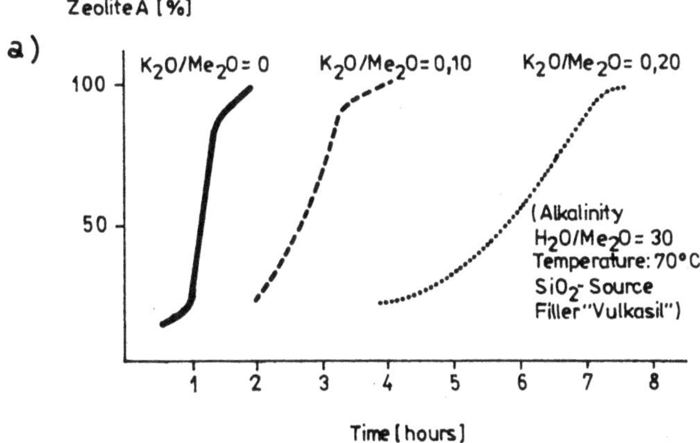



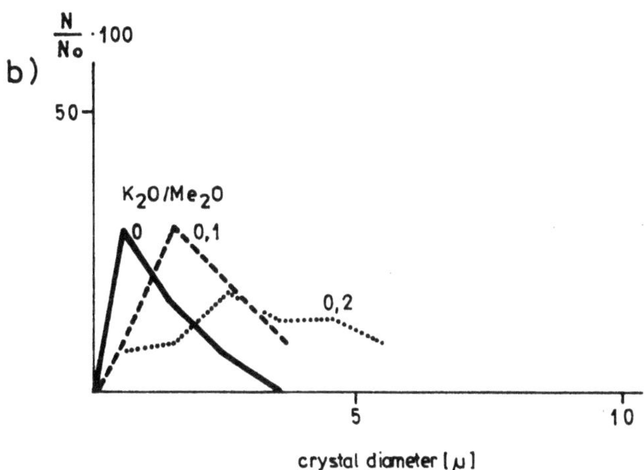



**Fig. 16** Influence of potassium ion on zeolite-A crystallization (a) and on crystal size distribution (b)

Beside strong "templating" and salting-out effects, the competition for the binding sites in gel and sorption of organic compounds both in the gel and in the crystals are most important. Sorption leads to a decrease of surface free energies, affecting induction and nucleation processes. The presence of different cations disturbs gel ⇆ solution cation equilibrium, leading to salt-out or salt-in. In both cases the mechanism can be switched from gel to solution. Sometimes both mechanisms can exist, one beside the other.

8) Seeding
There is a controversy about seeding of zeolite-A. However, at least there is a positive effect of seeding for X and hydrosodalite which always remain in small quantities in reactors, especially after wrong synthesis. It seems that the same is the case with amorphous aluminosilicate left on the reactor walls.

---

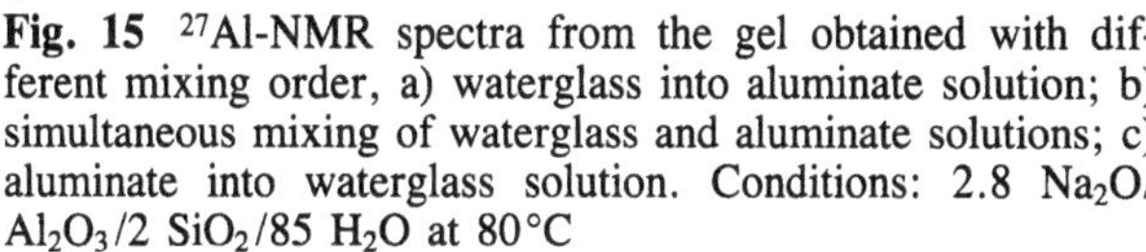

**Fig. 15** ²⁷Al-NMR spectra from the gel obtained with different mixing order, a) waterglass into aluminate solution; b) simultaneous mixing of waterglass and aluminate solutions; c) aluminate into waterglass solution. Conditions: 2.8 Na₂O/ Al₂O₃/2 SiO₂/85 H₂O at 80°C

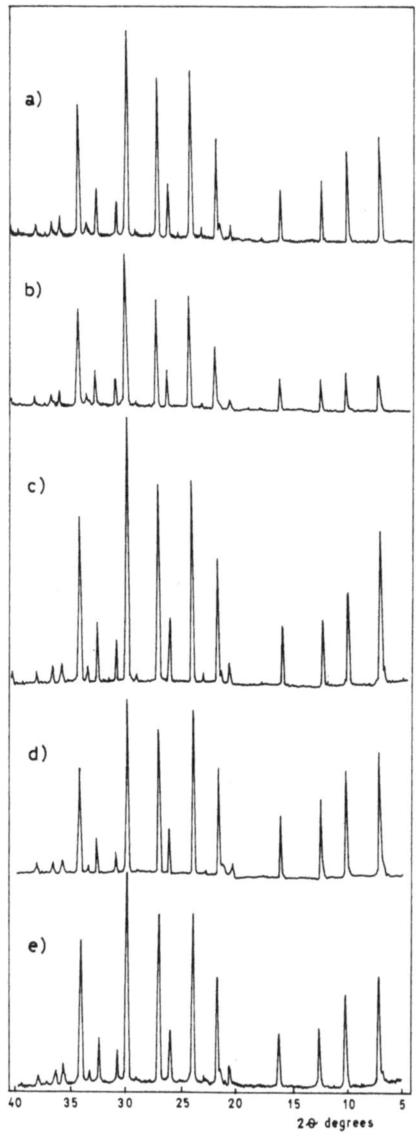



**Fig. 17** XRD of zeolite obtained from different Bayerish liquids. a) and c) Sherwin Plant, USA, courtesy of Reynolds; b) Worsely plant, Australia, courtesy of Reynolds; d) Birač, former Yugoslavia; e) Kaiser, Gramercy Plant, USA, courtesy of Kaiser

---

## Physico-chemical properties of zeolite-A affected by process or equipment design

For industrial synthesis, engineering design, large volume and real time can significantly change laboratory conditions and basic factors. The same is valid for the rest of the process. Environemental or cost restriction may also become prevailing. Optimization of basic and engineering factors are the secret state of art of all producers and only a small part of it can be found in patent literature. Nevertheless, some general factors are common to all hydrothermal processes.

### 1) $Na_2O/Al_2O_3/SiO_2/H_2O$ ratio

These affect gel consistency and environment. Gel is coarse in the beginning, but highly thixotropic at the end of the induction period, which creates a lot of problems for the pumps, stirrers, and valves. To eliminate these effects, increase of $Na_2O$, increase in $H_2O$, and overall low concentration of all components are desirable. The increase in $Na_2O$ is positive for reaction time and PSD/APS, but negative for production cost. Water increase is as much negative for the process as for the cost (longer reaction time, larger PSD and APS, more energy for heating, a small yield).

The ratio $SiO_2/Al_2O_3$ affects the mother liquid. High aluminium content is forbidden due to strong environmental restrictions (less than 1 ppm in EC waste waters). High silicate (to suppress Al) content is undesirable due to silicalization of energy units (evaporator).

### 2) Order and rate of mixing

Simultaneous mixing is time saving and leads to high quality. However, gel consistency causes a lot of problems for equipment. Addition of waterglass to aluminate solution or vice versa is easier for equipment but more costly, and special precautions must be taken to obtain a high quality product, if low quality starting materials are used. Today, the former process is widely accepted.

Mixing rate, which does not seem to be an important factor in laboratories or pilot-plants, is very important in production. Fast mixing is limited by dosing and pumping systems, so if reactors are of large volume (positive for production cost), a significant period of time (20—90 min) is needed. In the case of long mixing time, the first portion will crystallize before the last, leading to wide PSD and reverse transformation of zeolite-A to hydrosodalite can even start (usually seen as oscillation of IEC in a reactor). In the opposite case, short mixing time ($\sim 10$ min) produces a large mass of gel, cousing problems for pumps, stirrers, valves and, consequently, homogeneity.

### 3) Homogeneity

This is the main factor of reactor design which affects all the major properties of zeolite (IEC, PSD, morphology). Coarse gel in the beginning and a thixotropic one at the end both cause serious problems for large reactors. For good homogeneity high shear agitation is recommended [40]. Turbine stirrers, crown gear dissolvers, dispersing pumps, and centrifugal pumps or their combinations can be used.

From all the items presented above, it is clear that small reactors are better, but with high performance costs. Reactors of about 100 $m^3$ with yield of 120 $kg/m^3$ are typical [41, 42]. However, huge reactors of as much 1200 $m^3$ (Birač, former Yugoslavia) are in use, and so are small ones of 25 $m^3$ (Zeolite Mira, Italy).

### 4) Agglomeration

This is an important property for handling and washing. There are two sources of agglomeration processes. The primary (from reactor), and the secondary one, during finishing of already crystallized zeolite-A.

Factors influencing primary agglomeration are:

a) $Na_2O/Al_2O_3/SiO_2/H_2O$ ratio
b) mixing condition
c) homogeneity
d) foreign species (the most important being organics)
e) temperature
f) reactor profile

Because the surfaces of crystalline zeolites are negatively charged, pure crystal aggregation is not possible. Agglomeration of crystals are results of remnant silicate-anions (from solution or gel) and sorbed organic impurities (by electrostatic forces or hydrogen bond) are always involved.

Besides crystal agglomerates, amorphous agglomerates, with a high ability to grow are always present. Different factors influence agglomeration processes in different ways. These are also some of the key "know-how secrets" of all producers.

The secondary agglomeration occurs mainly at two places: filters and dryers. Effects are opposite. On the filters disagglomeration (if washing exists), and in the dryers fast growth of primary agglomerates take place. It is interesting that in both cases water seems to be the main force, although bringing opposite results. For filtration the role of water is to decrease pH and $Na_2O$ which is known in the presence of $OH^-$ to be a good bridge material. In drying, agglomeration through water hydrogen bridges, and not through Si-O-Si bridges, is desirable. Of course, what kind of bridges and what size of agglomerates will be formed depends on drying temperature, drying time, and dryer design.

### 5) Free flowing

This powder characteristic is completely tailored in a spray dryer, depending on the size/form and surface sorbed species of agglomerates only.

### 6) Sedimentation

Today, zeolite suspension has an advantage over powder only if a zeolite plant is directly linked to the detergent producing plant. Because zeolite crystals have a high tendency for sedimentation (making, if allowed, tough artificial sedimentary rocks in reactors and tubes), special precautions are needed; permanent stirring or addition of some suitable organic compounds (e.g., fatty alcohol polyglycol ethers [43]). Low particle size (1—2 μm) and all factors influencing it, seems, at present, to be the only way to decrease this undesirable zeolite crystal characteristic.

## Properties of zeolite-A important for cleaning process

Zeolites, including zeolite-A, have many properties, but only some of them are important for cleaning.

1) Particle size distribution (PSD);
   average particle size (APS)

   Both characteristics are important for two reasons:

   a) rate of ion exchange is linked directly to PSD and APS;
   b) large agglomerates incorporate within the fiber texture; instead of softening the fabric, they make it coarse and powdering.

In Fig. 18a typical PSDs of powders are shwon. The narrow PSD in Fig. 18a, with APS 43 μm, is a typical distribution for good powder quality. In contrast, although average particles are smaller, ~3 μm, the long tail between 6 and 20 μm represents an unacceptable powder, Fig. 18b. Finally, in Fig. 18c, a bimodal example is given. Generally, APS 4—5 μ and narrow PSD: 90% below 10 μm, 96% below 15 and 100% below 20 μm, respresents a typical high quality zeolite powder.

As was discussed above, practically all basic and engineering factors are involved in PSD, while APS is the result of basic factors only. Although several patents deal with good PSD in a crystallizer (for example [44]) to obtain such PSD in the powder as a final product is not easy. For handling of the powder it is important to avoid dusting and improve free flow. For these purposes a typical powder has PSD of agglomerates between 200—500 μm. The binding forces ought to be strong enough to prevent dusting during handling and storage, but weak enough to allow rapid dispersal in solution with PSD 2—20 μm. IF hydrogen bonds are responsible for secondary agglomeration, usually good results can be achieved.

2) Ion exchange capacity (IEC)

   Strong selectivity (binding) of zeolite-A towards divalent cations (calcium) is the basic function of zeolite-A as a builder.
   In simple form, the reaction may be represented by Eq. (4).

$$Na^+Z + M^{2+} \text{ (sol.)} \leftrightarrows M^{2+}Z + 2\,Na^+ \text{ (sol.)} \quad (4)$$

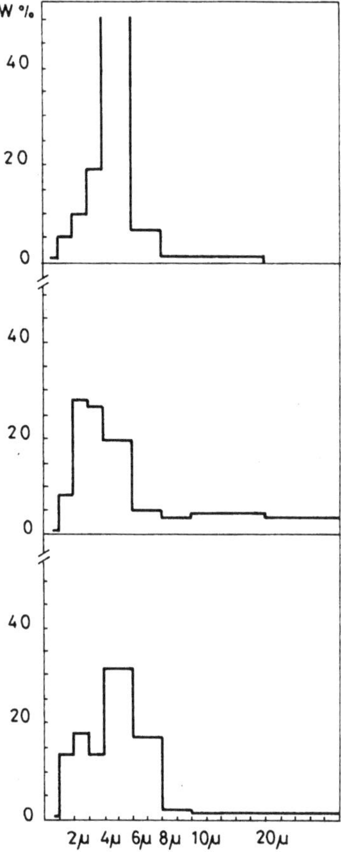


**Fig. 18** PSD from production: a) quality according to manufacturing standard; b) good APS, long tail does not meet standard; c) bimodal PSD due to reactor inhomogeneity

Classical thermodynamics can be used to obtain the equilibrium constant and free energy for this equation. The results for $Ca^{2+}$ and $Mg^{2+}$ (the only important ions in washing) in Table 1 show high affinity of zeolite-A to bind $Ca^{2+}$ and low for $Mg^{2+}$. High affinity for $Ca^{2+}$ is the basic builder characteristic most important for the washing process (see the classical paper from Schwuger and Smolka [45]).

For practical purposes, it is more convenient to deal with the amount or maximum amount, $Q_m$, ("exchange capacity") of exchangeable ions [46].

*$Ca^{2+}$-ion exchange capacity (IEC):*
This is the most important property of zeolite-A as a builder. In pure binary solutions of $Ca^{2+}$ and $Na^+$ it depends only on the concentration of $Na^+$ (the values of $C'_{Na+Cl}$ in solution are realistic because they reflect real washing conditions — all ionic surfactants are present in the form of sodium salts), and $Ca^{2+}$ ions in solution (Fig. 19).

Calcium binding increases with temperature (positive standard enthalpy $\Delta H^\varnothing$ in Table 1 [47, 48].

**Table 1** Thermodynamics and kinetics of ion exchange in zeolite-A

| Exchange [47] | | $T$ (°C) | $\Delta G^{\varnothing}$ | $\Delta H^{\varnothing}$ kJ (g equiv.)$^{-1}$ | $K_{c}$ |
|---|---|---|---|---|---|
| $Na^{+} \rightarrow 1/2\ Ca^{2+} \rightarrow$ | | 25 | −2.68 | 12.2 | 8.66 |
| | | 65 | −4.69 | | 27.8 |
| $Na^{+} \rightarrow 1/2\ Mg^{2+}$ | | 25 | 3.26 | 18.6 | 0.0717 |
| | | 65 | 1.20 | | 0.427 |

| * Diffusion coefficient | | Ca | Ca:Mg 2:1 | Ca:Mg 1:1 | Ca:Mg 1:2 | Mg ref. |
|---|---|---|---|---|---|---|
| $D \times 10^{15}$ at 25°C $(m^2/s)$ $Ca^{2+}$ | Crystal size, (μm) unknown | 1.67 | 1.70 | 1.38 | .62 | .18 [49]** |
| | 4.5 | 10 | | | | |
| | 3.4 | 14 | | | | |
| | 2.1 | 28 | this paper | | | |
| | 1.5 | 47 | | | | |
| | .9 | 44 | | | | |

* Measurement of $D$ depends strongly on the method and conditions used. Error of three orders of magnitude can be expected [50] although $D_{Ca^{2+}}$ were published in the range $10^{-10}$—$10^{-15}$ m²/s. However, when one method is used the results within one paper as a relative parameter are acceptable (e.g., row or column in the above table).

** Crystal size unknown.

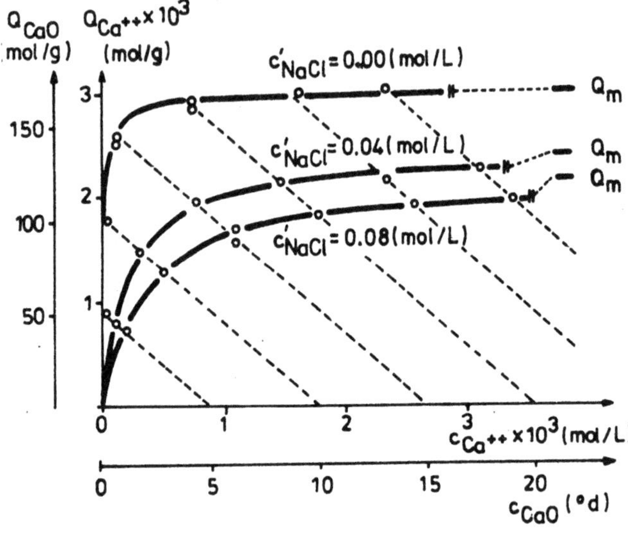


**Fig. 19** Ca$^{2+}$ exchange isotherms of zeolite-A in the presence of NaCl. Temperature, 22 ± 1 °C; time, 1 h. Drawn curves and $Q_m$ values are calculated [46]

From the results of standard free energy $\Delta G^{\varnothing}$ and equilibrium constants $K_c$ presented in Table 1 and in Fig. 19 one manufacturing standard is widely accepted: IEC at 22 ± 1 °C in hard water (30° d ~21 mg/l Ca$^{2+}$) after 10 min ought to be above 160 mg CaO/g of dry zeolite. this value is 95% of Q$_m$ (under these conditions Q$_m$ is 168 mg CaO/g) and 81% of total Na$^+$ in zeolite (Q$_{total}$ 197 mg CaO/g zeolite). Regardless of which of these values production IEC is

referred, it is a very high standard because 10 out of 12 Na$^+$ from each elementary cell have been exchanged.

IEC, being a thermodynamic constant for stoichiometric crystal, should not depend on any factor, basic or engineering. Only kinetic factors (e.g., if 10 min is not enough for equilibrium) could play some limited role. Agglomerates outside of the standard mentioned above or the occurrence of partial reverse transformation to hydrosodalite should be responsible for IEC decrease (reported decrease in IEC with APS 1—2 μ has never firmly confirmed).

*Mg$^{2+}$ exchange:*
The affinity for Mg$^{2+}$ is very low (Table 1). Expressed in practical terms (conditions as for Ca$^{2+}$ IEC), it is between 12 and 17 mg/g zeolite. The explanation for this low affinity seems to be of a kinetic nature. Although the pure ionic radius of bare Mg$^{2+}$ is smaller than that of Ca$^{2+}$, the hydrated one is larger, so Mg$^{2+}$ cannot pass through the channels. To let it pass through, stripping energy is needed. For Mg$^{2+}$ this energy is higher than for Ca$^{2+}$ (Table 1). Diffusion coefficients support this hypothesis [49, 50].

There is a controversy about the desirability of this low Mg$^{2+}$ IEC. For cleaning purposes it is beneficial [51], but in contrast, magnesium silicate precipitation on machine heaters is highly undesirable.

*Simultaneous Ca$^{2+}$ and Mg$^{2+}$ exchange:*
An exchange from pure Ca$^{2+}$ or pure Mg$^{2+}$ solutions is a rare event in practice. Usually, there is a mixture of

**Table 2A** IEC in hard water with 30°dH

| Sample | IEC in 30 dH Ca²⁺ (mg CaO/g) | IEC in 30 dH Mg²⁺ (mg MgO/g) | IEC in 30 dH Ca:Mg = 3:1 | | |
| --- | --- | --- | --- | --- | --- |
| | | | IEC total (mg CaO/g) | Ca²⁺ IEC (mg CaO/g) | Mg²⁺ IEC (mg MgO/g) |
| ZIB-5 | 160 | 17 | 158 | 153 | 4 |
| ZIB-6 | 169 | 35 | 155 | 145 | 10 |

**Table 2B** IEC in hard water with 16.8° dH

| Sample | IEC in 16.8 dH Ca²⁺ (mg CaO/g) | IEC in 16.8 dH Mg²⁺ (mg MgO/g) | IEC in 16.8 dH Ca:Mg = 3:1 | | |
| --- | --- | --- | --- | --- | --- |
| | | | IEC total (mg CaO/g) | Ca²⁺ IEC (mg CaO/g) | Mg²⁺ IEC (mg MgO/g) |
| ZIB-5 | 148 | 14 | 124 | 118 | 4 |
| ZIB-6 | 147 | 29 | 125 | 113 | 9 |

$Ca^{2+}$ and $Mg^{2+}$ ions in different ratios. In mid-European water the average hardness is 16.8°d with $Ca^{2+}/Mg^{2+}$ ratio between 3:1 and 2:1. Under these conditions, $Ca^{2+}$ IEC is about 20% depressed, due to the influence of $Mg^{2+}$, Table 2B. The effect is much less pronounced in hard water (30°d), Table 2A.

### 3) Exchange rate/depletion rate

Fast reaction of $Ca^{2+}$ in solution with anionic surfactants and inorganic components (silicate, carbonate) leads to undesirable non-soluble products in the washing process. The rate of exchange (rate of $Ca^{2+}$ concentration decrease in solution-depletion rate) ought to be competitive with these reactions. A typical high quality manufacturing standard demands, from 3 mM $Ca^{2+}$ solution (16.8°d) the following depletion rate:

| 1 min | less | 1.5 mM $Ca^{2+}$ | at 21 °C |
| 10 min | less | 0.5 mM $Ca^{2+}$ | |
| or | | | |
| 2 min | less | 1.0 mM $Ca^{2+}$ | at 30 °C |
| 10 min | | 0.5 mM $Ca^{2+}$ | |

Zeolite of 4 µm diameter, although matching this specification, due to slow diffusion through framework, is inferior to sodium triphosphate (STP) or $Na_2CO_3$ in $Ca^{2+}$ binding power, Fig. 20. Under real conditions, the diffusion coefficient of $Ca^{2+}$, Table 1, decreases ($Ca^{2+}/Mg^{2+}$ competition), decreasing (10—15%) the depletion rate as well. By decreasing APS, Fig. 21, it is possible to increase the depletion rate, even to surpass the depletion rate of $Na_2CO_3$.

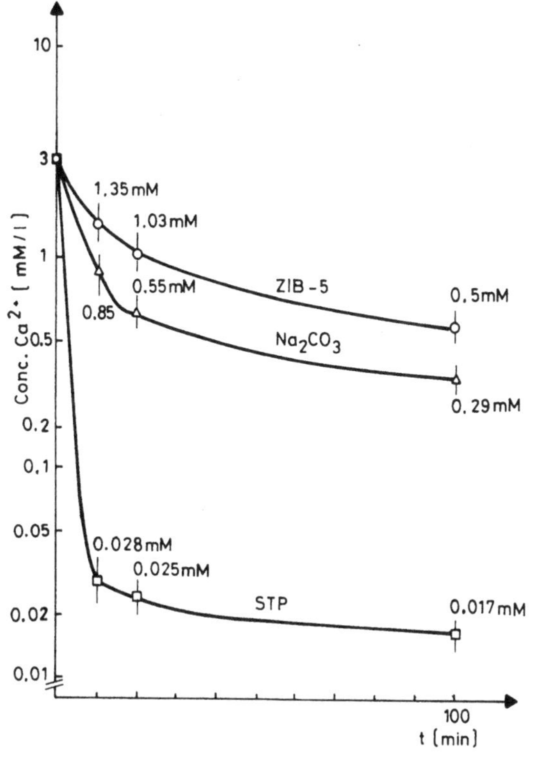


**Fig. 20** Competitive depletion rate between, STP, $Na_2CO_3$ and zeolite-A

### 4) Sublattice metastability

Up to now, the exchange properties of zeolites were regarded as transitions between one energy level in the zeolite and another in solution. However, $Na^+$-positions within zeolite cavities are not uniform. In dehydrated zeolite, 8 $Na^+$ are located in the plane of

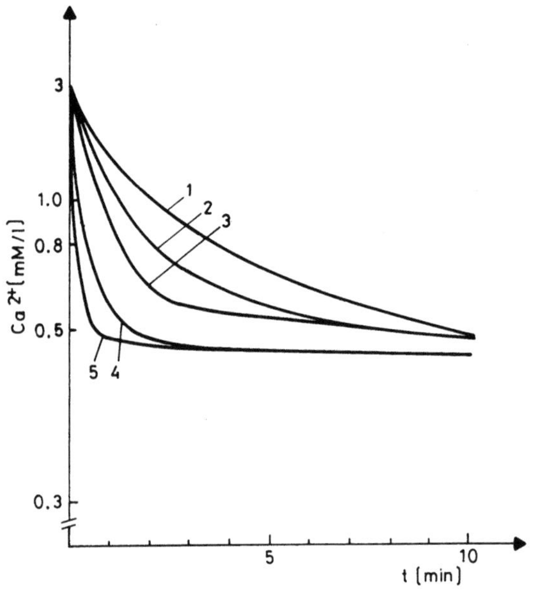

**Fig. 21** Dependence of depletion rate on zeolite-A APS. 1) ZIB-5 5.1 μm; 2) D-90 4.2 μm; 3) D-91 3.3 μm; 4) ZIB-A2 2.2 μm; 5) ZIB-A1 1.2 μm

six-membered rings displaced in a cage, 3 are in eight-membered rings, and 1 in four-membered rings displaced in a cage [52]. Hydration moves all of them into new positions. Water, through hydrogen bonds, is in the proximity of the framework, while all sodium is shifted from the rings. The 8 from six-membered rings are more mobile, while 4 (less shifted from eight- and four-membered rings) increase less in mobility [53]. The redistribution of sodium positions as a consequence of hydration is shown in Fig. 22. Reflection planes 110 and, to some extent, 100, 210, 311 and 321 are the most sensitive to water (for that reason these planes are better omitted for crystallinity determination). The loss of the water complex in the vicinity of six-rings occurs between 120 and 170 °C [25] followed by significant increase of all (except 100) water sensitive reflections. Loss of water from the eight-ring complex, at a temperature between 190° and 220 °C, has a small effect on the 311 and 321 planes only. Analogously, the quilibrium positions of $Ca^{2+}$ after ion exchange are: 3 in a cage hexacoordinated facing six-rings, 3 are four-coordinated, a little displaced from the eight-ring, in sodalite (β-cage) [54]. Besides these equilibrium positions many others exist (e.g., very mobile sites in four rings). If during synthesis sodium ions can be "frozen" in metastable positions

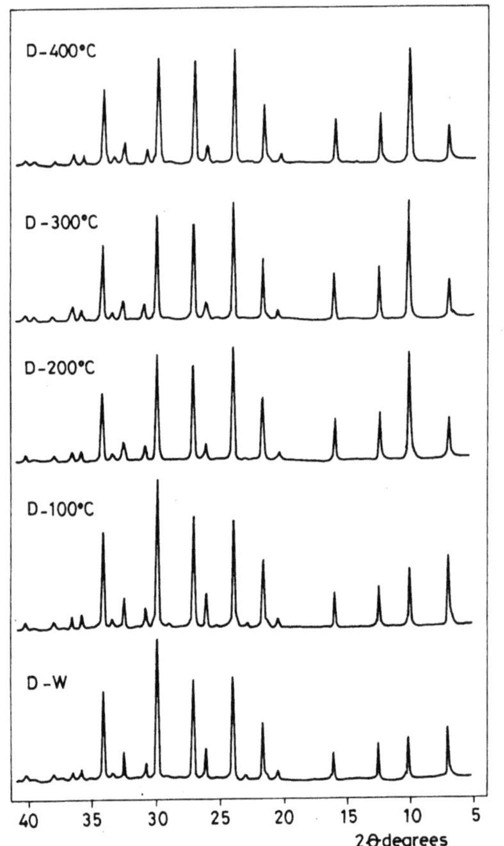

**Fig. 22** XRD dependence on temperature, ZIB-5

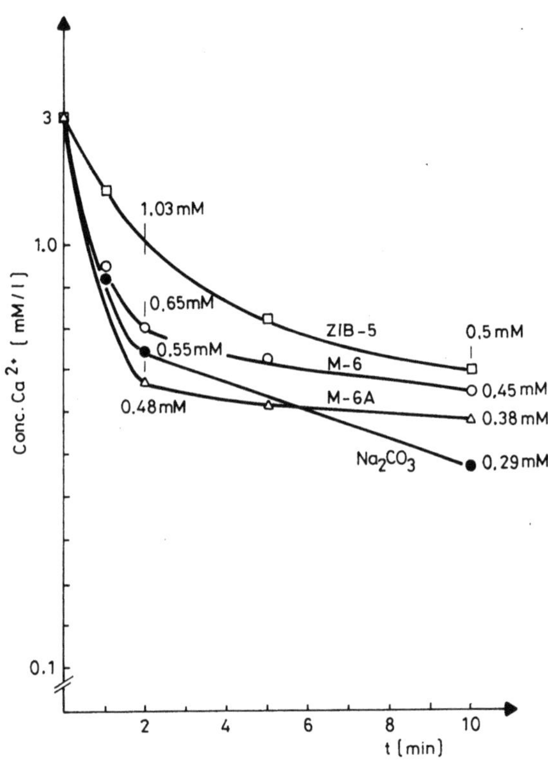

**Fig. 23** Depletion rate of zeolite in metastable cationic state. ZIB-5 5.1 μm, M-6 metastable 5.2 μm M-6A metastable 6.0 μm

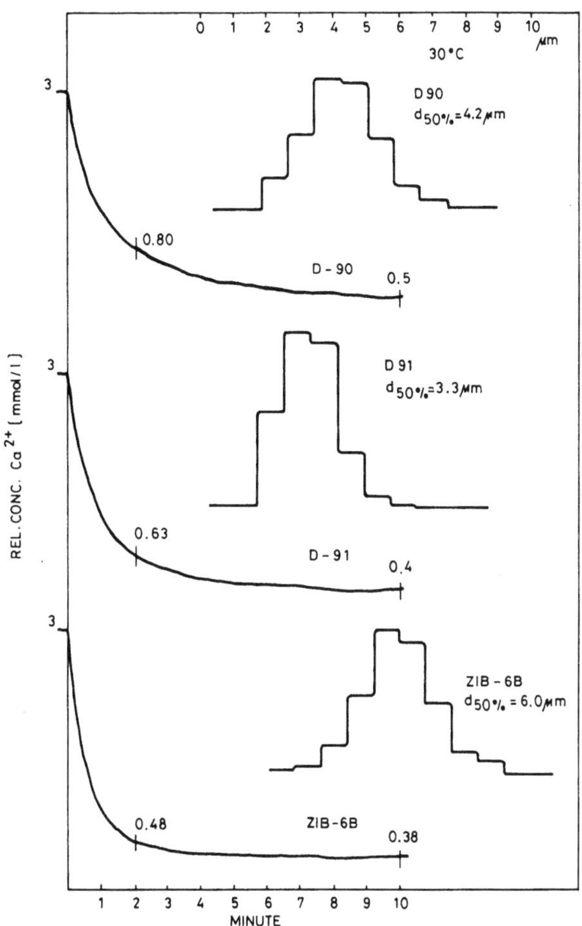



**Fig. 24** Increase of depletion rate with APS decrease (top and middle) and high depletion rate with zeolite in metastable cationic state (bottom)

(for example by a temperature jump) they will have higher IEC and rate of exchange independent of the crystal size. The typical examples are shown in Fig. 23. Although, the crystal size of M-6A is about 6 μm, $Ca^{2+}$ IEC's is 170—186 mg/g zeolite, and depletion rate increased two times (from 1 to 0.48 mM after 2 min). Obtained in this way, the crystals show significant IEC for $Mg^{2+}$, indicating not only new cation distributions, but also more open structures, Table 2 (series ZIB-6).

The mechanism responsible for this structure and behavior is not known. Very probably, the complex of water — aluminosilicate framework and sodium formed at the precursor stage gives rise to a sublattice with more mobile cations. The temperature jump does not allow molecules of water and sodium to settle down into their equilibrium positions. (A process based on a nonequilibrium cation distribution is realized in Zeolite Mira, Italy). The two different approaches, APS decrease or formation of a metastable cation sublattice, are illustrated in Fig. 24. As well as

avoiding the problem of dust in zeolite production (small APS), metabstable sub-lattice (although with APS of 6 μm) is as fast as D-zeolite of 3.2 μm.

## 5) Parameters depending on raw material impurities

The zeolite qualities in detergent standards include: color, amount of iron, heavy metals, and arsenic. All of them depend on impurities in waterglass, caustic soda, and aluminate. When these raw materials are of good quality, waterglass made from sand ($SiO_2$ 99% minimum, $Cr^{3+}$ below 10 ppm) and $Al(OH)_3$ (iron below 200 ppm, gibbsite 0.1% maximum, heavy metals below 5 ppm) zeolite quality standards can be achieved easily. However, when cheap and highly impure raw materials are used, special measures in the process have to be taken. For example, the simplest method is the addition of iron-complexing agents [55]. If synthesis is carried out under conditions such that the precursor gel has high porosity (large pores) impurity co-precipitation in the precursor stage can be avoided. In Birač, the production of ZIB zeolites from Bayerish liquid (containing impurities up to 22%, among them 1000 ppm of iron, 100 ppm of heavy metals) and sand, (96—98% of $SiO_2$) is based on this process. A second example process in Tallin, Estonia is based on sand with only 85—86% of $SiO_2$.

*Color:* Zeolite has to be white powder. The degree of whiteness depends on the method and standard used in the measurement (CIE LAB-Hunter; MgO, $BaSO_4$, CIBA as white standard). Although $BaSO_4$ as a secondary standard [56] is more stable than MgO, the latter, due to its diffuse reflectance, is better for powdered materials. The zeolite manufacturing standard demands the reflectance against MgO to be:

| Minimum 96% at 550 nm wavelength (yellow-green) | Minimum 95% 460 nm (blue) | Minimum 91% 375 nm (blue-purple) |
| --- | --- | --- |

This demand for whiteness is very stringent, especially at 375 nm, because zeolite reflectance decreases in this range, so this is almost an idealized value.

*Iron, heavy metals, arsenic:*
The above discussion is fully relevant to these impurities, because the color originates, mainly, from their presence.
For iron the manufacturing standard demands:

250 ppm as a total
 40 ppm chelatable

Heavy metals (expressed as Pb) below 10 ppm

and arsenic below 3 ppm.

## 6) Flowability

Zeolite powder has the lowest flowability of all detergent powder components. Bridging in the silos and screwers is a common event. An improvement can be obtained by modifying the surface by sorption of different additives (e.g., Silanes [57]), or modifying the surface as early as the induction period, as was discussed above.

## Detergent zeolite feed-back

Zeolite ion exchange properties had been known for a long time before their potential power for usage in the washing process was recognized. However, to be a detergent ingredient zeolite has to be cleaning active and compatible with other formulation components. Therefore, big changes in detergents formulation were necessary.

From these changes, feed-back demands for zeolites properties resulted. Four periods of this inter-influence can be recognized.

### The period 1973—1980

In this period, all major physicochemical parameters in complex surfactant — zeolite systems were obtained ([5] and refs. cited therein):

— ion exchange (selectivity: $Ca^{2+}$, $Mg^{2+}$) and counter ion effects;
— zeolite surfactant interaction. Cationic surfactant sorption;
— dye-stuff sorption (observed);
— zeolite cleaning properties (carrier mechanism, soil removal);
— pigment and perborate stabilization;
— first zeolite formulations;
— increase of incrustation as negative side-effect.

During this period, except for ion-exchange capacity, zeolite properties were accepted as they were.

### The second period (1980—1985)

Withtin this period, detergent production based on partial substitution of STP by zeolite started. It was found that [5, 58]:

— STP and zeolite interaction leads to a decrease in washing powder and to increased incrustation. Optimum for STP/zeolite detergent is 20/20 (in weight %);
— incrustation is as high as 3—5%, as compared to 0.1—1% for detergents without zeolite;
— redeposition effect is the same or a little better with zeolite.

Zeolite was used mostly in slurry forms and more attention was paid to preventing sedimentation than to the zeolite quality. For this reason standards for zeolite quality were simple: IEC around 150 mg CaO/g zeolite, crystallinity 100%, and the crystals to be rounded at the edges to avoid fiber damage during washing.

In this period the physico-chemical basis for cleaning performances was firmly established [5]:

— positive effects of $Mg^{2+}$ and $Na^{+}$ in combination with zeolite-A on detergent performances;
— zeolite electrophoretic mobility (negatively charged surface) decreases with pH decrease. (Practical consequences; positive heterocoagulation and anionic sorption on zeolite during rinsing phase of washing);
— adsorption of dyes and inhibition of pigment transfer;
— zeolite as nuclei for crystallization of sparingly soluble salts.

### The third period (1985—1990)

In this period, the main changes in detergent formulations happened: introduction of cobuilders [59—63], ZERO STP formulations [60, 64] and significant changes in overall detergent composition [65].

The introduction of co-builders (acrylic/maleic acid polymer) was shown to prevent incrustation of $Ca^{2+}$ salt, even if it was not added in stoichiometric quantities ("threshold effect") [62]. This automatically brought possibilities for 100% substitution of STP and involvement of $Na_2CO_3$ as an important pH washing regulator, Fig. 25 [66]. An average amount of these substances in modern detergents is:

Zeolite-A          20—25%
$Na_2CO_3$          10—15%

middle molecular weight polymers of acrylic and maleic acid 3—5%.

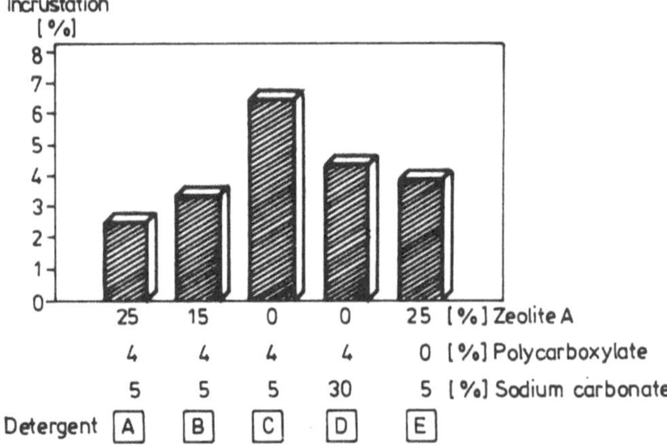


**Fig. 25** Incrustation dependence of zeolite, polycarboxylate, $Na_2CO_3$ ratio [66]

**Fig. 26** Changes in detergent composition and washing effects between (1970—1988) [65]

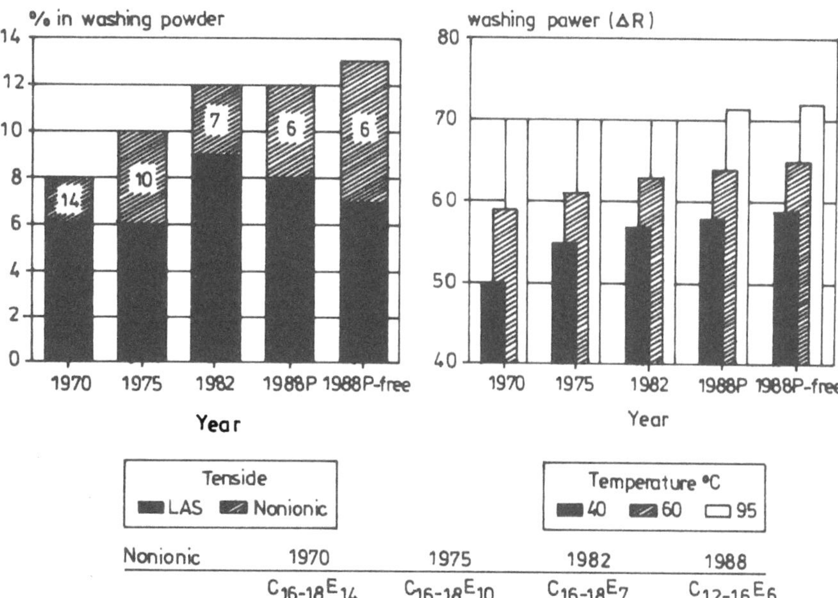


| Nonionic | 1970 | 1975 | 1982 | 1988 |
|---|---|---|---|---|
| | $C_{16-18}E_{14}$ | $C_{16-18}E_{10}$ | $C_{16-18}E_7$ | $C_{12-16}E_6$ |

## Surfactant composition:

Besides the involvement of zeolite, decrease in washing temperature, due to environmental energy saving appeal, was the main factor influencing surfactant changes. The percentages of both anionic and nonionic components increased. Significant increase of nonionics (Fig. 26; [65, 67]) especially with short ethoxy chain became very efficient, for the removal of oily and greasy soil. Silicates became incompatible with zeolites (in slurry causing phase instabilities, haziness or separation) decreased seriously to 1—4%.

Feed-back from these changes brought new demand for zeolite-A quality. Although copolymers decreased zeolite incrustation, the presence of $Na_2CO_3$ in high percentage brings undesirable competition for $Ca^{2+}$ followed by insoluble $CaCO_3$ formation. The increase of IEC and specially of the rate of exchange became more important. In many cases, the straightforward consequence is the need for narrow PSD and lower APS.

The consequence of increased nonionics is their partial or total postaddition. Therefore, zeolite post-addition increased and its sorption characteristics became important. If sorption is low, dusty products will be obtained, and if it is high, lumpy products will be obtained.

Finally, powder zeolite substitutes zeolite suspension almost completely. This brought a flowability standard.

## The fourth period (since 1990)

The main characteristic of this period is the increase of detergent bulk densities from 500—600 g/l up to 700—730 (compact) and in the second stage to 800—1000 g/l (ultra compact). Besides this, detergents without

bleaching agents (compact or normal) and liquid detergents with zeolites are already on the market.

Bulk density of dried detergent slurry is between 300 and 330 g/l in most spray towers. To increase bulk density the part of powder added without spraying (post addition) must increase substantially, too. Usually, water and nonionic are mixed with powders and agglomerate in a fluid bed (shugi-mix or similar) leading to the compact product ($\sim 700$ g/l). Without $Na_2SO_4$ filler (concentrate detergents) more and more powder is added in post addition. For ultracompactness dry neutralization (sulfonic acid (LAS/NaOH) is even necessary.

Dry mixing technology brought to center stage surface properties and powder particle interaction forces. The consequences were: zeolite surface charges and sorption properties became as important as, for example, IEC.

## 1) Sorption properties

This is an old story, starting with early papers by Schwuger [68—70]. High sorption was found, due to negatively charged zeolite surfaces, only for cationic surfactants. The sorption of nonionic is very low if the surface is not first pre-treated with cationic surfactant [70]. However, at the same time, it has been recognized that zeolite surfaces can be changed not only on a purely electrostatic basis. Sprayed zeolite from detergent slurry without any cationic surfactant can adsorb even negatively charged benzo purpurine dyes to some extent. Positively charged methylene bluc is sorbed much better (two orders of magnitude), indicating that the zeolite surface is still negatively charged. Besides charges, hydrophobic or hydrophilic properties and the value of specific surface area (BET) are important. The results for four different zeolites

**Table 3** Zeolite characteristics important for dry mixing

| Type | APS (μm) | BET (m²/g) | IEC (mg CaO/gz) |
|---|---|---|---|
| ZIB-5 | 5.1 | 1.25 | 163 |
| ZIB-5A | 3.7 | 2.60 | 165 |
| M-1 | 4.1 | 2.70 | 172 |
| M-2 | 3.5 | 9.40 | 183 |

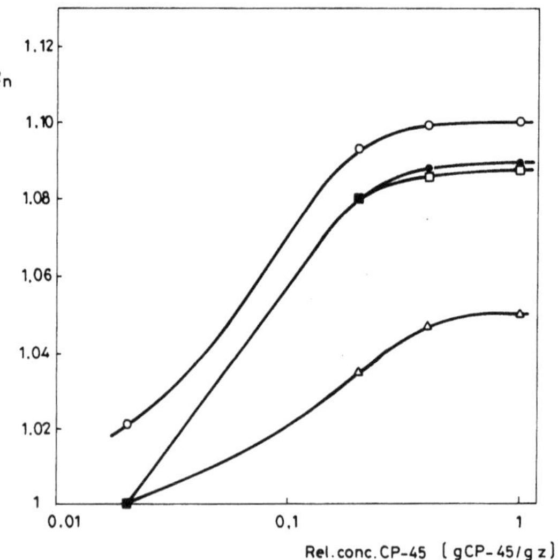

**Fig. 29** Sorption of maleic/acrylic copolymer CP-45 on different zeolites-A, △-ZIB-5, ●-ZIB-5A, □-M1, ○-M2

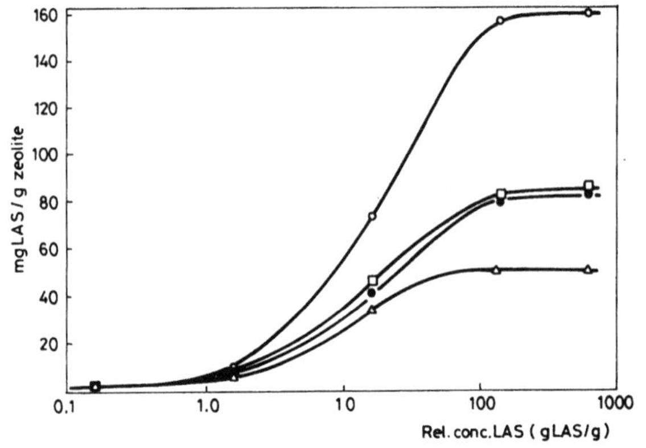

**Fig. 27** Sorption of LAS on different zeolites-A, △-ZIB-5, ●-ZIB-5A, □-M1, ○-M2

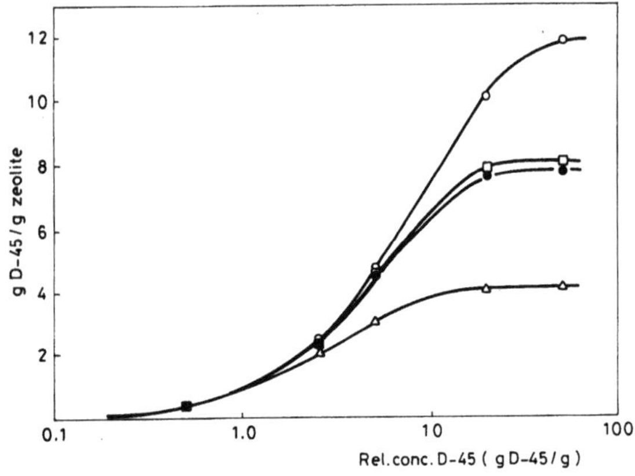

**Fig. 28** Sorption of nonionic Dobanol-45 on different zeolites-A, △-ZIB-5, ●-ZIB-5A, □-M1, ○-M2

(from production) are shown in Table 3, and Figs. 27, 28, and 29. Zeolites ZIB-5 and ZIB-5A are from Birač production and differ in APS and BET surface area only. M-1 and M-2 are from production in Zeolite Mira. M-2 contains cations in metastable states.

From the results presented in Figs. 27—29 the following general characteristics are obvious:

a) Sorption is much higher than was published for nonionics [69] or expected for a negatively charged surface. A possible explanation is that the present results were obtained under a simulation of mixing conditions, and not by sorption from solution [68, 69]. Under mixing conditions the pH is significantly lower (9—7, consequently increasing sorption).

b) BET surface area can be used as general indicator of sorption properties (higher sorption). However, APS reflects surface area only for the products obtained in the same process (ZIB-5 and ZIB-5A). Sample M-1 from a different process has the same BET although APS is higher than ZIB-5A. This is especially valid for zeolites M-2 with cations in metastates, which have a very high surface area (independent of APS close to ZIB-5A).

c) Neutral nonionic (Dobanol-45) has sorption three orders of magnitude higher than negatively charged LAS. However, relative to the others, LAS sorption on the M-2 is much higher than the corresponding increase for nonionic, indicating less negatively charged surface (closer proximity of the cation sublattice to the zeolite surface). In this respect ZIB-5 is the most negatively charged zeolite so far.

d) Large molecules, like maleic/acrylic copolymers (average m.w. 70 000) are sorbed only very slightly (CP-45) or not at all (CP-7). In the case of CP-45 selectivity towards acrylic copolymers was found. In Fig. 29 this affinity is represented through the coefficient $R_n$, referring to the ratio between acrylic and maleic in solution after sorption on the zeolite.

From the presented results for concentrated detergents, zeolite surface area ought to be a new imporant parameter. However, high surface area means a tendency towards bridging and a reduction in free flowing (in the process more lumps can be expected).

On the other hand, good flowability (low sorption, small surface area) leads to more dusting. Combinations of surface charges and BET surface are future items on the zolite manufacturing standard agenda.

High zeolite sorption affinity for the dyes lately found application in compact detergents without bleaching agents. Products are based on the known effect [71] that zeolite in combination with surfactant (dodecylsulphonate) and dye transfer inhibitor (polyvinylpyrrolidon) are such good binders of dye that perborate can be omitted and white and colored textiles can be washed together.

*2) Dry neutralization (ultracompact)*

The reactions of thin films on solid surfaces and the interactions of particles with and without thin films during mixing are extremely complex problems. A few basic facts are known at present:

— zeolite can be efficiently used (in combination with neutralizing agents) for dry neutralization of fatty or LAS acids. The main problem with fatty acids is phase separation, while with LAS it is caking in equipment during neutralization and mixing;
— in both cases the surfaces of zeolites and other powder particles present play a key role;
— important surface properties are: area, charges, water permeability, presorbed species and cation sublattice distribution;
— during neutralization, a significant part of crystal structure is destroyed, if zeolite surface pretreatment was not performed.

From all of these parameters, at present only surface area (BET) is ready for standardization.

*3) Liquid detergents*

Up to now zeolites were used almost exclusively for powders (the best known exception is Unilver's Italian "Biopresto"). Consequently, many standards were derived from powder. However, if liquids are to follow the zero phosphate tendency, the solution for liquid detergents must be zeolites of 2 μ size, to avoid sedimentation [72].

## Catalytic/inhibitory action of zeolite-A

It has been known for a long time that traces of heavy metals (Fe, Mn, Co) can catalyze perborate decomposition. Very early, due to the high affinity of zeolites for heavy metals, it was recognized that zeolite-A can play an inhibitory role for perborate decomposition both during storage and during washing, slowing oxygen release [68]. The tendency towards washing at lower temperature brought the demand for oxygen release at 40°C leading to the development of activators. Ethylene diamine tetraacetic acid salt (tetraacetyl-ethylene diamine, TAED) is widely used in Europe. The influence of zeolite on perborate — TAED system is very peculiar, Table 4. Zeolite blocks oxygen release, not only in pure perborate, but in perborate-activator systems as well. Of the two metal ions tested Fe ions in zeolite act as an inhibitor, but Mn-zeolite acts as a catalyst. If these two are combined, oxygen kinetics can be modified to obtain release during the whole washing process.

## Zeolite-A and the environment

It was shown, before huge production started, that crystalline zeolite-A is non-toxic and environementally benign [73, 74]. However, in the last decade soluble aluminum toxicity has been confirmed ([75] and refs. cited therein) leading to questions about zeolite itself. Maltoni's finding [76] in connection with carcinogenic effects was denied later by himself [77] and is not considered reliable.

**Table 4** Activity ($I_{O_2}$) of zeolite-A on catalytic decomposition of sodium perborate at 40°C

| No. | $BO_3^{3-}$ (mg) | $BO_3^{3-}$ TAED (1.4%) (mg) | Zeolite type | Zeolite (mg) | $I_{O_2}$ ($10^{-8}$ A/cm$^2$) |
|---|---|---|---|---|---|
| 1 | 5 | — | — | — | 23 |
| 3 | 5 | — | NaA | 5 | 5 |
| 4 | 5 | — | $Na_9Fe_{10}A$ | 5 | 3 |
| 5 | 5 | — | $Na_4Mn_4A$ | 5 | 28 |
| 2 | — | 5 | — | 5 | 5 |
| 6 | — | 5 | NaA | 5 | 5 |
| 7 | — | 5 | $Na_9Fe_{10}A$ | 5 | 5 |
| 8 | — | 5 | $Na_4Mn_4A$ | 5 | 25 |

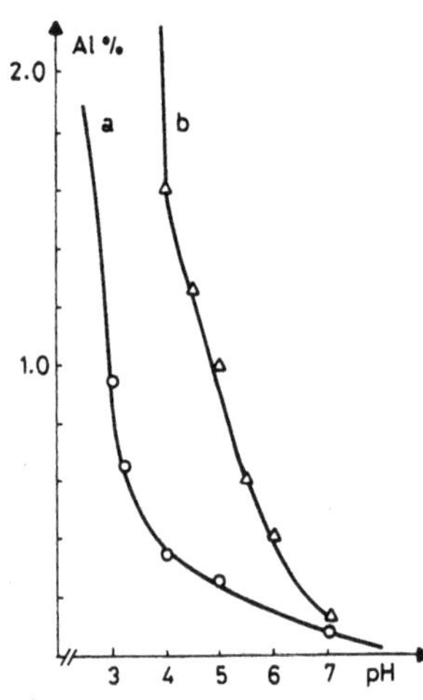


**Fig. 30** Dissolution of 0.1 (curve "a") and 0.01% (curve "b") suspensions of $Na_2Ca_5$ zeolite

## Dissolution of zeolite-5A

Sodium zeolite-A is not stable below pH 7 and dissolves to the $Al^{3+}$ forms and sodium silicate. However, in waste water it is not in the sodium but mainly in $Ca^{2+}$ form (zeolite-5A) which is more resistant to acid attack, Fig. 30. To simulate acid rains, even addition of water with pH = 3 in an amount to produce 0.1% zeolite suspension (pH of final suspension, due to alkalinity of zeolite, was 5.3) after 1 h releases only 0.95% of total Al in zeolites. This brings the value of free $Al^{3+}$ to ~0.2 ppm which is far below EC water standards. A part of the low solubility comes from the high buffering capacity of 0.1% zeolite suspensions, but results are similar with 0.01% of zeolite suspension (curve b in Fig. 30); "Acid rain" with pH = 4 will bring only 0.8 ppm of free aluminum in solution.

### Ecotoxicoloty

If free aluminum from zeolite in running water or in the ocean is not harmful, there is a question about its accumulation in coastal river water or in lakes. Toxicity of zeolite-5A up to 0.5% is presented in Table 5, with very sensitive eco-indicator algae *Chara gymnophylla*.

It is obvious that zeolite in solution affects membrane potential, decreasing it by 30—40% Cyclosis is not touched and mortality on the short term is unaffected. From these results it can be seen that no toxicity results in the short term from zeolite 5A accumulation. However, to see what kind of changes will appear due to membrane potential decrease, a few more years (seasons) are needed.

### Bacterial activity of zeolite-A

The sodium form of zeolite shows low activity or, in some cases, supports bacterial growth (*trichophyton mentagrophytes* and *Aspergillus fumigatus*), Table 6. Conversely, with small amounts of heavy metals, especially $Cu^{2+}$ and $Zn^{2+}$ (not shown) there is a tremendous inhibitory effect, Table 6. Except for *Candida albicans* at high microorganism concentration (MO), in all other cases depending on contact time, 100% sterility can be achieved. This is very important for people working with sewage sludge and farmers using it as a fertilizer. All microorganisms (except *Bacillus subtilis*) presented in Table 6 can cause disorders of different organs (respiratory, digestive or skin) in humans. In this way, zeolite in sewage tanks, through ion exchange, binds heavy metals, sterilizing microorganisms and behaving as a double natural filter.

**Table 5** Effect of zeolite-5A addition to aquarium of algae *Chara Gymnophylla*

| | $\Psi_m$ (mV) Membrane potential | Cyclosis (µm/s) | pH starting | pH fifth day | Mortality (%) |
|---|---|---|---|---|---|
| Control | −142 ± 20 n = 10 | 67 ± 5 n = 10 | 7.4 | 7.6 | 2 (100) |
| 0.1% Zeolite | −102 ± 17 n = 10 | 64 ± 11 n = 16 | 7.6 | 7.8 | 1.8 (80) |
| 0.5% Zeolite | −103 ± 7 n = 11 | 67 ± 8 n = 20 | 7.9 | 7.9 | 1.7 (60) |

**Table 6** Bacterial effect of zeolites (0.1% suspension 20 °C)

| Microorganism MO | Contact time (min) | Biocidal efficiency | | | |
|---|---|---|---|---|---|
| | | NaA zeolite | | $Na_{8.4}Cu_{1.8}$ | |
| | | $10^5$ MO/ml | $10^7$ MO/ml | $10^5$ MO/ml | $10^7$ MO/ml |
| *Escherichia coli* — ATCC 25922 | 15 | 0 | 0 | 0 | 0 |
| | 30 | 0 | 0 | 0 | 0 |
| | 60 | 99 | 50 | 50 | 0 |
| | 120 | 100 | 50 | 75 | 0 |
| *Staphylococcus aureus* ATCC 25913 | 15 | 60 | 0 | 40 | 0 |
| | 30 | 86 | 0 | 50 | 0 |
| | 60 | 93 | 50 | 50 | 0 |
| | 120 | 99 | 50 | 30 | 0 |

**Table 6** (Continued)

| Microorganism MO | Contact time (min) | Biocidal efficiency | | | |
|---|---|---|---|---|---|
| | | NaA zeolite | | $Na_{8.4}Cu_{1.8}$ | |
| | | $10^5$ MO/ml | $10^7$ MO/ml | $10^5$ MO/ml | $10^7$ MO/ml |
| *Bacillus* | 15 | 100 | 0 | 0 | 0 |
| *subtilis* | 30 | 100 | 0 | 0 | 0 |
| | 60 | 100 | 100 | 0 | 0 |
| | 120 | 100 | 100 | 0 | 0 |
| *Trichophyton* | 15 | 100 | 25 | −100 | 0 |
| *menta-* | 30 | 100 | 60 | −140 | 0 |
| *grophytes* | 60 | 100 | 76 | −100 | 0 |
| | 120 | 100 | — | −200 | 0 |
| *Candida* | 15 | 90 | −23 | 0 | 0 |
| *albicans* | 30 | 82 | −13 | 0 | 0 |
| | 60 | 100 | −23 | 50 | 0 |
| | 120 | 100 | −23 | 50 | 0 |
| *Aspergillus* | 15 | 90 | 75 | 6 | 0 |
| *fumigatus* | 30 | 80 | 75 | −80 | 0 |
| | 60 | 80 | 60 | −100 | 0 |
| | 120 | 80 | 70 | −110 | 0 |

# References


1. DE-OS 2412837 (1974) Henkel KGaA
2. DE-OS 2422655 (1974) US Patent 3852211, Procter & Gamble
3. Specker E (March 1993) Zeol Engineering Report, Zeol Engineering and Trading Ltd CH-6340 Baar, Switzerland
4. Milton RM (1959) US Patent, 2,882,243; Ibid 2,882,244
5. Schwuger M, Smulders E (1987) In: Gale Cutler W, Kissa E (eds) Detergency, Theory and Technology, Marcel Dekker, New York, pp 371—439
6. Derleth H, Walter L, Bretz K, Kurs A (1978) Ger Offen Patent, P 2705088.0
7. Ferris A (1977) GB Patent 23 049
8. Ettlinger M, Ferch H (1978) Manuf Chem Aerosol News 49:51—55
9. Yamane I, Nakazawa T (1986) In: Murakami Y, Lijima A, Ward JW (eds) New Developments in Zeolite Science and Technology, Kodansha/Elsevier, Tokyo/Amsterdam, pp 991—1000
10. Costa E, Lucas A, Uguina MA, Ruiz JC (1988) Ind Eng Chem Res 27: 1291—1296
11. Flaningen EM, Breck DW (1960) Abstracts, 137th National Meeting of the American Chemical Society, pp 33M, Cleveland, Ohio
12. Flanigen EM (1973) Adv Chem Ser 121:119—139
13. McNicol BD, Pott GT, Loos KR (1972) J Phys Chem 76:3388—3396
14. Khatami H, Flanigen EM Ref 12 pp 134
15. Barrer RM, Baynhan JW, Bultitude FW, Meier WM (1959) J Chem Soc 195:208
16. Kerr GT (1966) J Phys Chem 70:1047
17. Kerr GT (1968) J Phys Chem 72:1385
18. Barrer RM (1981) Hydrothermal Chemistry of Zeolites, Academic Press, London
19. Guth JL, Coullet P, Wey R (1980) In: Rees LVC (ed) Proc Fifth Int Conf on Zeolites, Heyden, London, pp 30—39
20. Roozeboom F, Robson HE, Chan SSh (1984) In: Ribeiro FR, Rodrigues EA, Rollmann LD, Naccache C (eds) Zeolites: Science and Technology, Martinus Nijhoff, Hague, pp 127—147
21. Bodart P, Nagy JB, Gabelica Z (1986) J Chim Phys 83:777—792
22. Veda S, Koizumi M (1979) Amer Mineral 64:23—32
23. Vučelić D, Juranić N (1976) J Inorg Chem 38:2091—2112
24. Vučelić D (1977) J Chem Physics 66:43—51
25. Vučelić V, Vučelić D, Karaulić D, Šušić M (1973) Thermochim Acta
26. Vučelić D, unpublished
27. Culfaz A, Sand LB (1973) Adv Chem Ser 121:140—151
28. Zhdanov SP, Samulevich NN (1980) In: Rees LV (ed) Proceedings of the Fifth International conference on zeolites, Heyden, London, pp 75
29. Kostinko J (1983) In: Stucky G, Dwyer F (eds) Intrazeolite Chemistry, Am Chem Soc 218:3—19
30. Meise W, Schwochow FE (1973) In: Meier W, Uytterhoeven J (eds) Molecular Sieves, Am Chem Soc 121:169—178
31. Harris RK, Knight CTG, Hull WE (1981) J Am Chem Soc 103:1577—1578
32. Cavill KJ, Masters AF, Filshier KG (1982) Zeolites 2:244—246
33. Engelhardt G, Hoebhel D (1984) J Chem Soc Chem Commun 5:4—516
34. Griffits L, Cundy SC, Plaisted JR (1986) J Chem Soc Dalton Trans 2256—2268
35. Knight CTG, Kirkpatrick RJ, Oldfield E (1986) J Chem Soc Chem Commun 66—67
36. Gabelica Z, Blom N, Derouane EG (1983) Appl Catal 5:227—339
37. Mostowicz R, Berak JM (1985) In: Drzaj B, Hocevar S, Pejovnik S (eds) Zeolites: Synthesis, Structure, Technology and Application, Elsevier, Amsterdam, pp 65—72
38. Lowe B, MacGilp N, Whittam T (1980) In: Rees L (ed) Proceeding of the Fifth International Conference on Zeolites, pp 85—93
39. Wieker W, Fahlke B (1985) In: Drzaj B, Hocevar S, Pejovnik S (eds) Zoelites, Synthesis, Structure, Technology and Application, Elsevier, Amsterdam, p 168

40. Strack H, Roebke W, Knietel D, Parr E (1981) US Patent 4,303,629
41. Vaughan DEW (1988) Chem Eng Prog 84:25—31
42. Roland E (1989) In: Karge HG, Weitkamp J (eds) Zeolites as Catalysts, Sorbents and Detergent Builders, Elsevier, Amsterdam, pp 645—659
43. Diehl M, Bergmann R, Stadtmuller G, Diener S (1987) US Patent 4,671,887
44. Wuest W, Guenther J, Berud P (1981) US Patent 4,271,135
45. Schwuger MJ, Smolka H (1979) Tenside Detergents 16:233—244
46. Schwuger M, Smolka H (1978) Colloid Polym Sci 256:1014—1028
47. Barri S, Rees L (1980) J of Chromatography 201:21—34
48. Barrer R, Rees L, Ward D (1963) Proc Roy Soc London, Ser A 273:180—196
49. Rees L (1989) In: Karge HG, Weitkamp J (eds) Zeolites as Catalysts, Sorbents and Detergent Builders, Elsevier, Amsterdam, pp 661—672
50. Barrer RM (1980) In: Rees L (ed) Proceeding of the Fifth International Conference on Zeolites, Heyden, London, pp 273—290
51. Schwuger M (1982) J Amer Oil Chem Soc 59:265—272
52. Pluth J, Smith J (1980) J Am Chem Soc 102:4704—4708
53. Gramlich V, Meier W (1971) Z Kristallogr 133:134—149
54. Deroy G, Vansant E, Mortier W, Uytterhoeven J (1980) In: Rees LVC (ed) Proceedings of the Fifth International Conference on Zeolites, Heyden, London, pp 214—222
55. Fitton R (1978) Ger Patent Appl 2743597
56. DIN standards (1982) Ger 50933, part a
57. Kittelmann V, Diehl M, Bergmann R, Stadtmuller G (1984) US Patent 4,454,056
58. Krings P, Verbeek H (1981) Tenside Deterg 18:250—262
59. Diehl M, Ettlinger M, Kuzel P (1982) Seifen, Oele, Fette, Wachse 108:451—460
60. Berth P, Berg M, Hachmann (1983) Tenside Deterg 20:276—283
61. Hettche A, Trieselt W, Diessel P (1986) Tenside Deterg 23:12—19
62. Kurzendörfer P, Liphard M, Rybinski W, Schwuger M (1986) Proc 7th International Conference of Zeolites. Tokyo, pp 17—22
63. Andree H, Krings P, Upadek H, Verbeek H (1987) In: Baldwin A (ed) Proceedings Second World Conference on Detergents, Amer Oil Chemists Soc, pp 148—152
64. Berth P, Krings P, Verbeek H (1981) Tenside Deterg 22:169—178
65. Krings P, Voght G (1988) Chimia 42:245—256
66. Upadek H, Krings P (1989) In: Karge H, Weitkamp J (eds) Zeolites as Catalysts, Sorbents and Detergent Builders, Elsevier, Amsterdam, pp 701—709
67. Schmulders E, Krings P (1988) Proc 2nd World Surfactant Congress (CESIO) Vol III, pp 430—448
68. Schwuger M, Smolka H (1976) Colloid Polym Sci 254:1062—1069
69. Schwuger M, Rybinski W, Krings P (1984) Prog Colloid Polymer Sci 69:167—173
70. Schwuger M (1982) J Amer Oil Chem Soc 59:265—272
71. Nüsslein H, Schumann K, Schwuger M (1979) Ber Bunsenges Phys Chem 83:1229—1237
72. Leonhardt W, Sax BM (1989) In: Karge HS, Weitkamp (eds) Zeolites as Catalysts, Sorbent and Detergent Builders, Elsevier, Amsterdam, pp 691—699
73. Gloxhuber Ch, Potokar M, Pitterman W, Wallat S, Bartnik F, Renter H, Braig S (1983) Fd Chem Toxic 21:209—220
74. Llenado R (1984) In: Olson D, Bisio A (eds) Proceedings of the 6th International Zeolite Conference, Butterworth, pp 940—965
75. Macdonald T, Martin B (1988) TIBS 13:15—19
76. Maltoni C (1988) May, Personal Report to Ausident Enimont group, Italy
77. Maltoni C (1988) May, Personal Report to Ausident Enimont group, Italy

Progr Colloid & Polym Sci (1994) 95:39—47
© Steinkopff Verlag 1994

M. J. Rosen

# Predicting synergism in binary mixtures of surfactants

**Abstract** Chemicals in the environment continue to cause concern. One method of decreasing the environmental impact of surfactants is to design new molecules that are more efficient and effective in their interfacial activity. New types of surfactants with two hydrophilic and two or three hydrophobic groups in the molecule, called gemini surfactants, exhibit these characteristics. When properly designed, they can be as much as three orders of magnitude more efficient in reducing the surface tension of aqueous solutions than comparable conventional surfactants. A second method is to use mixtures of known surfactants that exhibit synergism in their interfacial properties. This paper will discuss the second method.

The requirements, in quantitative terms, for the existence of synergy in mixtures of surfactants have been elucidated for several interfacial phenomena. They involve an interaction ($\beta$) parameter whose various values, for mixed monolayer formation at different interfaces and for mixed micelle formation in the solution phase, can be determined from surface, interfacial, or adhesion tension data on the individual interface-active components and at least one mixture of them. This, together with relevant data on the individual components, permits one to predict not only whether the mixture will exhibit synergy, but also the ratio of the two surfactants needed to reach the point of maximum synergism and the value of the investigated property at that point. Since synergy depends so greatly on the value of the $\beta$ parameter, this discussion will also cover the effect on the $\beta$ value of the chemical structures of the interface-active components of the mixture, the nature of the interface, and the molecular environment.

**Key words** Surfactants — Surfactant mixtures — Synergism — Surfactant interactions — $\beta$ parameters

M. J. Rosen
Surfactant Research Institute,
Brooklyn College,
City University of New York,
Brooklyn, New York 11210, USA

## Introduction

Chemicals in the environment continue to cause increasing concern. One method of addressing this problem is to design chemicals that are more efficient and effective at doing the job for which they are now used. Thus, chemists are synthesizing new insecticides and herbicides that are effective at much lower concentrations than products that were used previously. This means smaller amounts of chemical by-products and waste products to be removed from industrial plant effluents. In the field of surfactants, new materials with two hydrophilic groups and two or three hydrophobic groups in the molecule (gemini-type surfactants), when properly designed, are at least one order of magnitude and sometimes more than two orders of magnitude more surface-active than comparable conventional surfactants with the same hydrophile-lipophile balance.

Another approach to increasing the efficiency and effectiveness of surfactants is to use mixtures of them that exhibit synergy. When synergy exists, the mixture of two surfactants exhibits better interfacial properties than either surfactant by itself.

We have done considerable research on both these approaches to increasing the efficiency and effectiveness of surfactants and thus decreasing their environmental impact. We are currently intensely investigating the chemical structure/property relationships in gemini surfactants and for many years now have investigated synergy in mixtures of surfactants.

In this paper, I should like to cover the basic concepts involved in investigating synergy in binary mixtures of surfactants and to discuss some of the surfactant-surfactant molecular interactions that result in synergy. Some of the findings obtained from the study of surfactant-surfactant interactions can be used to suggest surfactant-pollutant interactions that can result in the enhanced removal of these contaminants from industrial wastes.

For a mixture of surfactants to exhibit synergism in its interfacial properties, the different types of interface-active molecules in the mixture must attract each other. Interactions between surfactants are measured by the so-called $\beta$ parameters: $\beta^\sigma$ for interaction in the mixed monolayer at the aqueous solution/air interface; $\beta^m$ for interaction in the mixed micelle in aqueous media; $\beta^\sigma_{LL}$ for interaction in the mixed monolayer at the liquid/liquid interface; $\beta^\sigma_{LS}$ for interaction in the mixed monolayer at the liquid/solid interface. The more negative the value of the $\beta$ parameter, the stronger the attractive interaction. Using the nonideal solution treatment (often called the "regular solution" treatment), $\beta^\sigma$ and $\beta^m$ values can be easily determined from surface tension-concentration curves of the individual surfactants and at least one mixture of them at a fixed molar ratio of the two surfactants; $\beta^\sigma_{LL}$, from interfacial tension-concentration curves; $\beta^\sigma_{LS}$ from surface tension- and contact angle-concentration curves. $\beta$ parameters can also be estimated from a published table containing values for more than 100 surfactant pairs [1].

The theoretical basis for this approach has been demonstrated previously [2]. From the thermodynamics of solutions, it can be shown that:

$$\alpha C_{12} = X_1 f_1 C_1^0 \tag{1}$$

$$(1 - \alpha) C_{12} = (1 - X_1) f_2 C_2^0 , \tag{2}$$

where $\alpha$ is the mole fraction of surfactant 1 in the total surfactant in the solution phase, $C_1^0$, $C_2^0$, and $C_{12}$ are the molar concentrations of individual surfactants 1 and 2 and their mixture at a given value of $\alpha$, respectively, required

to produce the same surface tension value of the solution, at a given temperature. $X_1$ is the mole fraction of surfactant 1 in the total surfactant in the mixed monolayer and $f_1$ and $f_2$ are the activity coefficients of individual surfactants 1 and 2, respectively, in the mixed monolayer.

The relationships

$$f_1 = \exp[\beta^\sigma (1 - X_1)^2] \tag{3}$$

$$f_2 = \exp[\beta^\sigma X_1^2] \tag{4}$$

are used for the activity coefficients. (This is the nonideal or regular solution assumption.) $\beta^\sigma$ is the interaction parameter for mixed monolayer formation.

Equations (1), (2), (3) and (4) yield

$$\frac{X_1^2 \ln(\alpha C_{12}/X_1 C_2^0)}{(1 - X_1)^2 \ln[(1 - \alpha) C_{12}/(1 - X_1) C_2^0]} = 1 \tag{5}$$

$$\beta^\sigma = \frac{\ln(\alpha C_{12}/X_1 C_1^0)}{(1 - X_1)^2} , \tag{6}$$

Equation (5) is solved numerically for $X_1$, which is then substituted in Eq. (6) to yield $\beta^\sigma$. Analogous equations (7), (8),

$$\frac{(X_1^m)^2 \ln(\alpha_1 C_{12}^m/X_1^m C_1^m)}{(1 - X_1^m)^2 \ln[(1 - \alpha_1) C_{12}^m/(1 - X_1^m) C_2^m]} = 1 \tag{7}$$

$$\beta^m = \frac{\ln(\alpha_1 C_{12}^m/X_1^m C_1^m)}{(1 - X_1^m)^2} \tag{8}$$

($X_1^m$ = mol fraction of surfactant 1 in total surfactant in the mixed micelle

$C_1^m$, $C_2^m$, and $C_{12}^m$ = critical micelle concs. of surfactants 1, 2, and their mixture, respectively.

$\beta^m$ = interaction parameter for mixed micelle formation in aqueous solution)

were derived by Rubingh [3] for evaluating $\beta^m$, the interaction paameter for mixed micelle formation, using the critical micelle concentrations (CMC) $C_1^m$, $C_2^m$, and $C_{12}^m$, of surfactants 1, 2, and their mixture at a given value of $\alpha$, respectively, instead of $C_1^0$, $C_2^0$, and $C_{12}$. In this case, Eq. (7) is solved for $X_1^m$, the mole fraction of surfactant 1 in the mixed micelle. The experimental evaluation of $\beta^\sigma$ and $\beta^m$ is shown in Fig. 1.

When interaction is strong $|\beta^\sigma| > 5$, it is necessary, in calculating $\beta^\sigma$ and $X_1$, to take into consideration the changes in the area per surfactant molecule at the interface. In that case, the following equations should be used [4]:

$$\frac{X_1^2}{(1 - X_1)^2}$$

$$\cdot \frac{\ln \dfrac{\alpha C_{12}}{X_1 C_1^0} - \dfrac{\gamma A_1^0}{RT} \left[ 1 - \dfrac{A_{av}}{X_1 A_1^0 + (1 - X_1) A_2^0} \right]}{\ln \dfrac{(1 - \alpha) C_{12}}{(1 - X_1) C_2^0} - \dfrac{\gamma A_2^0}{RT} \left[ 1 - \dfrac{A_{av}}{X_1 A_1^0 + (1 - X_1) A_2^0} \right]}$$

$$= 1 . \tag{9}$$

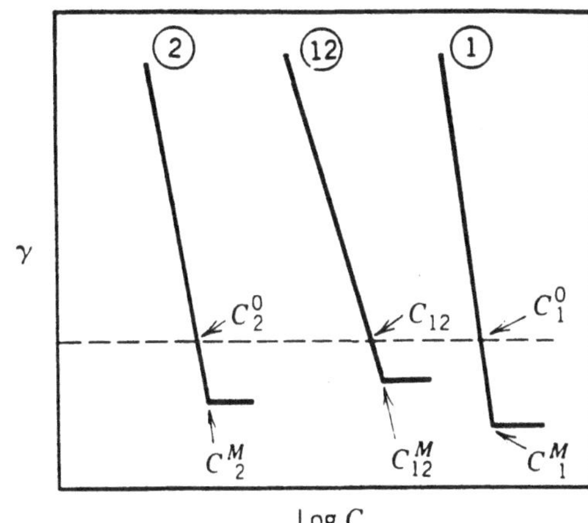



**Fig. 1** Experimental evaluation of $\beta^\sigma$ or $\beta^m$. ① Surfactant 1. ② Surfactant 2. ⑫ Mixture of surfactants 1 and 2 at mole fraction, $\alpha$, in the solution phase

$$\beta^\sigma = \frac{\ln \dfrac{\alpha C_{12}}{X_1 C_1^0} - \dfrac{\gamma A_1^0}{RT}\left[1 - \dfrac{A_{av}}{X_1 A_1^0 + (1 - X_1)A_2^0}\right]}{(1 - X_1)^2} . (10)$$

$A_1^0$, $A_2^0$, $A_{av}$ = surface area of surfactants 1, 2, and their mixture, respectively, at the air/water interface (from the slope of $\gamma$-$\log C$ plots).

We have shown these approaches to be valid for calculating $\beta^\sigma$ by comparing the value of $X_1$, obtained in the fashion, with the value of $X_1$ obtained by an entirely different method, using the Gibbs adsorption equation. Very good agreement was found [2, 4]. Having validated this approach to calculating $\beta^\sigma$, we then proceeded to investigate the phenomenon of synergism quantitatively.

There are various types of synergy in interfacial phenomena. The first type that I wish to discuss is:

## Synergism in surface tension reduction efficiency [5]

The efficiency of surface tension reduction by a surfactant is measured by its solution phase concentration required to produce a given surface tension (reduction). Synergism in this respect is present in a binary mixture of surfactants when a given surface tension (reduction) can be attained at a total mixed surfactant concentration less than that required of either surfactant by itself (Fig. 2).

The conditions for the existence of synergism of this type are obtaining by starting with the basic equations

$$\alpha C_{12} = X_1 f_1 C_1^0 \tag{1}$$

$$(1 - \alpha) C_{12} = (1 - X_1) f_2 C_2^0 \tag{2}$$

and

$$f_1 = \exp[\beta^\sigma (1 - X_1)^2] \tag{3}$$

$$f_2 = \exp[\beta^\sigma X_1^2] . \tag{4}$$

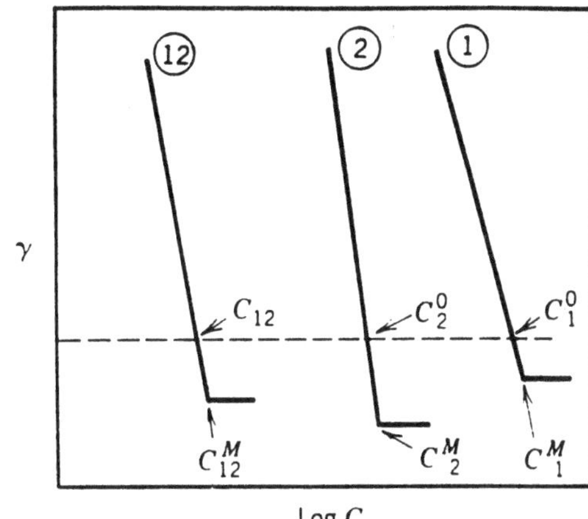



**Fig. 2** Synergism in surface tension reduction efficiency or in mixed micelle formation. ① Surfactant 1. ② Surfactant 2. ⑫ Mixture of surfactants 1 and 2 at mole fraction $\alpha$ in the solution phase. Synergism in surface tension reduction efficiency: $C_{12} < C_1^0$, $C_2^0$. Synergism in mixed micelle formation: $C_{12}^m < C_1^m$, $C_2^m$

From these, we obtain

$$\ln C_{12} - \ln C_1^0 = \ln X_1 - \ln \alpha + \beta^\sigma (1 - X_1)^2 . \tag{11}$$

The condition for synergism in this respect is:

$$C_{12} < C_1^0, C_2^0 .$$

Thus,

$$\ln X_1 - \ln \alpha + \beta^\sigma (1 - X_1)^2 < 0 . \tag{12}$$

When synergism exists, a minimum will be represent in the $C_{12}$ vs. $\alpha$ curve, i.e., $dC_{12}/d\alpha = 0$ at the point of maximum synergism.

From Eqs. (1) and (3),

$$\frac{dC_{12}}{d\alpha} = C_{12}\left[\frac{1}{X_1} \cdot \frac{dX_1}{d\alpha} - \frac{1}{\alpha} - \frac{2\beta^\sigma (1 - X_1)\,dX_1}{d\alpha}\right] .$$

Since $C_{12}$ can never be 0, $dC_{12}/d\alpha = 0$ when

$$\frac{1}{X_1} \cdot \frac{dX_1}{d\alpha} - \frac{1}{\alpha} - \frac{2\beta^\sigma (1 - X_1)\,dX_1}{d\alpha} = 0$$

or

$$\frac{dX_1}{d\alpha} = \frac{X_1}{\alpha[1 - 2\beta^\sigma X_1 (1 - X_1)]} . \tag{13}$$

From Eqs. (1) and (2), we obtain

$$\frac{\alpha}{1 - \alpha} = \frac{X_1 C_1^0 \exp\,]\beta^\sigma (1 - X_1)^2]}{(1 - X_1) C_2^0 \exp[\beta^\sigma X_1^2]} , \tag{14}$$

which becomes, upon rearrangement and conversion to logarithmic form,

$$\ln X_1 - \ln (1 - X_1) + \ln (1 - \alpha)$$
$$- \ln \alpha + \ln C_1^0/C_2^0 + \beta^\sigma (1 - 2X_1) = 0 \; .$$

This becomes

$$\frac{1}{X_1} \cdot \frac{dX_1}{d\alpha} + \frac{1}{(1 - X_1)} \cdot \frac{dX_1}{d\alpha}$$
$$- \frac{1}{(1 - \alpha)} - \frac{1}{\alpha} - \frac{2\beta^\sigma \, dX_1}{d\alpha} = 0$$

or

$$\frac{dX_1}{d\alpha} = \frac{X_1 (1 - X_1)}{\alpha (1 - \alpha) [1 - 2\beta^\sigma X_1 (1 - X_1)]} \; . \tag{15}$$

Combining Eqs. (13) and (15)

$$X_1 = \alpha^* \; , \tag{16}$$

where $\alpha^*$ is the mole fraction of surfactant 1 in the total surfactant in the solution phase at the point of maximum synergism.

That is, when $dC_{12}/d\alpha = 0$, the mole fraction of each surfactant in the mixed monolayer equals its mole fraction in the solution phase. Substituting (16) into the relationship (12) and Eq. (14), we obtain the following conditions for synergism of this type to exist:

1) $\beta^\sigma$ must be negative.
2) $|\beta^\sigma| > |\ln (C_1^0/C_2^0)|$ .

The first condition means that the two surfactants must have an attraction for each other; the second that this attraction must be stronger than the difference between the natural logarithms of the two concentrations needed to produce this same surface tension (reduction) of the solvent. Therefore, to maximize this type of synergism, two materials with strong attraction for each other (large negative $\beta^\sigma$ value) should be used.

On the other hand, if it is not possible to use two surfactants with strong attraction for each other, then the two surfactants should be selected with a $C_1^0$ value approximately equal to $C_2^0$. Under this condition, $\ln C_1^0/C_2^0 \approx 0$, and any negative $\beta^\sigma$ value will produce synergism.

To obtain maximum synergism in surface tension reduction efficiency, the two surfactants should be used at a mole fraction, $\alpha^*$ (of surfactant 1 in the total surfactant in the solution phase), given by the expression

$$\alpha^* = \frac{\ln (C_1^0/C_2^0) + \beta^\sigma}{2\beta^\sigma} \; . \tag{17}$$

This is also the value of $X_1^*$, the mole fraction of surfactant 1 in the total surfactant in the mixed monolayer at the interface at the point of maximum synergism. When $C_1^0 \approx C_2^0$, then $\alpha^*$ is $\approx 0.5$, meaning that equimolar amounts of the two surfactants should be mixed to give maximum synergism. When $C_1^0 \neq C_2^0$, $\alpha^*$ decreases with $\ln (C_1^0/C_2^0)$, meaning that the less surface-active

surfactant (with the larger value of $C^0$) must be used in smaller amounts to get maximum synergism.

When attractive interaction between the two surfactants is strong, and $\beta^\sigma$ is much larger than $\ln (C_1^0/C_2^0)$, then $\alpha^* \approx 0.5$ and equimolar amounts will give maximum synergism, irrespective of the $(C_1^0/C_2^0)$ ratio.

The minimum surfactant in the solution phase required to produce a given surface tension value is given by the expression:

$$C_{12,\text{min}} = C_1^0 \exp \left\{ \beta^\sigma \left[ \frac{\beta^\sigma - \ln (C_1^0/C_2^0)}{2\beta^\sigma} \right]^2 \right\} \; . \tag{18}$$

The larger the difference between the absolute value of $\beta^\sigma$ and $\ln (C_1^0/C_2^0)$ for a given value of $\beta^\sigma$, the smaller will be $C_{12,\text{min}}$. When $C_1^0 = C_2^0$, then $C_{12,\text{min}} = C^0 \exp \beta^\sigma/4$. In this case, the larger the negative value of $\beta^\sigma$, the smaller will be $C_{12,\text{min}}$.

Note that in all these relationships describing synergism, only the value of $\beta^\sigma$ and properties, $C_1^0$ and $C_2^0$ of the individual surfactants must be known. Some data are shown in Table 1.

## Synergism in mixed micelle formation [5]

Synergism in this respect is present when the CMC of the mixture is lower than that of either surfactant of the mixture by itself (Fig. 2). One might seek this type of synergism when one wishes to solubilize a pollutant in a micellar solution and consequently desires micelles to form at as low a surfactant concentration is possible.

By a treatment similar to that described for surface tension reduction efficiency, we have shown [5] that the condition for synergism in this respect are:

1) $\beta^m$ must be negative
2) $|\beta^m| > |\ln (C_1^m/C_2^m)|$ ,

where $\beta^m$ is the interaction parameter for mixed micelle formation, obtained from the CMC values $C_1^m$ and $C_2^m$, of the two individual surfactants and the CMC value, $C_{12}^m$, of at least one mixture of them at a given mole fraction, $\alpha$.

At the point of maximum synergism in mixed micelle formation, the mole fraction, $\alpha^{m,*}$, of surfactnat 1 in the solution phase equals its mole fraction in the mixed micelle, $X^{m*}$, and is given by the relationship:

$$\alpha^{m*} = X^{m*} = \frac{\ln (C_1^m/C_2^m) + \beta^m}{2\beta^m} \; . \tag{13}$$

The CMC at the point of maximum synergism, i.e., the minimum CMC, $C_{12,\text{min}}^m$, that the mixture can attain, is given by the relationship:

**Table 1** Synergism in surface tension reduction efficiency at various interfaces (21) N-octyl-2-pyrrolidinone/$C_{12}SO_3Na$ mixture: 0.1 M NaCl (aq.); 25°C; $\pi = \gamma_0 - \gamma = 32$ mN/m

| Interface | $\alpha_1$ ($\alpha_1^*$) | $C_1^0$ $\times 10^3$ M | $C_2^0$ $\times 10^3$ M | $C_{12,exptl.}$ $\times 10^3$ M | $C_{12,calcd.}$ $\times 10^3$ M |
|---|---|---|---|---|---|
| Air | 0.53 (0.46) | 1.91 | 1.51 | 0.78 | 0.77 |
| Hexadecane | 0.23 (0.30) | 1.32 | 0.55 | 0.46 | 0.47 |
| Teflon | 0.53 (0.45) | 3.31 | 2.59 | 1.58 | 1.55 |
| Parafilm | 0.53 (0.45) | 1.93 | 1.41 | 0.81 | 0.79 |

**Table 2** Synergism in mixed micelle formation in aqueous solution (in the absence or presence of a second liquid phase) N-dodecyl-N-benzyl-N-methylglycine/$C_{12}SO_3Na$ mixture (22) ($H_2O$, 25°C)

| Interface | $\alpha_1$ ($\alpha_1^*$) | $C_1^M$ $\times 10^4$ M | $C_2^M$ $\times 10^4$ M | $C_{12,exptl.}^M$ $\times 10^4$ M | $C_{12,calcd.}^M$ $\times 10^4$ M |
|---|---|---|---|---|---|
| Air | 0.69 (0.80) | 5.5 | 124 | 4.1 | 4.5 |
| n-Heptane | 0.95 (0.93) | 4.4 | 98 | 2.7 | 2.9 |
| n-Hexadecane | 0.95 (0.87) | 5.3 | 106 | 4.3 | 4.6 |

$$C_{12,min}^m = C_1^m \exp\left\{\beta^m \left[\frac{\beta^m - \ln(C_1^m/C_2^m)}{2\beta^m}\right]^2\right\} . \qquad (14)$$

Some data at various interfaces are shown in Table 2.

## Synergism in surface tension reduction effectiveness [6]

This exists when the mixture of surfactants at its CMC reaches a lower surface tension value than that obtained with either surfactant of the mixture at its CMC.

The conditions for this to occur are:

1) $\beta^\sigma - \beta^m$ must be negative.

2) $|\beta^\sigma - \beta^m| > \left|\ln\dfrac{C_1^{0,CMC} \cdot C_2^m}{C_2^{0,CMC} \cdot C_1^m}\right|$,

where $C_1^{0,CMC}$, $C_2^{0,CMC}$ are the molar concentrations of individual surfactants 1 and 2, respectively, required to yield a surface tension value equal to that of any mixture of the two surfactants at its CMC.

The first condition means that, for this type of synergism to exist, the two surfactants must have greater attraction for each other in the mixed monolayer at the interface than in the mixed micelle in the solution phase. This is usually obtained when the two surfactants are oppositely charged and have approximately equal hydrophobic chain lengths. From the second condition, it can be shown [6] that when both surfactants have approximately the same surface tension values at their CMCs, then almost any negative value of $(\beta^\sigma - \beta^m)$ will yield synergism of this type. Some data at various interfaces are shown in Table 3.

Since all of these relationships describing synergism involve the values of the relevant $\beta$ parameters, it is instructive to see how these values are affected by the structures of the two surfactants in the system and the microenvironment (pH, ionic strength) in which they exist. Determination of the $\beta$ values of mixtures containing a variety of structural types of surfactants also provides very useful information on the interactions of various structural groupings at interfaces and in mixed micelles. Such information provides insights into possible interactions of surfactants with pollutants adsorbed at interfaces or solubilized in micelles, and may suggest new methods for enhancing the removal of pollutants from waste streams by surfactants.

Values of the $\beta$-parameter, both for mixed monolayer formation at an interface and for mixed micelle formation in aqueous media, indicate that the interactions between the two different surfactants are dominated by electrostatic forces [1]. Attractive interactions are strongest when each surfactant has a charge of opposite sign. When one surfactant is charged and the other has no net charge but is capable of acquiring a charge of opposite sign (e.g., by accepting or losing a proton), then attractive interaction is somewhat weaker. When the second surfactant has no net charge and is incapable of acquiring a charge of opposite sign, then attractive interaction is even weaker, and when the two surfactants have charges of the same sign, then attractive interaction is weakest. Some data are given in Table 4.

The greater negative $\beta$ values for the $C_{12}SO_4^-Na^+ - C_{12}N^+(CH_3)_3Br^-$ system, compared to those for the $C_8SO_4^-Na^+ - C_8N^+(CH_3)_3Br^-$ systems are due to closer packing of the oppositely-charged head groups in the former system. The strong interaction in the $C_{12}SO_3^-Na^+ - C_{14}N(CH_3)_2O$ system is due to acquisition of a

**Table 3** Synergism in interfacial tension reduction effectiveness at various interfaces (22) N-dodecyl-N-benzyl-N-methylglycine/$C_{12}SO_3Na$ mixture: $H_2O$, 25 °C

| Interface | $\alpha_1$ ($\alpha_1^*$) (mN/m) | $\gamma_1^{CMC}$ (mN/m) | $\gamma_2^{CMC}$ (mN/m) | $\gamma_{12,expt}^{CMC}$ (mN/m) | $\gamma_{1min,calc.}^{CMC}$ |
|---|---|---|---|---|---|
| Air | .028 (.047) | 32.8 | 39.0 | 28 | 29.9 |
| n-Heptane | .051 (.042) | 1.8 | 7.7 | 1.1 | 0.6 |
| n-Hexadecane | .051 (.047) | 3.5 | 9.9 | 1.4 | 1.4 |
| Isooctane | .051 (.042) | 2.0 | 8.3 | 1.0 | 1.3 |
| Heptamethylnonane | .051 (.048) | 2.8 | 9.6 | 1.3 | 1.1 |

**Table 4** Effect of charges of head groups on $\beta$ parameter values (25 °C; pH 5.8—5.9)

| Mixture | $\beta^\sigma$ | $\beta^m$ | Ref. |
|---|---|---|---|
| $C_8SO_4^-Na^+$—$C_8N(CH_3)_3^+Br^-$ ($H_2O$) | −14.2 | −10.2 | 16 |
| $C_{12}SO_4^-Na^+$—$C_{12}N(CH_3)_3^+Br^-$ ($H_2O$) | −27.8 | −25.5 | 16 |
| $C_{12}SO_3^-Na^+$—$C_{14}N(CH_3)_2O$ (0.1 M NaCl) | −10.3 | −7.0 | 13 |
| $C_{12}SO_3^-Na^+$—$C_{12}N^+(Bz)(Me)CH_2COO^-$ ($H_2O$) | −5.7 | −5.0 | 11 |
| $C_{12}SO_3^-Na^+$—$C_{12}N^+H_2(CH_2)_2COO^-$ (0.1 M NaBr) | −4.2 | −1.2 | 11 |
| $C_{12}SO_3^-Na^+$—$C_8$Pyrrolidinone (0.1 M NaBr) | −3.1 | − | 11 |
| $C_{12}SO_3^-Na^+$—$C_{12}(OC_2H_4)_8OH$ (0.1 M NaCl) | −2.6 | −3.1 | 17 |
| $C_{12}SO_3^-Na^+$—LAS (0.1 M NaCl) | −3.0 | −0.3 | 18 |
| $C_{12}SO_3^-Na^+$—$C_7F_{15}COO^-Na^+$ ($H_2O$) | +2.0 | − | 15 |

**Tab. 5** Ionic-zwitterionic interactions (25 °C; pH 5.8—5.9)

| Mixture | $\beta^\sigma$ | $\beta^m$ | Ref. |
|---|---|---|---|
| $C_{12}SO_3^-$—$C_{12}N^+H_2(CH_2)_2COO^-$ (0.1 M NaBr) | −4.2 | −1.2 | 11 |
| $C_{12}Pyr^+Br^-$—$C_{12}N^+H_2(CH_2)_2COO^-$ (0.1 M NaBr) | −4.8 | −3.4 | 11 |
| $C_{12}SO_3^-Na^+$—$C_{12}N^+(Bz)(Me)CH_2COO^-$ ($H_2O$) | −5.7 | −5.0 | 19 |
| $C_{12}N^+(CH_3)_3Br^-$—$C_{12}N^+(Bz)(Me)CH_2COO^-$ ($H_2O$) | −1.3 | −1.3 | 19 |
| $C_{12}SO_3^-Na^+$—$C_{10}N^+(Bz)(Me)(CH_2)_2SO_3^-$ ($H_2O$) | −2.5 | − | 19 |
| $C_{10}SO_4^-Na^+$—$C_{10}S^+(CH_3)_2O^-$ ($H_2O$) | −4.3 | −4.3 | 20 |
| $C_{10}N^+(CH_3)_3Br^-$—$C_{10}S^+(CH_3)_2O^-$ ($H_2O$) | −0.6 | −0.5 | 20 |
| $C_{12}SO_3^-Na^+$—$C_8$ Pyrrolidinone (0.1 M NaCl) | −3.1 | − | 11 |
| $C_{12}Pyr^+Br^-$—$C_8$ Pyrrolidinone (0.1 M NaBr) | −1.6 | − | 11 |

positive charge by the amine oxide as a result of protonation of the oxygen in the presence of an anionic surfactant, even at pH 5.8—5.9 [7]. In similar fashion, the zwitterionics, $C_{12}N^+(Bz)(Me)CH_2COO^-$ and $C_{14}N^+H_2CH_2CH_2COO^-$, accept protons (although to a lesser degree than the amine oxide), in the presence of an anionic surfactant. This is also the reason for the attractive interaction (although weak) in the $C_{12}SO_3^-Na^+$—$C_8N$ system and in the $C_{12}SO_3^-Na^+$—$C_{12}(OC_2H_4)_8OH$ system (see below). It is noteworthy that when hydrocarbon chain and fluorocarbon chain surfactants with charges of the same sign are mixed, there is repulsive reaction, as indicated by the positive value of the $\beta$ parameter. This produces demixing in the monolayer or separate micelle formation [8—10].

The ease with which a zwitterionic surfactant can accept or lose a proton at a fixed pH in the presence of a second, charged surfactant determines the magnitude of its interaction with anionic and cationic surfactants, as measured by the value of the $\beta$ parameter [11]. Some data are given in Table 5. The ampholyte, $C_{12}N^+H_2CH_2CH_2COO^-$, which is capable of both gaining and losing a proton, interacts almost equally with an anionic or a cationic surfactant. On the other hand, the betaine $C_{12}N^+(Bz)(Me)CH_2COO^-$, which is capable of accepting a proton but incapable of losing one, reacts much more strongly with an anionic surfactant than with a cationic. The sulfobetaine, $C_{10}N^+(Bz)(Me)CH_2CH_2SO_3^-$, which is a much weaker base than $C_{12}N^+(Bz)(Me)CH_2COO^-$, interacts more weakly with the anionic $C_{12}SO_3^-Na^+$. The sulfoxide, $C_{10}S(CH_3)_2O$, which, like a

**Tab. 6** Anionic-zwitterionic interactions: effect of pH (25°C)

| Mixture | pH | $\beta^\sigma$ | $\beta^m$ | Ref. |
|---|---|---|---|---|
| $C_{12}SO_3^-Na^+$—$C_{12}N^+(Bz)(Me)CH_2COO^-$ ($H_2O$) | 9.3 | −2.9 | −1.7 | 19 |
| $C_{12}SO_3^-Na^+$—$C_{12}N^+(Bz)(Me)CH_2COO^-$ ($H_2O$) | 6.7 | −4.9 | −4.4 | 19 |
| $C_{12}SO_3^-Na^+$—$C_{12}N^+(Bz)(Me)CH_2COO^-$ ($H_2O$) | 5.0 | −6.9 | −5.4 | 19 |
| $C_{12}SO_3^-Na^+$—$C_{14}N(CH_3)_2O$ (0.1 M NaCl) | 5.8 | −10.3 | −7.0 | 13 |
| $C_{12}SO_3^-Na^+$—$C_{14}N(CH_3)_2O$ (0.1 M NaCl) | 2.9 | −13.5 | − | 13 |
| $C_{12}SO_3^-Na^+$—$C_{12}(OC_2H_4)_8OH$ (0.1 M NaCl) | 10.1 | −2.47 | − | 17 |
| $C_{12}SO_3^-Na^+$—$C_{12}(OC_2H_4)_8OH$ (0.1 M NaCl) | 3.1 | −2.64 | − | 17 |
| $C_{12}SO_3^-Na^+$-LAS (0.1 M NaCl) | 5.8 | −0.3 | −0.3 | 18 |
| $C_{12}SO_3^-Na^+$-LAS (0.1 M NaCl) | 2.9 | −0.3 | −0.3 | 13 |

betaine, is capable only of accepting a proton, shows a similar tendency only to interact significantly with an anionic surfactant. This is true also for the N-alkyl pyrrolidione, $-C_8N$, which, although formally uncharged, has a zwitterionic resonance form similar to that on an amine oxide [11]).

Since protonation plays so significant a role in ionic-zwitterionic interactions, it is to be expected that these interactions will be greatly influenced by the pH of the system. Some data on anionic-zwitterionic interactions are shown in Table 6, together with data on some anionic interactions with non-zwitterionics for comparison.

For the betaine-containing mixtures, the carboxylate group of the betaine has significant basicity, and attractive interaction with an anionic surfactant, which itself has little basicity, increases with decrease in the pH. This reflects the increasing cationic character of the betaine as a result of increased protonation of the carboxylate group. Similar results are obtained with the amine oxide-containing mixtures. On the other hand, the $C_{12}SO_3^-Na^+$-LAS mixtures, in which neither surfactant has significant basicity, show no change in $\beta$ values when the pH is lowered, while the $C_{12}SO_3^-Na^+$-$C_{12}(OC_2H_4)_8OH$ mixtures show only a very small increase in the absolute values of the $\beta^\sigma$ values, reflecting the weak basicity of the polyoxyethylene chain.

Recently, we have been investigating [12] surfactant-surfactant interactions where one of the surfactants has two hydrophilic groups and two or three hydrophobic groups in the molecule (gemini surfactants), specifically, anionic surfactants of this type. We have found that anionic surfactants of this type interact much more strongly at the aqueous solution/air interface than comparable surfactants with only one hydrophilic group and one hydrophobic group in the molecule, with both nonionic and zwitterionic surfactants. This is not unreasonable since, as we have seen, attractive interactions are dominated by electrostatic forces and one would therefore expect a doubly-charged molecule to interact more strongly. However, interaction at the aqueous solution/air interface is stronger also for molecules having two ionic heads and two or three hydrophobic groups than for comparable molecules containing two ionic heads and only one hydrophobic group. Data are shown in Table 7. In all the cases investigated, the absolute values of $\beta^\sigma$ were in the order: two heads and two tails > two heads and one tail > one head and one tail.

On the other hand, in mixed micelle formation, the presence of more than one hydrophobic group in the surfactant molecule tended to inhibit micellization and decreased the interaction in the mixed micelle (decreased the negative value of $\beta^m$). We feel that this is because of the difficulty of incorporating a surfactant with multiple hydrophobic groups into a spherical or cylindrical micelle.

An interesting by-product of this study was some insights into the effect of multiple ether linkages in the molecule on interfacial properties and interactions with other surfactants [13]. The molecules under investigation have the structures shown in Fig. 3. Those with multiple ether linkages in the molecule showed a smaller increase in their interaction with an amine oxide with a decrease in the pH of the system than molecules without multiple ether linkages. Data are shown in Table 8.

In addition, surfactants with multiple ether linkages in the molecule showed much weaker interactions than expected with polyoxyethylenated nonionics. Data are shown in Table 9. We have suggested [13] that the double

**Table 7** Effect of number of hydrophilic and hydrophobic groups on $\beta^\sigma$ and $\beta^m$; 0.1 M NaCl, 25°C (12)

| Mixture | $\beta^\sigma$ | $\beta^m$ |
|---|---|---|
| $C_{10}MADS$—$C_{12}EO_7$ | −1.8 | −0.9 |
| $C_{10}DADS$—$C_{12}EO_7$ | −6.9 | −0.8 |
| $C_{10}MAMS$—$C_{14}N(CH_3)_2O$ | −4.7 | −3.2 |
| $C_{10}MADS$—$C_{14}N(CH_3)_2O$ | −8.3 | −7.5 |
| $C_{10}DADS$—$C_{14}N(CH_3)_2O$ | −9.9 | −2.4 |

$C_{10}MADS$ — monodecyl diphenylether disulfonate;
$C_{10}DADS$ — didecyl diphenylether disulfonate;
$C_{10}MAMS$ — monodecyl diphenylether monosulfonate.

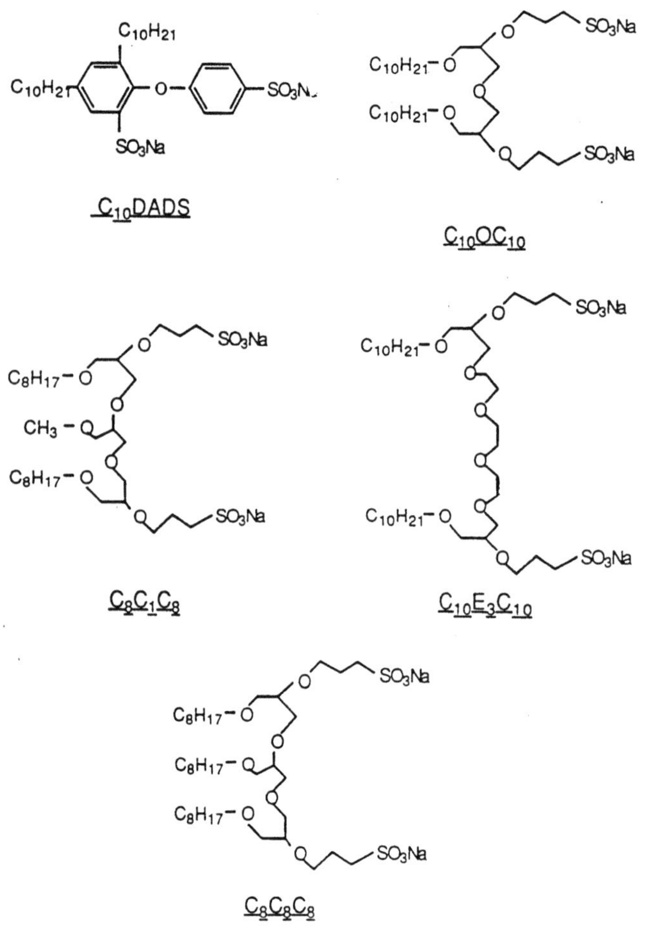

Fig. 3 Structures of gemini surfactants investigated

**Table 8** Effect of pH on molecular interactions of some geminis at the a/w interface; 0.1 M NaCl, 25 °C (13)

| Mixture | pH | $\beta^\sigma$ |
|---|---|---|
| $C_{10}DADS-C_{14}N(CH_3)_2O$ | 6.0 | −7.3 |
| $C_{10}DADS-C_{14}N(CH_3)_2O$ | 2.9 | −9.2 |
| $C_8C_1C_8-C_{14}N(CH_3)_2O$ | 5.8 | −6.8 |
| $C_8C_1C_8-C_{14}N(CH_3)_2O$ | 2.9 | −7.8 |
| $C_8C_8C_8-C_{14}N(CH_3)_2O$ | 5.8 | −4.9 |
| $C_8C_8C_8-C_{14}N(CH_3)_2O$ | 2.9 | −5.5 |

negative charge in these molecules increases the tendency of some of the oxygen atoms in the molecule to acquire a proton, thereby decreasing the net negative charge of the anionic surfactant. With decrease in pH of the system, this protonation is enhanced. Consequently, these molecules interact less strongly than expected with amine

**Table 9** Molecular interactions of some geminis with polyoxyethylenated nonionics; pH = 5.8, 0.1 M NaCl, 25 °C (13)

| Mixture | $\beta^\sigma$ | $\beta^m$ |
|---|---|---|
| $C_{10}DADS-C_{12}EO_7$ | −5.9 | −0.8 |
| $C_8C_1C_8-C_{12}EO_7$ | −1.5 | −0.2 |
| $C_8C_8C_8-C_{12}EO_7$ | −3.2 | +0.7 |
| $C_{10}OC_{10}-C_{12}EO_8$ | −1.5 | −0.6 |
| $C_{10}E_3C_{10}-C_{12}EO_8$ | −1.6 | −0.4 |
| $C_{12}SO_3Na-C_{12}EO_8$ | −2.6 | −3.1 |

**Table 10** Anionic-anionic molecular interactions: 0.1 M NaCl, 25 °C (13)

| Mixture | pH | $\beta^\sigma$ | $\beta^m$ |
|---|---|---|---|
| $C_8C_1C_8-C_{12}SO_3Na$ | 5.9 | −0.7 | +0.7 |
| $C_8C_1C_8-C_{12}SO_3Na$ | 2.9 | −2.7 | −0.9 |
| $LAS-C_{12}SO_3Na$ | 5.8 | −0.3 | −0.3 |
| $LAS-C_{12}SO_3Na$ | 2.9 | −0.3 | −0.3 |

oxides at decreased pH. Even at pH 5.8—5.9, there is sufficient protonation of the ether oxygens to decrease the molecular interaction with polyoxyethylenated nonionics.

Finally, in contrast to the very weak interaction between two anionics without multiple ether linkages in their molecules, irrespective of pH, a mixture of an anionic surfactant of the gemini type with multiple ether linkages in the molecule and an anionic surfactant without multiple ether linkages shows an increase in attractive interaction between the two different types of surfactants when the pH of the system is decreased. Data are shown in Table 10. We feel that this protonation of some of the ether oxygens may explain why anionic surfactants containing a few oxyethylene groups adjacent to the ionic group are considerably more surface-active (lower CMC, larger $pC_{20}$-log of the concentration of surfactant to reduce the surface tension of the solvent by 20 mN/m-values) than the corresponding surfactants without those oxyethylene groups [14], in spite of the water-solubilizing character of the oxyethylene group. It may also explain why anionics containing a few oxyethylene groups adjacent to the ionic group interact synergistically with other anionics containing no oxyethylene groups in themolecule [15].

## References

1. Rosen MJ (1989) Surfactants and Interfacial Phenomena. 2nd edition, John Wiley and Sons, New York, pp 398—402

2. Rosen MJ, Hua XY (1982) J Colloid Interface Sci 86:164

3. Rubingh DN (1979) Solution Chemistry of Surfactants, Mittal KL (ed) Vol 1, Plenum, new York, pp 337—354

4. Gu B, Rosen MJ (1989) J Colloid Interface Sci 129:537
5. Hua XY, Rosen MJ (1982) J Colloid Interface Sci 90:212
6. Hua XY, Rosen MJ (1988) J Colloid Interface Sci 125:730
7. Rosen MJ, Friedman D, Gross M (1964) J Phys Chem 68:3219
8. Zhao G-X, Zhu B-Y, Zhou Y-P, Shi L (1984) Acta Chimica Sinica 42:416
9. Carlfors J, Stilbs P (1984) J Phys Chem 88:4410
10. Zhao G-X, Zhu BY (1986) In: Scamehorn JF (ed) Phenomena in Mixed Surfactant Systems. ACS Symp Series 311, Amer Chem Soc, Washington DC, pp 184—198
11. Rosen MJ (1991) Langmuir 7:885
12. Rosen MJ, Zhu ZH, Gao T (1993) J Colloid Interface Sci 157:254
13. Rosen MJ, Gao T, Nakatsuji Y, Maxuyama A, Colloids and Surfaces, in press
14. Dahanayake M, Cohen AW, Rosen MJ (1986) J Phys Chem 90:2413
15. Schwuger MJ, In: Rosen MJ (ed) Structure/Performance Relationships in Surfactants. ACS Symp Series 253, Amer Chem Soc, Washington DC, pp 1—26
16. Zhu BY, Rosen MJ (1984) J Colloid Interface Sci 99:435
17. Rosen MJ, Zaho F (1983) J Colloid Interface Sci 95:443
18. Rosen MJ, Zhu ZH (1989) J Colloid Interface Sci 133:473
19. Rosen MJ, Zhu BY (1984) J Colloid Interface Sci 99:427
20. Zhu D, Zhao G-X (1989) Acta Phys Chem Sin 4:129
21. Rosen MJ, Gu B, Murphy DS, Zhu ZH (1989) J Colloid Interface Sci 129:468
22. Rosen MJ, Murphy DS (1989) J Colloid Interface Sci 129:208

Progr Colloid & Polym Sci (1994) 95:48—60
© Steinkopff Verlag 1994

H. W. Dürbeck
E. Klumpp
J. D. Schladot
M. J. Schwuger

# Environmental specimen bank of the Federal Republic of Germany — Significance of surfactants

J. D. Schladot (✉) · H. W. Dürbeck
E. Klumpp · M. J. Schwuger
Institute of Applied Physical Chemistry,
Research Centre Jülich (KFA),
P.O. Box 1913,
52425 Jülich, FRG

**Abstract** Environmental contaminants in water, soil, and air, as well as changes in their concentration with respect to space and time may be quite effectively detected by the analysis of appropriate indicators (biomatrices from different levels of the food chain, sediments, sludges, dusts) which accumulate these chemicals by several orders of magnitude.

However, regular monitoring of the environment should not be restricted to presently known substances. The suitable storage of representative indicator specimens allows also the retrospective analysis of chemicals which are not detectable at present or which have not been regarded as environmental pollutants so far.

Therefore, the German Environmental Specimen Bank (ESB) of the Federal Government was established in 1985 with the purpose of storing representative samples from the terrestrial and aquatic environment as well as from human beings for future decades without any change in chemical composition.

In spite of its relatively limited time of operation, the ESB has already obtained a variety of promising results which support not only the success of legislative regulations (introduction of unleaded fuels, ban of pentachlorophenol), but demonstrate also the decrease of pollutants in rivers due to reduced industrial or municipal discharges. Moreover, the effectivity of new technologies with respect to environmental protection may be traced back by a specific and retrospective characterization of suitable indicator samples.

In addition to its routine program, the Institute of Applied Physical Chemistry of the Research Centre Jülich (KFA) is performing basic studies on the speciation of selected elements such as arsenic, mercury, and tin, and the determination of new compounds. In this respect, surfactants play an important role because they can influence the immobilization or remobilization of pollutants in soils or sediments. The mobility of other chemicals in such matrices can be estimated, if surfactant concentrations are known. With that, a prediction of possible contaminations of ground or surface waters is possible. Selected examples will be discussed in detail.

**Key words** bioaccumulation — biomonitoring — environmental specimen bank — mobilization/immobilization of pollutants — retrospective analysis

## Introduction

One of the most urgent tasks of the environmental protection policy consists in the regular monitoring of pollutants in water, soil, and air, as well as in various stages of important food chains (plants, animals), finally leading to man. However, this monitoring should not be restricted to the determination of stationary pollutant loads. On the contrary, it requires, in particular, information about their time- and space-dependent behavior under natural environmental conditions. Therefore, the following questions should be answered as a first priority:

— Where do these materials persist and where do they possibly accumulate?
— What chemical form are they present in?
— How mobile are they in the environment?
— Why are they mobilized or immobilized?
— What short- or long-time effect do they have on man and the environment?
— When and how do new pollutants appear in the environment?
— What new substances are environmental chemicals converted into and in what time?
— Do toxic subtances — possibly — result during this process and how stable are they?

With the knowledge and methods previously available, possible influences and impacts of environmental chemicals produced by man and intentionally or unintentionally released into the environment on natural or semi-natural ecosystems, and also on the basic necessities of life and health of mankind can only be inadequately forecast or their potential dangers assessed [1].

According to the "European Inventory of Existing Commerical Substances" **EINECS** (after GDCh/BUA) [2], there are currently approx. 100 000 different chemical substances whose behavior and action in the environment are still largely unknown [3].

It is therefore all the more important that even the smallest changes in the environment and in the various trophic stages and food chains should be detected by specific observation and monitoring. Their origins have to be elucidated and if necessary their further development should be halted. Corresponding scientific and technical activities place particular emphasis on the protection of mankind and the environment against anthropogenic and geogenic pollutants, as well as their systematic and continuous detection in soil, water, air and selected biological specimens.

On the basis of their specific accumulation potential for certain pollutants, biological samples have the advantage that changes in the local or regional pollution situation can be recognized much more easily. They thus have a special indicator function in the determination of pollutant trends or the early recognition of new chemicals in the environment (see Fig. 1).

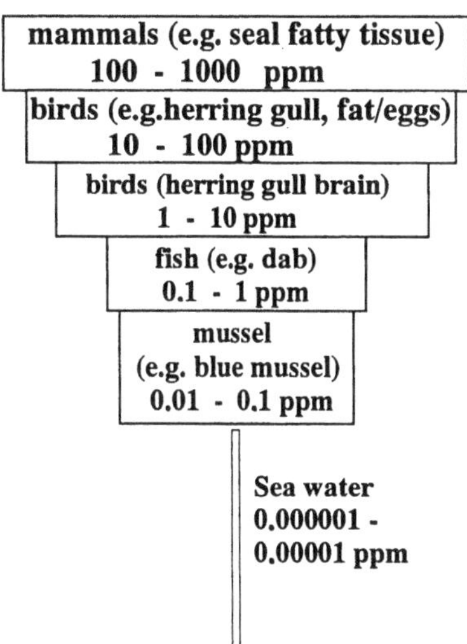


**Fig. 1** Accumulation of polychlorinated biphenyls (PCB) in various trophic stages of the marine environment

## Tasks of the environmental specimen bank

The characterization and evaluation of environmental and human samples — in their **actual state**, but also in their development over time — creates important prerequisites within the framework of a precautionary policy for the following:

— recognition of impending undesirable developments;
— estimation of nature and extent of undesirable developments already in evidence and their consequences;
— obtaining insights to set priorities for political activities, and
— compiling basic principles for the precautionary policy of the Federal Government in the field of nature conservation and environmental protection and also for human health.

As early as 1973, these insights led to the development of a strategic concept of providing the necessary scientific basis for realizing this precautionary policy by establishing an environmental specimen bank. Special guidelines (Standard Operation Procedures, SOP) were developed for all the operating steps necessary for implementation including the analytical work [4].

The long-term usefulness of the stored samples, particularly for retrospective analysis, is ensured by the fact that all materials are deep frozen at temperatures below −150 °C immediately after collection. In this way, any possible changes in chemical composition are completely prevented or reduced to a minimum [5, 6]. Futher sample

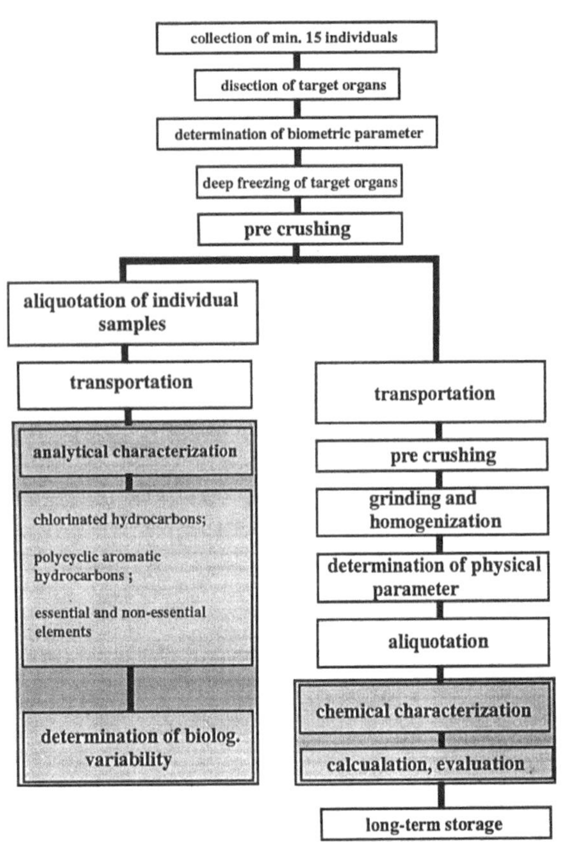



**Fig. 2** Flow chart of the preparation of environmental specimens

treatment (processing, homogenization and aliquotization into standardized sub samples) (see Fig. 2) is similarly carried out in the above-mentioned temperature range, and strict compliance with these conditions is monitored by continuous controls.

After preliminary analytical studies in 1976—1978, the *logistic* and *technical* prerequisites for the feasibility of an environmental specimen bank were thoroughly investigated between 1979 and 1984 in the pilot project "Environmental Specimen Bank" financed by the BMFT (Federal Ministry for Research and Technology) [7].

## Participating institutions

Since January 1, 1985, the Environmental Specimen Bank has been established as a permanent institution under the responsibility of the BMI (later the *BMU, Federal Ministry for the Environment, Nature Conservation and Reactor Safety*). The *Federal Environmental Agency (UBA)* is responsible for co-ordination. Two specimen banks are subsumed under the general heading of the *German Environmental Specimen Bank:*

— the *Specimen Bank for Environmental Specimens* — at the Institute of Applied Physical Chemistry of the Research Centre Jülich (KFA), and
— the *Specimen Bank for Human Organ Specimens* — at the Institute of Pharmacology and Toxicology of the Universiy of Münster.

These institutes also take part in the specimen characterization by analyzing heavy metals, metalloids, and essential elements in environmental samples at the Research Centre Jülich, whereas the corresponding activities at Münster comprise both the inorganic and also the organic (primarily chlorinated hydrocarbons) analysis.

Furhtermore, the work of the Environmental Specimen Bank in the field of ecology and in the selection of representative areas and specimen species is supported and guided by

— *The Institute of Biogeography at the University of the Saarland.*
— The *Institute of Ecological Chemistry at the GSF Neuherberg,* and
— *the Biochemical Institute for Environmental Carcinogens, Grosshansdorf*

are responsible for the analysis of organochlorine compounds and polycyclic aromatic hydrocarbons (PAH) in the stored terrestrial and aquatic environmental samples (see Fig. 3).

ENVIRONMENTAL SPECIMEN BANK
of the
FEDERAL GERMAN GOVERNMENT

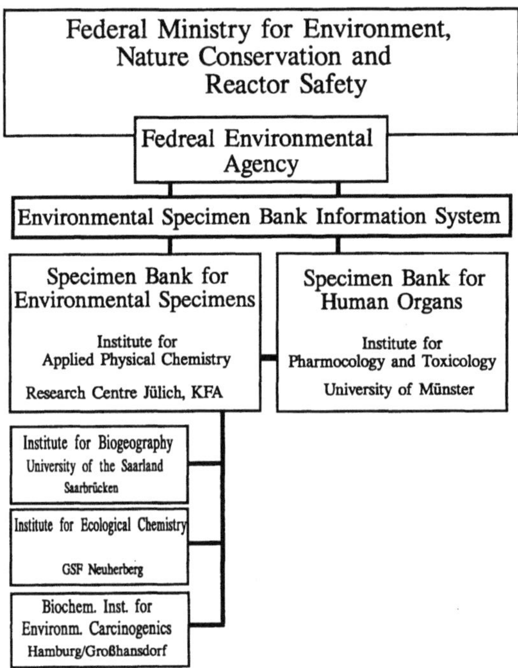



**Fig. 3** Organizational chart of the project "Environmental Specimen Bank" of the Federal Republic of Germany

## Representatively selected ecosystems

With the operation of the German Environmental Specimen Bank [8—10] as part of the environmental monitoring and ecosystem research network currently in the process of development biological objects in the context of their ecosystem will now be preferentially observed now in addition to the monitoring program for air and water which exists already. An overall concept has been developed for this monitoring and thus also for the environmental specimen bank by a committee of experts under the auspices of the BMU, taking into consideration different types of ecosystems with corresponding representative sampling areas. In these areas (see Fig. 4) continuous sampling has been carried out to some extent since 1985 (up to now, every 2 years). However, this will be converted into a 1-year sampling frequency in accordance with an overall concept to be realized by the year 2000 in order to utilize the analytical, biometric and meteorological data much more effectively in the sense of real-time monitoring. Since the reunification of Germany in 1989, representative areas in the former East Germany have also been increasingly studied with the support of local institutions.

## Representative specimen species

The selection and assignment of representative specimen species from the terrestrial and aquatic environment, and also from the human field, was similarly undertaken by a committee of experts. This selection is regularly reviewed on the basis of new insights and, if necessary, corrected. However, very serious reasons would be required to eliminate a specimen species that has been collected and stored for a considerable period of time. At present, the following specimen species are collected on a routine basis:

— from the *terrestrial* field: soil samples, tree samples (leaves and shoots), and animal samples (pigeons' eggs, roe deer liver, earthworms);
— from the *limnic* field: sediments, fish (muscle and liver), and freshwater mussels;
— from the *marine* environment: sediments, algae, mussels, fish (muscle and liver), and sea bird eggs (see Fig. 5).

The specimen species from the *human field* (fatty tissue, liver, kidney, blood, urine, hair, saliva) are stored in the specimen bank for human organ samples at the University of Münster. The range of analyzed elements and organic pollutants is largely identical with that of the environmental specimens in order to detect any possible connections between the pollution situation in man and his immediate environment.

As part of a supplementary research and development at the Institute of Applied Physical Chemistry, investigations are currently been carried out on materials of the en-

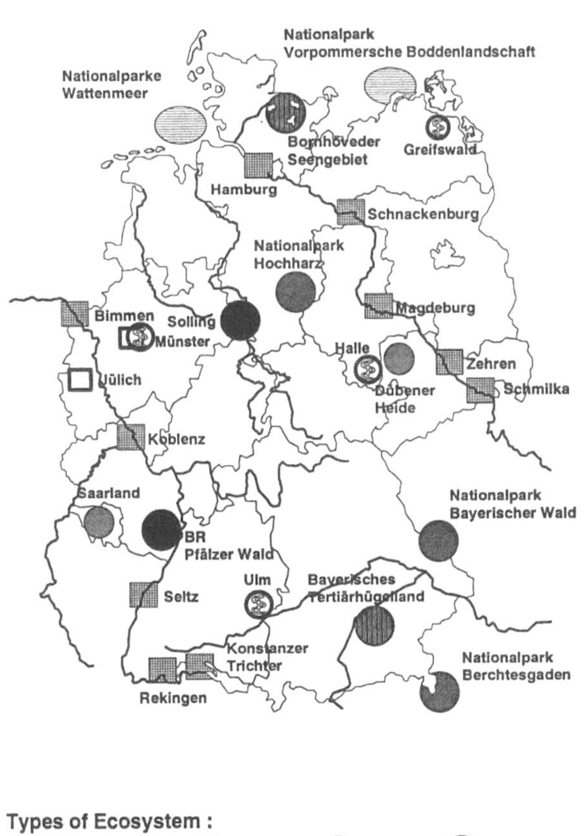


**Fig. 4** Geographical map of the representative sampling areas for the Environmental Specimen Bank

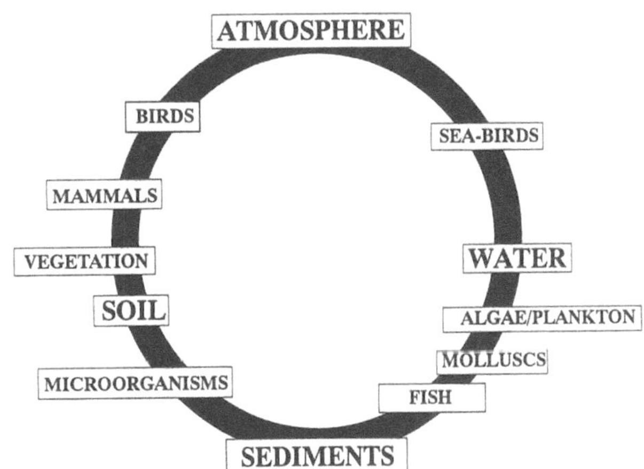


**Fig. 5** Various selected sample types from different sectors of the environment

vironmental specimen bank with the objective of clarifying the possible influences of surface-active substances (surfactants) on the transport behavior and bioavailability of inorganic and organic environmental chemicals in soils and sediments. Particular attention is being paid to the determination of synergistic or antagonistic effects as well as to questions of the immobilization (bound residues) and remobilization of pollutants.

## Results

Terrestrial environment — atmospheric input

The initial characterization of environmental specimens from the terrestrial environment — with four complete data sets to date — as well as the regular examinations of human blood from selected parent populations carried out at 6-month intervals has already led to reliable and thus credible results about the ubiquitous decrease in lead pollution. Making use of the Institute's own measuring network to determine pollutant deposition by precipitation, it can be demonstrated without any doubt that the legal measures to reduce lead emissions (introduction of unleaded fuels in 1984) have achieved significant success. The data collections carried out continuously since 1980 — with weekly sampling — indicate, for example, that in rural regions close to industrial centers lead deposition decreased by more than 70% in 1921 in comparison to 1980. A comparable trend is apparent in so-called "background regions," not directly polluted, in which wet lead deposition can be almost exclusively attributed to atmospheric long-distance transport (see Fig. 6).

The tree specimens (spruce shoots, beech leaves) collected and stored since 1985 from the regions of the environmental specimen bank also provide convincing documentation of the decrease in lead pollution (see Fig. 6). The same is also true for the human field, as confirmed by the results of the continuous lead studies of blood from student populations at the University of Münster. On the whole, the results provide clear evidence of the positive effects of the introduction of unleaded fuels.

Cadmium contents in tree specimens (leaves, needles) from various regions display higher concentrations at an agricultural site than in an urban industrialized area. Since the plant uptake of this element essentially takes place via the roots, it must be assumed that the different contents depend on the bioavailability or mobility of this heavy metal in the soil.

The reasons for a varying bioavailability are very complex and extensive site-related studies are necessary to clarify them. For example, differences in humus content or permanent soil moisture are conceivable. A further possibility results from soil acidification caused by acidic rain. Moreover, ubiquitously occurring surface-active substances (surfactants) should also be taken into consideration with respect to a change in mobility. It was thus demonstrated in laboratory experiments that cationic surfactants are basically capable of releasing both essential trace elements ($Ca^{2+}$, $Mg^{2+}$, $Na^+$) as well as toxic heavy metals ($Cd^{2+}$) from degraded loess soil (orthic luvisol — approx. 25% clay mineral fraction). Furthermore, investigations on the clay mineral montmorillonite demonstrated that this release of cations occurs quantitatively and stoichiometrically [11]. In Fig. 7 this change in the balance of nutrients and trace elements is demonstrated by the example of a layer silicate loaded with $Na^+/Cd^{2+}$. The addition of the cationic surfactant initially leads to the replacement of relatively weakly bound $Na^+$ ions. Only after this displacement process has been completed a quantitative release of $Cd^{2+}$ into the soil solution takes place. However, if the monovalent $Na^+$ ion is replaced by divalent, essential cations (e.g., $Ca^{2+}$) in the silicate layer, a simultaneous displace-

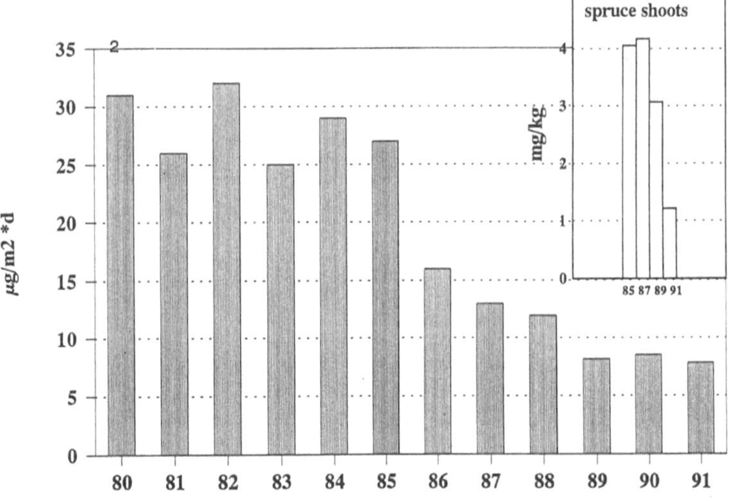



Fig. 6 Decline in lead pollution from wet deposition in rural areas (site: Jülich) and in spruce shoots from a "background area" (site: Berchtesgaden National Park)

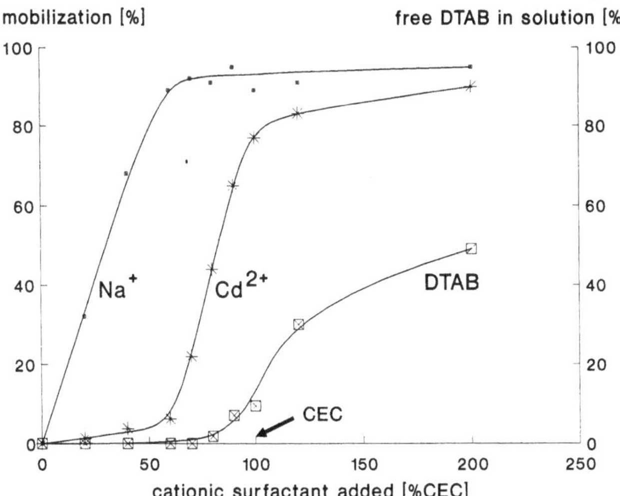



**Fig. 7** Displacement of Na$^+$ and Cd$^{2+}$ ions from a layer silicate by cationic surfactant [11]; (DTAB = dodecyl trimethyl ammonium bromide, CEC = cation exchange capacity)

ment by the cationic surfactant takes place because both Ca$^{2+}$ and also Cd$^{2+}$ are bound to the exchange sites with comparable strength [11].

## Limnic environment — inputs into river water systems

Considerable pollution with heavy metals and chlorinated hydrocarbons was detected in sediments from the River Elbe collected in 1991 and 1992 in co-operation with the Federal Hydrology Agency (BfG, Berlin office). Thus, for example, the salt contents and also the concentrations

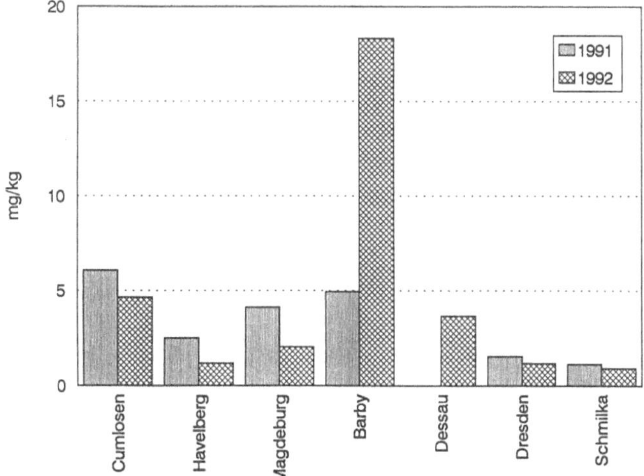



**Fig. 8** Distribution of mercury concentrations in Elbe sediments from 1991 and 1992 along the Elbe at sampling sites of the International Commission for the Protection of the Elbe; Schmilka — border Czech Republic/Germany; Barby — Saale mouth; Cumlosen — former boundary between the two German states

of some heavy metals such as mecury, cadmium, lead, and chromium are extremely high in the region of the Saale estuary. Even if the sediment samples from 1992 display declining mercury concentrations for almost all sampling sites, nevertheless, the drastic rise in the region of the Saale mouth (Barby site) is alarming (see Fig. 8).

Likewise, investigations of the Elbe sediments collected in 1991 for chlorinated hydrocarbons and dioxins point to particular pollution problems [12]. However, since these surveys have only recently been included in the analytical program, reliable conclusions about the time sequence and origin of this pollution can only be obtained by the retrospective analysis of corresponding stored samples.

The convention concluded by the *International Commission for the Protection of the Elbe (IKSE)*, the closure of industrial plants, and the improvement in quality of municipal sewage in the upper regions of the Elbe and its tributaries will undoubtedly lead in the future to an improvement of water quality particularly in this region. The same expectations can also be placed in the extension of new environmental technologies. In the same way, an experimental verification of this forecast can only be obtained from a retrospective analysis of different sample types from the Environmental Specimen Bank, for example sediment, freshwater mussels and fish (bream).

In contrast to the systematic investigations of the Elbe, which have started only recently, surface sediments from the Rhine have been analyzed by the Institute of Applied Physical Chemistry since 1978. Samples taken at the Emmerich border station demonstrate a clear reduction in lead and cadmium pollution from 1978 to 1990 (see Fig. 9). The effects of the convention concluded by the *International Commission for the Protection of the Rhine* (IKSR) are particularly apparent in the same way as the decline in the biological oxygen demand (BOD$_5$) or the rise in the oxygen concentration, which is one reason for a clear increase in the variety of fish species in the Rhine [13].

As part of the continuous expansion of the range of elements characterized in the initial analysis, the thallium concentration in various limnic samples has been determined for some time. The comparison of current data for bream muscle from two different sampling areas (Rhine/Lake Constance and lake land district of Bornhöved, Schleswig-Holstein) show that the thallium contents of the samples from Lake Constance are higher than the reference samples from the North German area by approximately a factor of 8 (see Fig. 10). More detailed investigations of stored material from the past 8 years (retrospective monitoring) furthermore revealed that this is a permanent, excessive pollution of Lake Constance. An acute thallium contamination, initially assumed, can thus be ruled out.

Surface-active substances, in particular, cationic surfactants and also the metabolites of the alkyl phenol

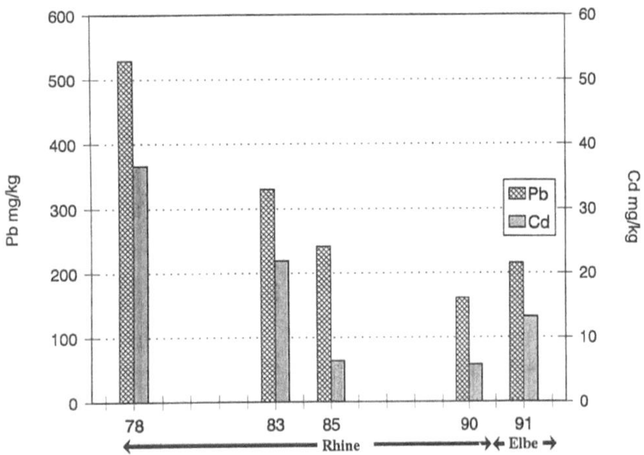

Fig. 9 Time curve of the lead and cadmium concentration in surface sediments of the Rhine (site: Emmerich); for comparison, values in Elbe sediments from 1991 at the Cumlosen site

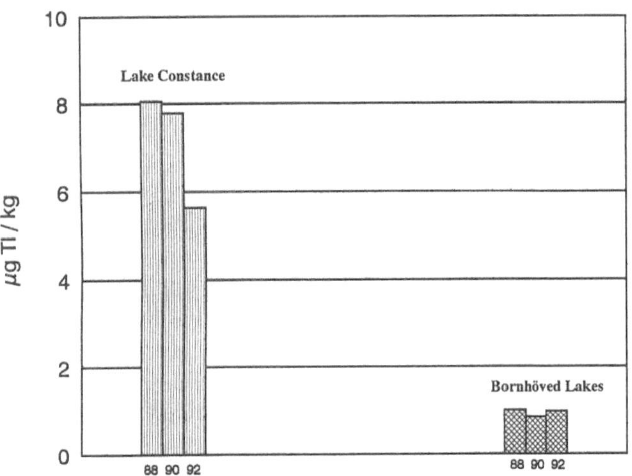

Fig. 10 Distribution of the thallium concentration in bream muscle from Lake Constance and the Bornhövel Lake district

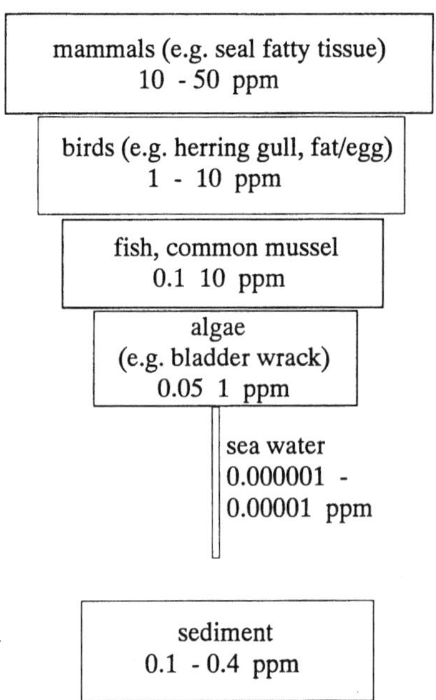


Fig. 11 Accumulation of mercury in various trophic stages of the marine environment

ethoxylates, possibly also have certain toxic (environmentally toxic) effects in the aquatic environment (surface sediments from rivers, lakes and estuary regions) in the same way as in the terrestrial field. It may also be assumed that a remobilization of heavy metals becomes possible under the influence of surfactants and thus additional quantities of these elements become available for bioaccumulation in various organisms of the limnic or marine food chains.

## Marine environment —
impacts on estuary and coastal areas

Both sea water and also sediment analyses have as yet only rarely been implemented with respect to the biological availability of pollutants. However, since only the

available pollutant fraction has toxicological relevance and therefore causes "*pollution*", it is necessary to use increasingly biological indicators to monitor the pollution and pollution capacity of surface waters.

The special significance of such indicators consists in their accumulation potential, which is described in the literature for various marine organisms [14] (see Fig. 1 and 11). For this reason, mainly sediments, brown algae, common mussels, fish (viviparous eelpout) and herring gull egs are investigated within the framework of the Environmental Specimen Bank program to characterize a marine sampling area. It is planned to sample seal organs (muscle, fatty tissue, liver) as the final link in the marine food chain.

Brown algae (bladder wrack) and common mussels from the same sampling site display quite different accumulation behavior with respect to certain elements. As shown in Fig. 12, As, Ba, Mn, S and Zn are, for example, preferentially accumulated in bladder wrack, whereas an increased accumulation of Cu, Fe and Hg is found in common edible mussels. Furthermore, significant deviations arise for both sample species in a comparison of the North Sea and Baltic Sea (see Fig. 12), which differ considerably in their salt content. Both North Sea specimens conform in displaying a higher accumulation potential for As and Hg, whereas Ba, Mn, and Zn are accumulated to a greater extent in the Baltic Sea specimens. No site-specific allocations of this kind can be made for Cu, Fe, and S. On the whole, apart from the matrix-dependent parameters, in particular the salinity and the temperature seem to play a certain part here.

Progr Colloid Polym Sci (1994) 95:48—60
© Steinkopff Verlag 1994



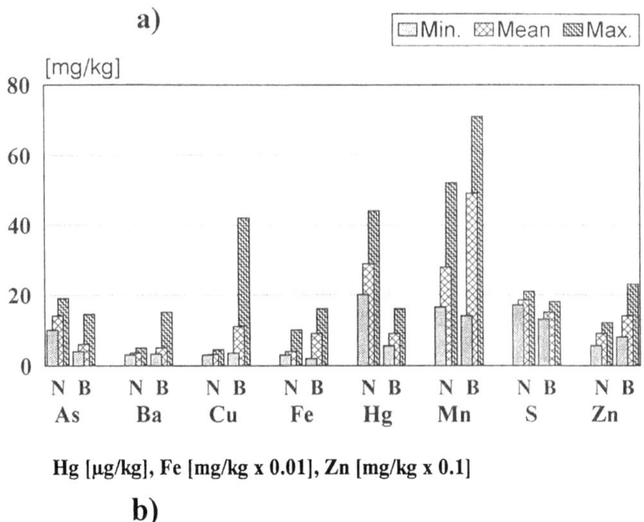

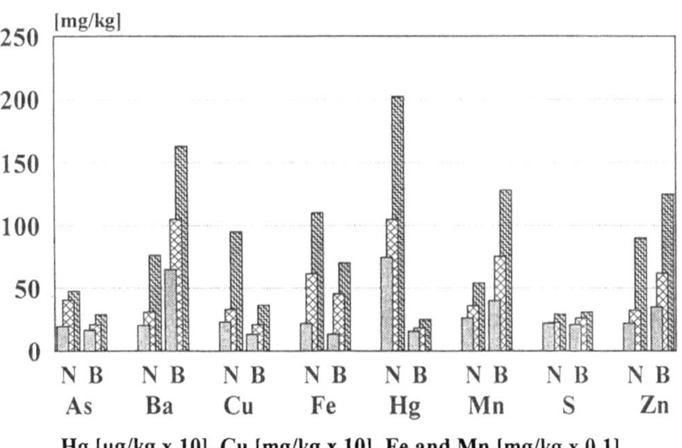



**Hg [µg/kg], Fe [mg/kg x 0.01], Zn [mg/kg x 0.1]**



**Hg [µg/kg x 10], Cu [mg/kg x 10], Fe and Mn [mg/kg x 0.1]**

**Fig. 12** Accumulation of various elements in (a) edible mussels and (b) bladder wrack from the North Sea (N) and the Baltic Sea (B)

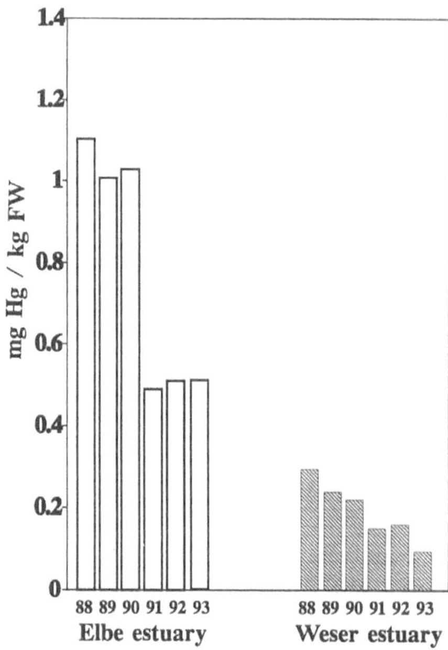



**Fig. 13** Decline in mercury pollution in the estuary regions of the Elbe (Island of Trischen) and Weser (Island of Mellum) represented by the pollution values of herring gull eggs; FW = fresh weight

Thus, for example, in the literature a relationship is described in certain marine organisms between increased cadmium uptake at high temperature and low salinity [15].

A clear decline in mercury pollution in the estuary region of the Elbe in the past few years can be documented on the basis of samples of herring gull eggs collected in the "Trischen" bird sanctuary and in which up to more than 90% of the mercury was present in the form of the highly toxic methyl mercury. In comparison to the period before reunification of Germany (1988/89) a reduction in pollution to less than half can be determined for the subsequent years 1991/93 (see Fig. 13). Similar results are also shown by an investigation of polychlorinated biphenyls (PCB) [12]. These findings are very probably associated with the closure of industrial plants and municipal sewage disposal in the upper regions of the Elbe and its tributaries. On the other hand, the reduced run-off of the Elbe in the past few years of low

rainfall may also have contributed to these results. This urgently requires continuous investigations in the coming years to monitor the new environmental technologies coming into operation. Realistic and reliable statements about the development of the pollution situation however can undoubtedly be made only by a systematic, retrospective comparison with past samples.

Due to the sedentary habits of this species, the determination of the mercury concentration in herring gull eggs also permits a clear distinction between adjacent sampling areas with different degrees of pollution. Thus, for example, samples from the bird sanctuary of "Mellum" (Weser estuary) display considerably lower mercury contents than herring gull eggs from the bird sanctuary of "Trischen" (Elbe estuary) (see Fig. 13).

Additional research activities

Apart from continuous monitoring of known pollutants and operation of the Environmental Specimen Bank, the tasks of prognostic environmental precautions also include, in particular, the investigation of chemicals which, according to the present state of knowledge, do not cause any direct ecotoxicological effects, but whose environmental impacts or possible effects on man and the environment by synergistic or antagonistic effects are as yet completely unknown.

The Institute of Applied Physical Chemistry is therefore involved in investigating those substances which, due to their surface-active properties, may cause an immobilization, remobilization or change of mobility of pollutants — heavy metals and organic substances such as polychlorinated biphenyls (PCB), PAH's or pesticides. The possible surfactant-induced remobilization of pollutants fixed in a "sink" (soils, sediments) is of particular importance since this would make bound residues biologically available again for plant or animal organisms.

Surface-active substances alter the properties of soil minerals (e.g., layer silicates) by adsorption and hydrophobization of the surface. Even slight quantities of surfactants cause significant effects with respect to the transport and biological availability of pollutants in soils and sediments [16]. This may result in the accumulation of pollutants in the trophic stages of the food chain and the elimination of essential trace elements. Cationic surfactants, for example, cause a drastic alteration in the electrolyte equilibrium between the soil and soil solution due to ion exchange. Both essential cations ($Ca^{2+}$, $Mg^{2+}$, $Na^+$) as well as toxic heavy metal ions (e.g., $Cd^{2+}$) are quantitatively displaced [11].

Hydrophobization of the surface causes basic changes in the adsorption properties of the soil minerals. Neutral (hydrophobic) molecules may be taken up into the surfactant layer. On the basis of this knowledge, alkyl ammonium montmorillonites are being investigated and recommended for sealing landfill sites [17, 18]. Similar sorption processes may take place in soils and sediments. Probably, the hydrophobic character of soil minerals will increase due to the accumulation of surfactants, thus promoting the adsorption of non polar less water-soluble substances. That would immobilize them and exclude them from further natural transport or degradation processes.

The following examples are therefore concerned with the impact of surfactants on the interaction between environmental chemicals and the most active clay mineral components (montmorillonite, illite) of the mineral soil horizon or the natural soil structures (e.g. degraded loess soil). Firstly, the influence of surfactants on non-essential elements will be discussed. In examining the interactions of organic substances with surface-active substances, the examples are oriented towards the three surfactant classes in the sequence: cationic, non-ionic and anionic surfactants. Details of the materials and analytical methods used, as well as the nomenclature, are described in the literature [11, 19, 20].

## Surfactant interactions with inorganic pollutants

The influence of various classes of surfactants on the transport behavior of $Cd^{2+}$ in degraded loess soil (B horizon) is shown in Fig. 14 [21]. It is apparent that only cationic surfactants are capable of mobilizing sorbed $Cd^{2+}$ by an ion exchange mechanism, in the course of which the surfactants are quantitatively bound to the

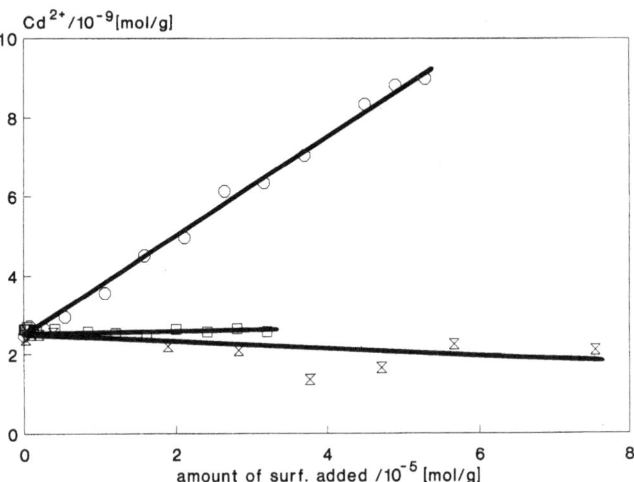


**Fig. 14** $Cd^{2+}$ mobilization from degraded loess soil by various classes of surfactant: o—o didodecyl dimethyl ammonium bromide; ⊠—⊠ sodium dodecyl sulphate; □—□ alkylphenyl polyethylene glycol ether (Triton-X-100) with 9.5 EO

mineral surface and are not detectable in the equilibrium solution. Non-ionic surfactants (e.g. Triton X 100) are adsorbed much more weakly on degraded loess soil, which may be attributed partially to the van der Waals forces. Accordingly, no $Cd^{2+}$ mobilization takes place. Anionic surfactants form an only slightly soluble salt with $Cd^{2+}$, namely $Cd(DS)_2$. Therefore, $Cd^{2+}$ immobilization occurs due to precipitation.

## Surfactant influences on organic pollutants

### Cationic surfactants

In principle, the adsorption of organic environmental chemicals on layer silicates proceeds extremely slowly so that it may take several weeks to establish an equilibrium state [22]. In contrast, if the layer silicate surface is coated with cationic surfactants in different quantities [22] or of different hydrophobicity, e.g., with DTAB (dodecyl trimethyl ammonium bromide) or DDDMAB (didoceyl dimethyl ammonium bromide) [23] then equilibrium in both systems is reached within 30 min, i.e., a drastic increase in the adsorption rate of organic pollutants results. Their residence time or probability of residence in the soil electrolyte is therefore drastically reduced and their bioavailability thus also drops.

Figure 15 shows the influence of added surfactants on the equilibrium states in the form of adsorption isotherms for 2-naphthol. In this concentration range the isotherms can be described by straight lines, which points to a constant distribution of the naphthol between the solid and liquid phases [22, 24]. Its adsorption rises with increasing surfactant coverage — relative to the cation exchange

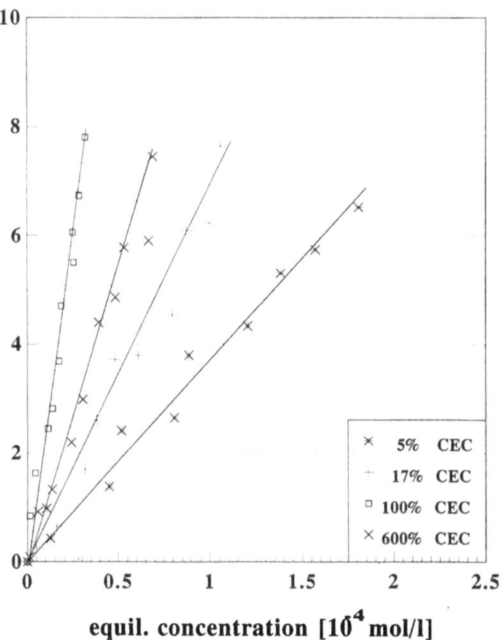



**Fig. 15** Adsorption isotherms of 2-naphthol on cetyltrimethyl ammonium ($CTA^+$)-illite with increasing surfactant coverage (related to the CEC)

capacity (CEC) — until the exchange capacity has been reached. With the addition of even higher quantities of surfactant (600% of the CEC), the adsorbed amount of 2-naphthol declines again considerably. Micelles are formed in the solution which then compete with the adsorbed surfactant films for the pollutant molecules, which leads to a decrease of the incorporated, adsorbed pollutant molecules (soil washing effect) [22].

Detailed information about the binding site of the organic compounds and the structure of the adsorbate layer can be obtained by x-ray diffractometry. For example, nitrophenols are intercalated in the interlayers of the swellable, hydrophobed bentonites [23], which may essentially influence their further fate (transport, degradation behavior).

Calorimetric measurements provide information about the binding forces of the pollutant adsorbates. Whereas the low adsorption heat of the phenols on hydrophilic clay surfaces can hardly be measured, clearly exothermal enthalpies were determined for the adsorption of 4-nitrophenol on surfactant-bentonite complexes [23]. Accordingly, the hydrophobic interactions (hydrophobic bond) between the surfactant alkyl chains and the aromatic part of the pollutant molecules are of major importance for these adsorption processes.

On the whole, the displacement of adsorbed environmental chemicals by cationic compounds seems to be a process of general environmental relevance. Thus,

for example, the pesticide atrazine adsorbed on layer silicates can be quantitatively displaced by the cationic plant growth inhibitor chlormequat [25], which may lead to a contamination of the ground water with atrazine. Furthermore, as a consequence of similar processes surfactants may also reach deeper layers of the soil and initiate further secondary reactions there.

### Non-ionic surfactants

Apart from the cationic surfactants, non-inonic surfactants also cause an increased hydrophobization of the layer silicate surface with increasing concentration, thus promoting, for example, the adsorption of the hydrophobic fungicide biphenyl (see Fig. 16) [26]. The increased biphenyl adsorption also measured at a very weakly hydrophobized bentonite (10 µmol non-ionic surfactant/g bentonite) indicates that these effects also become active even with very small (environmentally relevant) surfactant concentrations.

### Anionic surfactants

In the case of clay minerals pretreated with small amounts of anionic surfactants (e.g., sodium dodecylsulphate (SDS), c < CMC (critical micelle concentration)), for example, the adsorption of biphenyl is not significantly influenced in comparison to pure layer silicates [26]. These findings may be explained, among other aspects, by the low adsorption potential for anionic surfactants of the negatively charged layer silicate surface (cf. Fig. 17, curve A). Quite different conditions exist in acid soils, in which the soil minerals may be positively charged. As

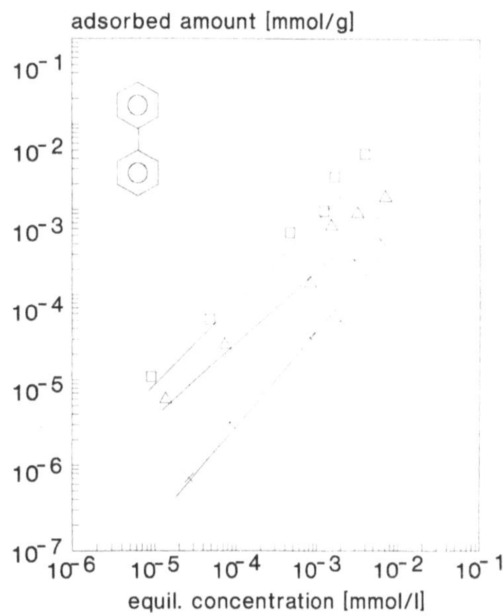



**Fig. 16** Adsorption isotherms of biphenyl on bentonite in the presence of the non-ionic surfactant dodecyl octaethylene glycol ether ($C_{12}E_8$): (×—× without surfactant; △—△ with 0.013 mmol/g $C_{12}E_8$; □—□ with 0.3 mmol/g $C_{12}E_8$) [26]

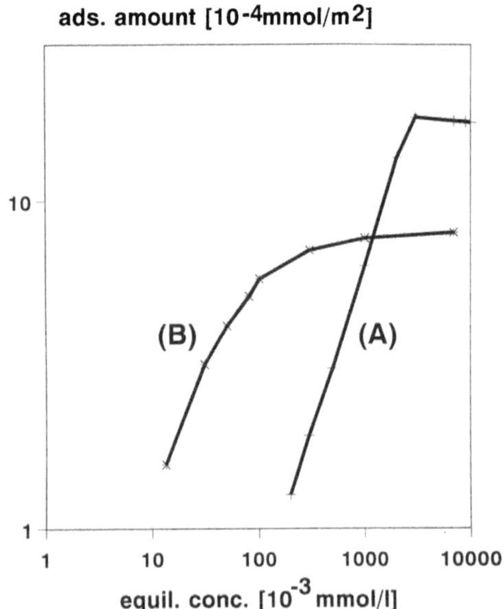



**Fig. 17** Adsorption isotherms of sodium dodecylsulphate on kaolinite. (A): without non-ionic surfactant; (B): with the addition of dodecyl octaethylene glycol ether in the ratio 1:1 [28]

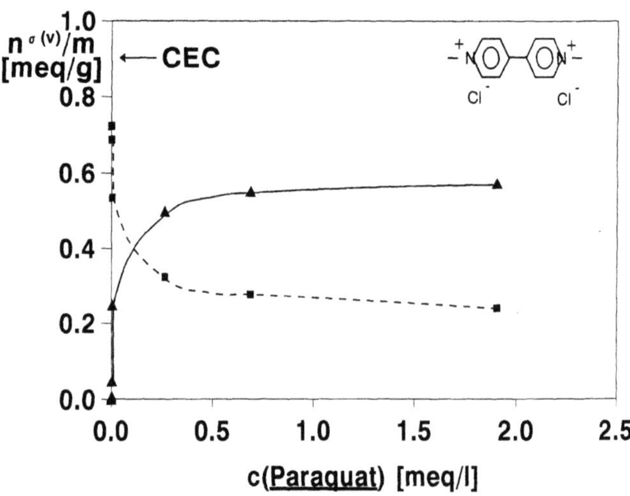



**Fig. 18** Adsorption of paraquat on Ca-bentonite pretreated with cationic surfactant: △—△ adsorption isotherm of paraquat, □—□ displacement of dodecyl trimethyl ammonium bromide (DTAB). (DTAB pretreated 0.73 meq/g) [20]

expected, higher adsorbed quantities of anionic surfactants were found, for example, on aluminium oxide at pH 3–7 [27]. Even in the neutral pH range, increased anionic surfactant adsorption on soil minerals is possible if cationic or non-ionic surfactants are also present in the system. Adsorption from a mixture of anionic and non-ionic surfactant on kaolinite thus shifts the isotherms of both components towards smaller equilibrium concentrations [28], as illustrated by curve B in Fig. 17 for SDS in the presence of the non-ionic surfactant $C_{12}E_8$. At a mixing ratio of 1:1, a SDS concentration in the bulk phase lower by one order of magnitude is already sufficient to achieve the same SDS adsorption as without the addition of non-ionic surfactant. Since in real systems surfactant mixtures can almost always be expected, the processes shown may play a significant role in the adsorption of organic pollutants on inorganic soil constituents.

## Pollutant influences on surfactant adsorption

Just as surfactants may determine the mobility behavior of environmental chemicals in the soil, surfactant transport may also be influenced by other accompanying substances. For example, surfactants adsorbed on layer silicates may be exchanged by ionic pesticides, as illustrated in Fig. 18 for a clay mineral precoated with DTAB [20]. The preferentially adsorbed organic dication paraquat (PQ) displaces the single charged DTAB (the substance quantities are given in equivalent quantities (eq) to clarify the comparison). For each equilibrium concentration, the associated adsorbed amounts of PQ

and DTAB approximately correspond to the CEC. On the basis of these experimental findings, the following exchange reaction may be assumed:

$$(DTA^+)_2\text{-clay} + PQ^{2+} = PQ^{2+} \text{ clay} + 2\ DTA^+ .$$

Since in contrast to cationic surfactants, non-ionic surfactants are largely physisorbed on clay mineral surfaces, they can be relatively easily replaced by monocationic organic substances. If, for example, the monocationic pesticide cyperquat (CQ) is present in the soil electrolyte then apart from the essential cations ($Ca^{2+}$, $Na^{2+}$) it also displaces the physisorbed non-ionic surfactant $C_{12}E_8$ [26]. In order to illustrate the associated structural changes at the layer silicate, Fig. 19 shows the basal spacing of an Na montmorillonite pretreated with $C_{12}E_8$ as a function of the adsorbed amount of CQ. The initial basal spacing of the clay-surfactant complex (1.8 nm) caused by the intercalation of $C_{12}\ E_8$ decreases with increasing amount of adsorbed CQ, and at 1.5 nm finally reaches the basal spacing of pure CQ-montmorillonite.

To summarize, the following conclusions may be drawn as provisional results of these model studies:

— Cationic surfactants mobilize essential elements and heavy metals in stoichiometric relations and are quantitatively adsorbed.
— Only cationic surfactants are capable of mobilizing $Cd^{2+}$, whereas non-ionics do not display any effects and anionic surfactants cause an immobilization.
— The originally hydrophilic surface of the layer silicates is hydrophobized by the adsorption of cationic and non-ionic surfactants resulting in an expanded interlayer of the swellable clay minerals.

Progr Colloid Polym Sci (1994) 95:48–60
© Steinkopff Verlag 1994



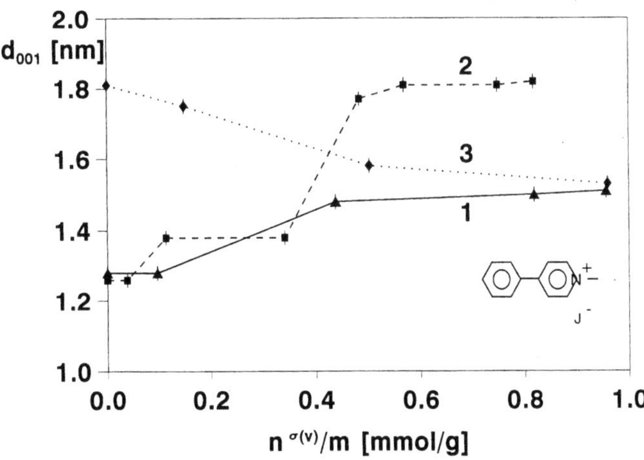



**Fig. 19** Change in the basal spacing of Na montmorillonite with the addition of cyperquat (1), dodecyl octoethylene glycol ether (2) as well as cyperquat to montmorillonite precoated with $C_{12}E_8$ (3) [26]

— The adsorption of hydrophobic environmental chemicals is enhanced and accelerated at surfactant-clay mineral complexes. The extent of this enhancement depends on the degree to which the surface is hydrophobized.
— The inorganic soil horizon represents an effective sink for organic pollutants, whose effectiveness may increase considerably in the presence of small quantities of surfactant.
— Surfactants adsorbed on clays may be displaced by ionic pesticides, i.e., surfactant transport can also be influenced by other accompanying substances.

## Conclusions

The results presented as examples of the operation and accompanying research program of the Environmental Specimen Bank represent an initial attempt to indicate perceptible developments in the chemical pollution of our environment on the basis of reliable data. Based on these results, it is already apparent that, in concert with the "Ecological Environmental Monitoring" and the "Environmental Precaution Policy" of the Federal Government, the Environmental Specimen Bank may be used as a most effective instrument to determine pollutant trends and to monitor the success of legislative measures. Even after the relatively short period of permanent operation, it has been possible to document the success of the introduction of unleaded fuels both in the environment and also in the human sector by the element characterization of representative samples. In the future, it will furthermore also be possible to monitor the impacts of modern environmental technologies on improving the quality of water and air by means of appropriate indicator systems. It is, moreover, of special significance that, in particular, those substances which cannot be determined now with the necessary reliability or whose ecotoxicological or human toxicological effects are not yet known can be studied by retrospective monitoring with authentic samples from the past.

As a supplement to this work directed towards individual elements or substances, in future those investigations will have to be intensified which are related the synergistic or antagonistic effects of multicomponent systems. Major emphasis is placed on elucidating basic relations responsible for the mobilization or immobilization of pollutants under the influence of extraneous chemicals. Typical examples are the effects of surface-active substances on the bioaccumulation or bioavailability of environmental chemicals in soils or sediments.

Finally, on the basis of its application-related aims, the Environmental Specimen Bank offers the possibility of illustrating the significance and necessity of forward-looking environmental research to a broad spectrum of the population. The Environmental Specimen Bank thus also makes a vital contribution towards providing an objective perspective for public discussions of environmental problems.


## References

1. Lewis RA (1989) Forecasting, assesment and the nature of real systems. What are the limits?, Environmental Monitoring and Assessment
2. GDCh/BUA (1987) Altstoffbeurteilung S 32
3. SRU (Rat der Sachverständigen für Umweltfragen) Umweltgutachen 1987, Deutscher Bundestag, Drucksache 11/1569 und Verlag Kohlhammer, Stuttgart/Mainz 1987
4. Schladot JD, Backhaus FW (1993) Probenahmerichtlinien für Blasentang (*Fucus vesiculosus*), Miesmuscheln (*Mytilus edulis*), Silbermöveneier (*Larus argentatus*) und entsprechende Richtlinien für die Aufarbeitung aller Umweltproben bei tiefen Temperaturen, Umweltbundesamt (in press)
5. FAO (Food and Agriculture Organization of the United Nations) (1976) Manual of Methods in Aquatic Environment Research, Part 2 — Guidelines for the Use of Biological Accumulators in Marine Pollution Monitoring, FAO Fisheries Technical Paper No 150
6. Schladot JD, Backhaus FW (1988) Preparation of Sample Material for Environmental Specimen Banking Purposes — Milling and Homogenization at Cryogenic Temperatures, in S. A. Wise, R. Zeissler, G. M. Goldstein (eds) Progress in Environmental Specimen Banking, NBS Special Publication No. 740, 184—193

7. BMFT (Bundesministerium für Forschung und Technologie) (1988) Umweltprobenbank — Bericht und Bewertung der Pilotphase, Springer-Verlag Berlin Heidelberg

8. Stoeppler M, Schladot JD, Dürbeck HW (1989) Umweltprobenbank in der Bundesrepublik Deutschland, Teil 1: Realisation von Umweltprobenbanken, GIT Fachz Lab 20:1017

9. Stoeppler M, Schladot JD, Dürbeck HW (1989) Umweltprobenbank in der Bundesrepublik Deutschland, Teil 2: Betrieb der Umweltprobenbank seti 1985, GIT Fachz Lab 11:1017

10. Schladot JD, Stoeppler M, Kloster G, Schwuger MJ (1992) The Environmental Specimen Bank — long term storage for retrospective studies, Analusis 20:3, 45

11. Narres HD, Rützel H, Dekany I, Schwuger MJ (1992) 3rd Cesio Int Surfactants Congress, Proceedings section E, London S 88

12. Oxynos K, Schmitzer J, Kettrup A (1992) Private communication

13. BMU (1992) Umwelt 8

14. Steinhagen-Schneider G (1981) Fucus vesiculosus als Schwermetall-Bioindikator, Bericht Nr 93 aus dem Institut für Meereskunde, Christian-Albrechts-Universität Kiel

15. Vernberg WB (1974) Multiple environmental factor effects on physiology and behaviour of the fiddler crab Uca pugilator. In: Vernberg FJ, Vernberg WB (eds) Pollution and physiology of marine organisms. Academic Press, London

16. Klumpp E, Struck BD, Schwuger MJ (1992) Nachr Chem Tech Lab 40:428

17. Weiss A (1989) Appl Clay Sci 4:193

18. Nüesch R (1991) Tonmineralogie und Geotechnik 1:17

19. Klumpp E, Heitmann H, Lewandowski H, Schwuger MJ (1992) Progress Colloid Polym Sci 89:181

20. Rheinländer T, Klumpp E, Rossbach M, Schwuger MJ (1992) Progress Colloid Polym Sci 89:190

21. Gonzales J, Pohlmeier A, Narres HD, Schwuger MJ (1993) Mitteil der deutschen Bodenkundl Gesellschaft 68:235

22. Klumpp E, Heitmann H, Schwuger MJ (1991) Tenside Surf Det 28:6

23. Heitmann H, Dissertation, Univ. Dortmund (in preparation)

24. Klumpp E, Heitmann H, Schwuger MJ (1993) Colloids and Surfaces 78:93

25. Weiss A (1992) private communication

26. Rheinländer T (1993) Dissertation, Univ Düsseldorf

27. Fuerstenau DW (1971) IN: Hair ML (ed) The Chemistry of Biosurfaces, Vol 1, Marcel Dekker, New York

28. Xu Q, Vasudevan TV, Somasundaran P (1991) J of Colloid and Interface Sci 142(2):528

Progr Colloid & Polym Sci (1994) 95:61—72
© Steinkopff Verlag 1994

G. Lagaly

# Bentonites: adsorbents of toxic substances

Prof. Dr. G. Lagaly
Institute of Inorganic Chemistry
University of Kiel
Olshausenstraße 40
24098 Kiel, FRG

**Abstract** The use of bentonites as adsorbents results from the reactivity of montmorillonite which is the main mineral in this clay-like material. Montmorillonite, a $2:1$ clay mineral, impresses by a diversity of reactions in the interlayer space and at the external surfaces, which are the cause of strong adsorption of heavy metal ions and organic compounds. Bentonite adsorbents are used as crude bentonite, in soda-activated form, after degradation to bleaching earths, after modification by organic cations, or in form of polyhydroxometal or polyoxometal derivatives ("pillared clays"). The ease with which the bentonites are modified allows an optimization of the properties so that the requirements of numerous practical applications are fulfilled.

**Key words** Bentonites — clay minerals — clay-organic interactions — environment technology — pesticides — pollutants

## Introduction

The bentonites are outstanding adsorbents for heavy metal ions and organic compounds. A large variety of organic materials is bound. The actual or possible uses are advanced by the ease with which the clays are modified. Modified bentonites are produced which fulfill the requirements for an amazing variety of applications in different fields and at different scales.

Bentonite rocks are composed of natural, earthy, fine-grained argillaceous materials [1]. The main $2:1$ clay mineral is montmorillonite which belongs to the group of smectites (Table 1) (structure see [2—5]).

The montmorillonite content of bentonites varies considerably from locality to locality and within the deposits. Determination of the montmorillonite content is a very tedious procedure which in general gives approximate values only (see chapter 1 in [2]). In industrial test laboratories the montmorillonite content is obtained by methylene blue adsorption. This comfortable but rough method is very approximative (see below) (cf. [6]; chapter 3.3.6 in [2]). It is recommended to derive the montmorillonite content from the C, N content of the

**Table 1** Classification of planar $2:1$ clay minerals (non-planar modulated structures (e.g. sepiolite, palygorskite) see [123])

| Interlayer material | Group | Octahedral character | Species |
|---|---|---|---|
| none $\xi \approx 0^{1)}$ | talc-pyrophyllite | tr$^{2)}$ <br> di | talc <br> willemseite <br> pyrophyllite |
| hydrated exchangeable cations, $\xi \approx 0.2—0.6$ | smectite | tr <br><br> di | saponite <br> hectorite <br> sauconite <br> montmorillo-nite <br> beidellite <br> nontronite |
| hydrated exchangeable cations, $\xi \approx 0.6—0.9$ | vermiculite | tr <br> di | trioctahedral vermiculite <br> dioctahedral vermiculite |
| non-hydrated monovalent cations, $\xi \approx 0.6—1.0$ | mica | tr <br><br> di | biotite <br> phlogopite <br> lepidolite <br> muscovite <br> illite |

**Table 1** (continued)

| Interlayer material | Group | Octahedral character | Species |
|---|---|---|---|
| | | | glauconite |
| | | | celadonite |
| | | | paragonite |
| non-hydrated divalent cations, $\zeta \approx 1.8{-}2.0$ | brittle mica | tr di | clintonite margarite |
| hydroxide sheet ($\zeta$ variable) | chlorite | tr di di-tr | clinoclore donbassite sudoite |

[1] $\zeta$ = charges/formula unit (= eq/mol), formula unit = $\{M^{3+}, M^{2+}, M^+)_{2-3}[(Si, M^{3+})_4O_{10}(OH)_2]^{\zeta-}$.
[2] di, tr = di-, trioctahedral

alkylammonium derivatives and the interlayer cation density determined by the alkylammonium ion exchange [6, 7]. The mineralogical composition of two high quality bentonites is given in Table 2.

## Clay mineral-organic interactions

The layer structure of 2:1 clay minerals provides different types of surfaces and different sites for the interaction with organic compounds (Fig. 1): the external surfaces (with distinct differences between the basal plane and edge surfaces) and internal surfaces. The organic molecules can interact with the exchangeable interlayer cations, the cations or anions residing at the edges, the silanol and aluminol groups on the edges, and the siloxane oxygen atoms of the layers. The oxygen atoms of the —Si—O—Si— bonds are weak acceptors for hydrogen bonds; the basicity is strongly increased by $Al^{3+}$ substitution, i.e., for —Si—O—Al— bonds [8]. Thus, isomorphous substitution in the tetrahedral layer (hectorite, saponite → montmorillonite, beidellite) intensifies the interaction of the silicate layer with water and organic molecules.

Many reactions are related to the interlayer cations (Fig. 2): hydration and solvation of the interlayer cations, complex formation, and exchange reactions with inorganic and organic cations. In solvates, a certain fraction of the interlayer molecules interact with the cations, the remaining molecules are filling the space between the solvation shells around the cations. When complexes form, the number of molecules directly coordinated to the interlayer cations is small and can be the same as in homogeneous solution.

The principles of clay mineral-organic interactions are reported by Mortland [9a]. This paper still provides an excellent survey. For pesticides see [9b, c, d].

**Table 2** Mineralogical composition and properties of a Wyoming bentonite (MX-80 from Bentonite International, Duisburg Meiderich) und Montigel (Süd-Chemie, Bavaria) [124, 125]

| | MX-80 | Montigel |
|---|---|---|
| fraction <0.2 μm | | |
| total | 73.5% | 71.3% |
| quartz | 2% | 1% |
| kaolinite | — | 2% |
| muscovite | — | 5% |
| montmorillonite in fraction >0.2 μm | 2% | 1% |
| loss during separation | 2% | 2% |
| total montmorillonite content in the bentonite | 75.5% | 66.3% |
| montmorillonite content, density separation [1] | 75.4% | ≈71% |
| density (bentonite), g/cm³ | 2.76 | 2.85 |
| cation exchange capacity (bentonite), meq/g | 0.76 | 0.62 |
| molecular mass M (montmorillonite), g/mol | 373 | 375 |
| interlayer cation density $\zeta$ [2], eq/mol | 0.30 | 0.28 |
| cation exchange capacity (montmorillonite), $C_t$, meq/g | 1.01 | 0.94 |
| interlayer exchange capacity [3] (montmorillonite), $C_i$, meq/g | 0.80 | 0.75 |
| ratio $C_i/C_t$ | 0.79 | 0.80 |

chemical composition (calculation based on the interlayer cation density)

MX-80
$Na_{0.30}\{(Al_{1.55}Fe^{3+}_{0.20}Fe^{2+}_{0.01}Mg_{0.24})(Si_{3.96}Al_{0.04})O_{10}(OH)_2\}$

Montigel
$Ca_{0.14}\{(Al_{1.36}Fe^{3+}_{0.31}Fe^{2+}_{0.01}Mg_{0.35})Si_{4.0}O_{10}(OH)_2\}$

[1] Separation of octadecylammonium montmorillonite as fraction <1.88 g/cm³ in mixtures of tetrabromoethane and ethanol.
[2] alkylammonium method.
[3] $C_i$ = 1000 $\zeta$/M.

## Neutral guest molecules

The interlamellar water in 2:1 clay minerals is displaced (not always quantitatively) by enough polar molecules, e.g., alcohols, ethylene glycol, glycerol and complex forming agents. When it is desorbed, smectites rehydrate reversibly or adsorb organic molecules which solvate the interlayer cations. Usually, the guest molecules are offered in liquid state. They can also be adsorbed from gaseous state or by solid state reactions (e.g., acrylamide [10]; alkylamines [10]; p-amino azobenzene [12]). Some-

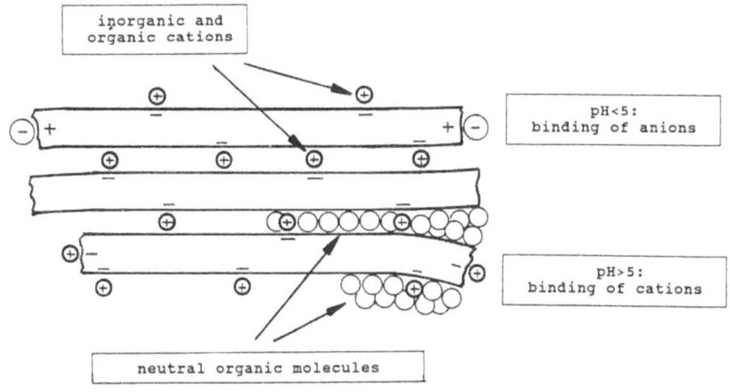


**Fig. 1** The different types of reacting sites on 2:1 clay mineral particles

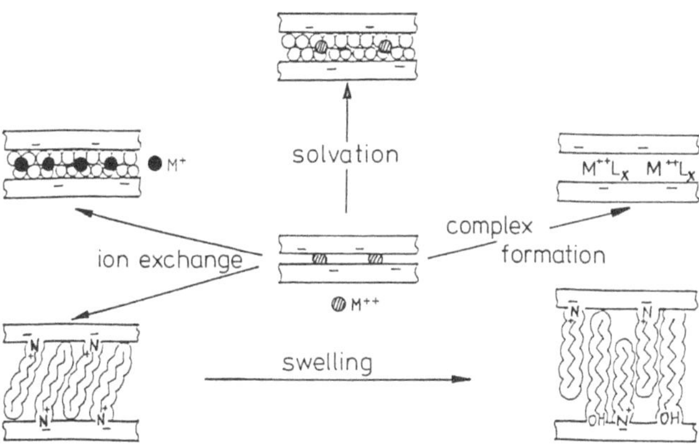


**Fig. 2** The interlamellar reactions of 2:1 clay minerals

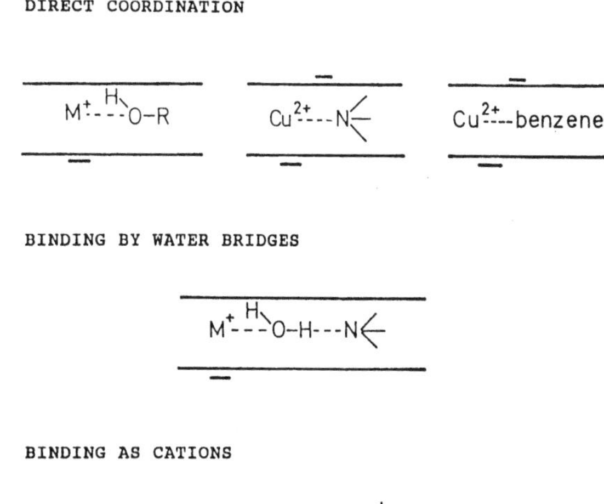


$$\geqslant\!N + H_2O \rightleftharpoons \geqslant\!\overset{+}{N}H + OH^-$$

$$\geqslant\!\overset{+}{N}\!\!<^H \qquad \left[\geqslant\!N\text{--}H\text{--}N\!\leqslant\right]^+$$

**Fig. 3** Binding of neutral molecules in the interlayer space of 2:1 clay minerals

times, pronounced selectivity is observed in solid-state reactions which is absent for intercalation from solution. Maleic and fumaric acid are adsorbed from solution but only maleic acid is intercalated by solid state reaction [13].

Interlamellar adsorption of dried smectites is often kinetically hindered. Organic derivatives are then obtained by pre-adsorbing a suitable compound which subsequently is displaced by the desired compound ("propping open" [14, 15]).

Many toxic substances which are strongly bound by montmorillonite are nitrogen-containing basic compounds. These molecules can interact directly with soft interlayer cations (e.g., copper ions) but not with hard cations (e.g., sodium ions). The amine molecules are bound by water bridges to the hard cations (Fig. 3). (Concept of hard and soft acid and bases see textbooks of chemistry; see also [16]).

The amines are often protonated and held as cations in the interlayer space, even if they are unprotonated in solution.

Important is binding of pairs consisting of the base and its protonated form ("semi-salt formation" [9a, 17]).

The enhanced protonation of bases in the interlayer space is a consequence of the increased acidity of the interlamellar water molecules and is further promoted by the electrostatic interaction of the protonated form with the surface charges. Intercalated bases can also be protonated by proton transfer from a protonated species already present (e.g., protonation of methylamine, pyridine, 3-aminotriazole by $NH_4^+$ ions [18]).

Enhanced levels of adsorption may be attained by surface induced aggregation. Organic molecules (e.g., parathion [19]) or certain cations (paraquat, diquat, thionine [20]) are bound at the surface and enhance the adsorption of further molecules or act as nuclei for stacking arrangements. The surface aggregates may have a lower solubility than the free compound.

The contrary effect is also observed. It is known in pharmaceutical science that the bioavailability of scarcely soluble compounds can be increased by adsorption [21]. Adsorption of a drug from a non-polar solution at the solid interface can produce an arrangement different from that of the pure, crystalline drug and with increased solubility. Another mechanism is solubilization by colloidally dispersed solids. The solubility of griseofulvin in water is remarkably increased in the presence of sodium montmorillonite [22]. The drug is adsorbed at the external surface of the smectite from acetone solution (Fig. 4). When this complex is dispersed in water, the smectite delaminates and the silicate layers with the drug attached act as carriers. The maximum solubilization is obtained for loadings <0.2 mmol/g montmorillonite. Solubilization at a somewhat reduced level is also observed for

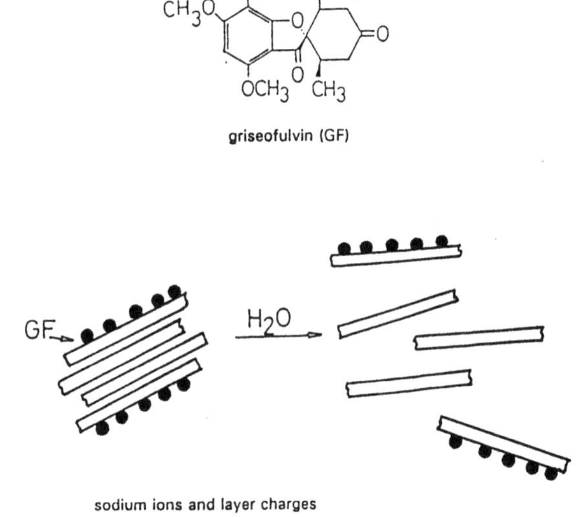


**Fig. 4** Solubilization of the drug griseofulvin (GF) by sodium montmorillonite crystals delaminating in water [22]

physical mixtures of the drug and the clay. The dispersed sodium smectite adsorbs the drug and carries it into the solution.

To my knowledge, increased solubility in the presence of clays or solubilization by clays are not reported for pesticides but may occur in certain cases. The effect may have consequences for the use of clays in environmental technology as the solubility of a toxicant can be increased in the presence of bentonite and sodium ions.

Usually, the main factors governing the adsorption on smectites are concentration or vapor pressure, pH, interlayer cation, layer charge and temperature. Many other factors can also be operative: the layer charge distribution [23], pretreatment of the smectites, degree of dispersion [24], presence of salts: not only concentration or ionic strength, but also the very nature of the ions [20, 23, 25].

Elaborating synergistic and competitive adsorption is a current challenge. Well documented are synergistic effects for the adsorption of nuclein bases [23, 26].

Organic cations

Various types of organic cations are exchanged for the inorganic interlayer cations. Smaller organic cations intensively studied are the herbicides diquat (I) and paraquat (II), which contain the cations

The interaction of diquat and paraquat with clay minerals was extensively studied [27—31]. The

heterocyclic rings arranged in a monolayer lie flat in the interlayer space of montmorillonite. Montmorillonite absorbs paraquat preferentially over diquat. Micromolecular concentrations of paraquat displace inordinately large amounts of diquat. Enhanced ionic strength increases the surface aggregation of both herbicides, but decreases the degree of cation exchange as the salt cations are competitive with the herbicide cations [20].

A strong adsorption of the major part of a biocide on the soil clay minerals reduces its bioavailability and must be compensated by addition of increased amounts. On the other hand, dangerous biocide cations adsorbed on the clay minerals constitute a long-term health hazard when slowly released to the ground water. Coadsorption of the pesticide together with a non-phytotoxic compound or displacement of the pesticide (for instance diquat by 4′-pyridyl pyridinium chloride [32]) reduce this threat and increase the bioavailability. As shown by Narine and Guy [20], the divalent cations paraquat and diquat are more easily displaced by washing with salt solutions than monovalent cations such as methylene blue and thionine.

Another group of organic cations of interest comprises dye molecules such as methylene blue, crystal violet, and methyl green. Methylene blue adsorption is used for estimating the montmorillonite content of bentonites. This is a very approximate procedure because the amount adsorbed depends on the exchange capacity and is also influenced by formation of dimers and trimers on the external surfaces [33, 34]. Rather, the adsorption of methylene blue and crystal violet may be used for determining the exchangeable cations and the cation exchange capacity of montmorillonites [35]. The metachromasy of crystal violet after adsorption on laponite is related to the degree of flocculation of the clay mineral. The adsorbed dimers are accumulated in the interparticle pores of the flocs [36, 37].

Competitive binding of different types of cations was recently described by Margulies et al. [38]. The studies are based on Nir's model of cation exchange processes [40, 41].

When adsorbed together with the biocide on montmorillonite, methyl green, thioflavin or naphthylammonium ions may be used for photoprotection of insecticides.

The interaction of smectites with cationic surfactants is important for numerous practical applications of bentonites ([1] (chapter 9)). Important surfactants used in modifying bentonites are mono- and dialkylammonium chlorides:

For scientific studies primary alkylammonium ions $RNH_3^+$ and alkylpyridinium ions are often used.

The strong binding of cationic surfactants makes the amount of surfactant bound by cation exchange less dependent on the experimental conditions. Generally, this amount is equivalent to the exchange capacity. During the exchange reaction, ion pairs (surfactant cation + gegen ion) are also intercalated between the layers and are accumulated at the external surfaces [43]. The total amount of surfactant adsorbed is more sensitively dependent on the solution properties (concentration, pH, presence of salts, temperature).

Ion pairs adsorbed on the external surfaces recharge the particles and cause peptization. The surfactant clay mineral complexes behave as hydrophobic materials only when the ion pairs are removed by washing.

Long chain surfactants are more strongly bound than shorter chain cations [44]. The longer chain cations are preferentially adsorbed from surfactant mixtures. However, the adsorption of long chain cations such as dimethyl dioctadecylammonium is strongly reduced and restricted to a mono- or bilayer arrangement of flat-lying surfactant ions when tetramethylammonium or tetraethylammonium cations are present [45]. The reason is the strong tendency of tetramethylammonium ions to hold the silicate layers at close distances [46].

Surfactant cations are difficult to displace from the clay mineral surface. Quaternary alkylammonium ions are easier to exchange than primary alkylammonium ions. Primary alkylammonium ions are removed by hydrolysis into the amine in alkaline medium [47]. The mutual exchange of surfactants in non-aqueous dispersions was studied by Mc Atee [48, 49]. Solvent was a mixture of isooctane and isopropyl alcohol. The alcohol served two functions: as a solvent for the surfactant chloride salts and the dispersing agent for the clay mineral surfactant particles. An effective agent to displace quaternary alkylammonium ions (dodecyl trimethylammonium ions) is paraquat [50].

## Anions

Anions are bound at the edges when, at lower pH, protons are adsorbed by the aluminol groups (Fig. 1) (see also [51]). The amount of anions which is held as gegen ions is low but influences the particle-particle interactions and rheological properties [52—54]. Anions can also replace hydroxyl groups at the edges (ligand exchange) and become directly coordinated to octahedral cations, mainly aluminum ions. Ligand exchange should predominate around pH = 5—7 when the number of positive and negative edge charges is small. By ligand exchange reactions higher amounts of anions may be bound than by gegen ion binding. Typical examples are oxo-anions of acids (e.g., ortho-, oligo-, and polyphosphates) (see also [55]). The pH dependence of anion adsorption by both processes is governed not only by the edge charges of the silicate layers, but also by the degree of dissociation of the acids.

Anions can also be bound by calcium bridges. The cation exchange is then equimolar [56]:

$$\left.\begin{matrix} \\ - \\ \\ \\ \end{matrix}\right\} Na^+ + Ca^{2+} + X^- \rightleftharpoons \left.\begin{matrix} \\ - \\ \\ \end{matrix}\right\} Ca^{2+}X^- + Na^+$$

Anionic surfactants are adsorbed by smectites in acidic medium and, at pH near neutral, when inert salts (e.g., NaCl) are added to compress the diffuse ionic layers [53]. They can also be attached to the mineral surface by calcium bridges.

Anion adsorption is more sensitively dependent on the type of particle aggregation and the degree of dispersion than cation exchange reactions because formation of edge/face contacts blocks edge sites for anion adsorption. The adsorption of adenosine triphosphate on smaller sized fractions of montmorillonite increased after addition of traces of phosphate which break the edge/face contacts and make additional adsorption sites available [24].

The amount of anions adsorbed cannot be directly measured by the depletion in solution. Repulsion of the anions by the high density of negative surface charges (negative adsorption [57—59]) increases the anion concentration in the bulk and reduces the depletion by the (positive) adsorption.

In the presence of di- and trivalent exchangeable cations many anions are precipitated as insoluble salts (e.g., calcium dodecylsulfate). When this possibility is not considered, adsorption data are misinterpreted. (It was reported [60] that dodecylsulfate anions were adsorbed in the interlayer space of calcium montmorillonite. However, the (001) reflections observed and ascribed to the calcium montmorillonite surfactant complex ($d$ (001) = 30 Å) are the reflections of calcium dodecylsulfate. Anionic pesticides may be partially inactivated by precipitation as insoluble $Ca^{2+}$, $Al^{3+}$, $Fe^{3+}$ etc. salts (e.g., acifluorfen [61]).)

## Bentonites as adsorbents

Bentonite adsorbents are or may be used as

— crude bentonite
— soda activated bentonite
— acid activated bentonite
— organically modified bentonite and
— pillared bentonite.

## Crude and sodium bentonites

Crude bentonite and sodium bentonite (soda activated bentonite) bind heavy metal ions by cation exchange. Usually, the preference is (see also [70])

$$Cu^{2+} > Pb^{2+} > Zn^{2+} > Cd^{2+} > Mn^{2+} \, .$$

The separation factor $a_s$ or selectivity coefficient $K_s$ depend not only on the layer charge, but change with proceeding exchange, too [62—64].

The separation factor

$$a_s = \overline{C_{Cu^{2+}}} C_{Mn^{2+}} / \overline{C_{Mn^{2+}}} C_{Cu^{2+}}$$

(bars: concentration at the surface) for a mixture of $Cu^{2+}$ and $Mn^{2+}$ ions on 2:1 clay minerals is highest for illite (Fig. 5) [65]. Correctly, the data are related to the ternary system comprising $Cu^{2+}$, $Mn^{2+}$ ions, and the $Ca^{2+}$ or $Na^+$ ions of the clay mineral. The preference for $Cu^{2+}$ decreases with increasing molar fractions of $Cu^{2+}$ ions in solution, that is, with increasing coverage by $Cu^{2+}$ (Fig. 5). A high selectivity for heavy metal ions at low equivalent fractions in solution is often observed. Van Bladel et al. [66] report this behavior for cadmium and zinc ions on calcium vermiculite, illite, and bentonite from Wyoming. (No preference of zinc and cadmium ions over calcium ions is found for montmorillonite from Camp Berteau, Morocco). A very high selectivity at a very low degree of coverage was also observed by Brouwer et al. [67]. Illite shows a high preference of cesium ions even over rubidium ions as long as only 0.5% of the exchangeable sites are occupied by cesium ions.

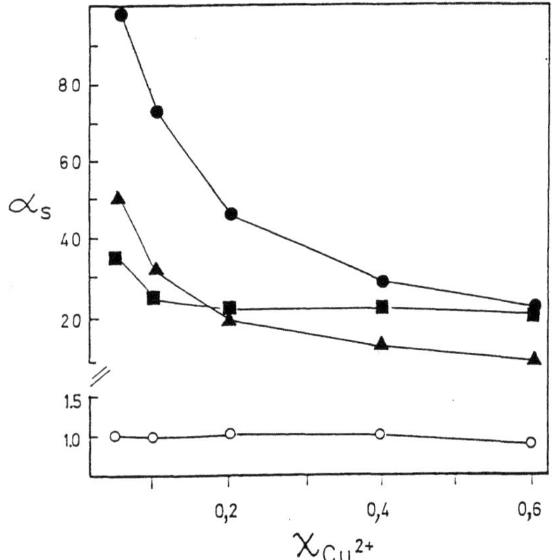


**Fig. 5** Selectivity of copper ions over manganese ions on clay minerals. $a_s$ is the separation factor related to $Cu^{2+}$ and $Mn^{2+}$ ions; $\chi_{Cu^{2+}} = C_{Cu^{2+}}/(C_{Cu^{2+}} + C_{Mn^{2+}})$ in solution [65]
● illite (<74 μm; green shale, New York, Ward's, Rochester)
▲ vermiculite, Toc. Mine, Transvaal, South-Africa
○ sodium bentonite (Clay Spur, Wyoming, USA)
■ calcium form of the bentonite

Sodium montmorillonite binds copper and manganese ions without any preference. The restriction of the interlayer expansion by calcium ions seems to be one of the prerequisites for selectivity of montmorillonite. Sposito and LeVesque [68] illustrated that the adsorption of sodium ions by illite eliminated the preference of the clay minerals for $Ca^{2+}$ over $Mg^{2+}$ ions.

The selectivity for heavy metal ions can be increased by modifying the clay mineral with compounds which offer suitable ligands for complexation of the metal ions. Examples of modifiers are 2-ammonio ethyldithio carbamate (I) or the sodium salt of 1,3,5 triazine 2,4,6-trithiol (sym. trimercaptotriazine) (II). (The triazine is used for adsorbing heavy metal ions from waste water in the presence of complexing agents).

The adsorption of heavy metal ions can be largely influenced by the presence of organic compounds. Complexation may increase or reduce adsorption of heavy metal ions. A calcium bentonite (Montigel, Süd-Chemie, Bavaria) adsorbs 71% $Zn^{2+}$ from 1100 ppm $Zn^{2+}$ in solution (2 g bentonite/100 ml solution) (Fig. 6). In the presence of diethyl ketone and phenol 89% and 92% of the $Zn^{2+}$ ions are adsorbed [69]. Zinc ions also promote the adsorption of diethyl ketone (from 5 to 28%) and phenol (from 2 to 6%). Both organic compounds form weak complexes or solvation shells around the zinc ions. As the concentration of the organic compounds is very low (1%), enrichment of the zinc ions on the surface due to the decreased dielectric constant is not decisive.

When complexation holds the gegen ions preferentially in solution, the ratio of the cations in bulk solution to the cations in the diffuse and Stern layer increases. Even chloride ions are strong enough to hold an increased amount of cadmium ions in bulk solution (Fig. 7). The calculated amount of cadmium ions in the Stern and diffuse layer agrees with the measured amounts of cadmium ions adsorbed [41]. The fraction of cadmium ions in solution increases from 3.6% in 0.01 M NaCl to 56% in 0.05 M NaCl; the adsorption in the Stern layer decreases from 90.4% to 43% (calculated values, total concentration of $Cd^{2+}$ ions 0.13—1.07 μM). In the presence of $ClO_4^-$ which does not form complexes with $Cd^{2+}$, the fraction of $Cd^{2+}$ in solution is 1.9% in 0.01 M $NaClO_4$ and 30.7% in 0.05 M $NaClO_4$.

These examples reveal that as well solvation by organic molecules as direct complexation by inorganic and organic ligands strongly influences the adsorption of heavy metal ions. These effects must be considered when

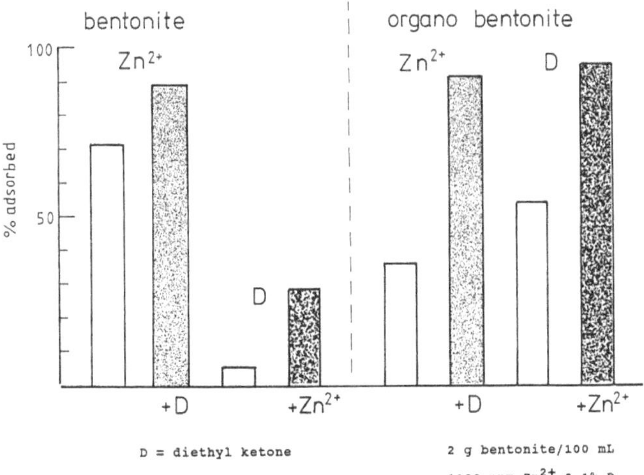



**Fig. 6** Synergistic effects for the adsorption of heavy metal ions and organic compounds:

Adsorption of zinc ions on a crude bentonite (Montigel, a calcium bentonite from Bavaria, Süd-Chemie, FRG) and an organo-bentonite (Thixogel VP, dimethyl dioctadecylammonium bentonite, Süd-Chemie, FRG) in the absence and presence of diethyl ketone. Also is shown the influence of zinc ions on the adsorption of diethyl ketone [69]

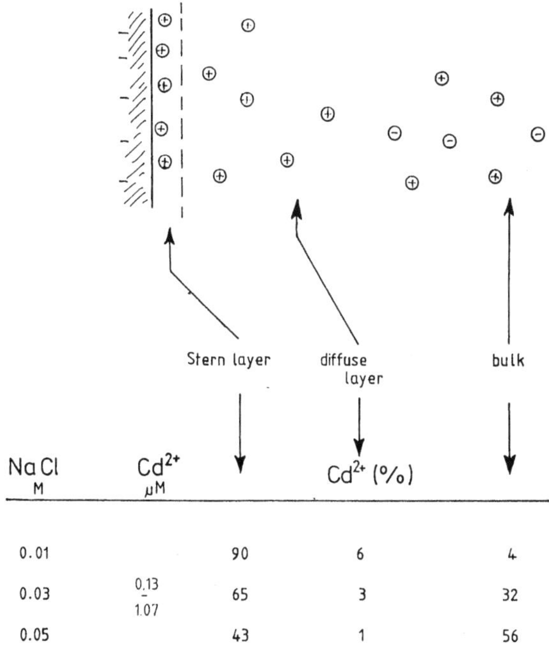



| NaCl M | Cd²⁺ μM | | Cd²⁺ (%) | |
|---|---|---|---|---|
| 0.01 | | 90 | 6 | 4 |
| 0.03 | 0.13 – 1.07 | 65 | 3 | 32 |
| 0.05 | | 43 | 1 | 56 |

**Fig. 7** Distribution of cadmium ions between the Stern layer, the diffuse ionic layer and the bulk for sodium montmorillonite in NaCl solutions [41]

the immobilization of metal ions within clay barriers is discussed (see also [70]).

Bentonites are excellent adsorbents for cationic surfactants which are strongly held at the clay mineral surface. However, stability against biocides cannot generally be presumed. As mentioned above, dodecyl trimethylammonium ions are displaced by paraquat [50].

Bentonites also remove compounds from contaminated water which, at first sight, should not interact strongly with smectites. An example are halogenated hydrocarbons (chloroethanes, m-dichlorobenzene and several bromo hydrocarbons) which are irreversibly bound by sodium montmorillonite. (The adsorption on fumed silica is reversible). The irreversibility of the process indicates chemisorption. Certain aromatic halogeno hydrocarbons (2 and 4-bromotoluene and 1,2 dichlorobenzene) are, at least partially, catalytically transformed into higher molecular weight products [71].

In environmental technology, adsorption of organic compounds, which is an outstanding advantage of smectites, can be accompanied by serious disadvantages. Contraction of the interlayer space after adsorption can produce fissures and cracks in the clay layer of barriers which destroy the sealing function [70, 72]. Organic substances may also change the particle aggregation which not only influences the permeability, but also the rheological behavior of the clay. Hydrocarbons and less polar compounds lead to break-down of voluminous networks as a consequence of reduced electrostatic repulsion. On the other hand, fixation of the interlayer separation by adsorption of organic cations may be used to increase the stability of smectitic clay layers against shearing [73].

Adsorption of permanent gases and vapors depends very sensitively on the texture of the smectites, which is mainly determined by the type of the exchangeable cation and by the way the samples were dried after cation exchange [74, 75]. Centrifugation or sedimentation and air-drying or freeze-drying of the dispersion produces different types of particle-particle aggregations which strongly influence the gas adsorption properties, in particular when adsorption occurs at the external surfaces only. (The effect of texture is of strong importance also in catalyst design [76, 77]).

Desorption and adsorption phenomena also influence distribution and transport of volatile compounds in soils. Adsorption of dibromomethane and trichloroethylene on pyrophyllite, kaolinite, illite, and montmorillonite proceeds very slowly (within hours) because these compounds are hampered in penetrating the aggregates. The desorption of these compounds from soil aggregates is also retarded [78].

## Bleaching earths

Bleaching earths (acid activated bentonites) are produced by boiling bentonites in hydrochloric acid (up to 17 mmol HCl/g bentonite at 90—100°C) [1]. This procedure removes a large part of the octahedral cations. After washing and drying, an excellent adsorbent is obtained. The material is x-ray amorphous but highly disordered tetrahedral layers still persist, which no longer lie parallel to each other. Adsorption of organic substances can cause a certain degree of parallel orientation of these layers [79, 80].

On an industrial scale, acid-activated bentonite is used for decolorization of mineral, vegetable, and animal oils, fats, waxes, and beverages (wine, beer, juices). In the paper industry acid-activated bentonite serves as color developer for carbonless copying paper [80]. The degraded bentonite is used as a carrier of fungicides and insecticides. It should be a suitable adsorbent for particular organic compounds, but actually, it is only used to regenerate organic fluids for dry cleaning.


Organo-montmorillonites

Comprehensive studies of gas adsorption by organically modified montmorillonites were reported by Barrer and coworkers [81, 82]. Non-polar compounds such as saturated hydrocarbons are adsorbed without distinct basal spacing expansion. (Small changes, however, may be observed.) The amount adsorbed is mainly determined by the interlamellar porosity. A pronounced selectivity against mixtures of hydrocarbons is observed [83, 84].

The interlamellar pores have the dimensions of micropores (diameters up to 20 Å). In such pores the adsorption energy is increased but kinetic effects become more pronounced. When long chain surfactant cations are adsorbed on montmorillonite, gas adsorption is strongly reduced because the long alkyl chains clog the entrance to the interlayer spaces [85, 86].

Aromatic hydrocarbons [46] and more polar compounds [87, 88] are adsorbed with appreciable basal spacing expansion. The layer charge is of minor influence as long as extreme values are avoided [89]. The way the samples were prepared has a strong influence on the partition coefficients [90]. This effect can be more important than the choice of the surfactant cation. Technical organo-bentonites differ largely in their gas adsorption properties because they are often prepared without removing an excess of surfactants or even with (designedly) incomplete exchange of the inorganic cations.

It is somewhat surprising that organo-bentonites (dimethyl dioctadecylammonium bentonite, Fig. 6) do retain considerable amounts of heavy metal ions (together with the gegen ions) from aqueous solutions. The adsorption of zinc ions is increased in the presence of phenol and diethyl ketone (from 36% to 87% (phenol) and 91% (ketone)). The synergetic effect is very strong: zinc ions increase the retention of the ketone from 54% to 95%. Phenol is almost quantitatively adsorbed even in the absence of zinc ions [69].

The adsorption of ion pairs may also be used to bind undesirable anions such as radio iodide [91, 92]. Anions appear to be more strongly attracted by the water molecules surrounding the alkyl chains than cations [93]. Thus, the adsorption of cations on organo-bentonites may be promoted by the adsorption of their anionic gegen ions in the region around the alkyl chains.

Adsorption studies of scarcely soluble neutral organic compounds from aqueous solutions were neglected in the past but are presently of interest. A large diversity of organic compounds is adsorbed from aqueous solution by octadecyl trimethylammonium bentonite [94]. The interlamellar space created by the large increase of the basal spacing is only partially filled by the organic compounds because water is also adsorbed. In other systems (e.g., with amines) the interlamellar space is mainly occupied by the organic molecules and the amount of coadsorbed water is small. The distribution equilibria of hexanol and octanol between the bulk phase and the interlamellar phase in n-alkylammonium montmorillonites were determined by Stul and Bock [95].

When comparable amounts of water and organic compounds are adsorbed, the depletion of the organic compound in solution does not give the correct amounts adsorbed, but rather the surface excess $n_1^{\sigma(n)}$ (see [2] pp. 150; [95a]).

An outstanding advantage of bentonites is that adsorption selectivity and capacity can be optimized by varying the kind and procedure of organic modification. The organic cations may be short chain cations such as tetramethylammonium or cationic surfactants. A tetramethylammonium smectite behaves like a zeolite with narrow micropores of rather hydrophilic character (Fig. 8). Long chain cations make the interlayer space hydrophobic. The amount adsorbed depends sensitively on the polarity of the adsorptive and the solvent and the degree of hydrophobization of the interlayer space. (The silicate layer itself is hydrophilic only when the tetrahedral charge, i.e., $Al^{3+}$ for $Si^{4+}$ substitution, is high; [8]). Generally, the adsorption is accompagnied by an appreciable interlayer expansion.

The balance between the polarity of the adsorptive, the solvent, and the interlamellar environment is instructively illustrated by the adsorption of phenol and trichlorophenol from water and hexane on tetramethylammonium (TMA) and hexadecyl trimethylammonium (HDTMA) montmorillonite (Fig. 9) [96]. Most striking is the adsorption of trichlorophenol from water: it is very low for TMA but high for HDTMA cations. Both smectites adsorb comparable amounts from hexane solutions. Evidently, the pronounced hydrophobic character of the interlayer space of HDTMA montmorillonite favors adsorption of the less hydrophilic trichlorophenol from water. The adsorption of pentachlorophenols from water is also distinctly higher on montmorillonite modified with long chain cations than on montmorillonite with short chain cations [97].

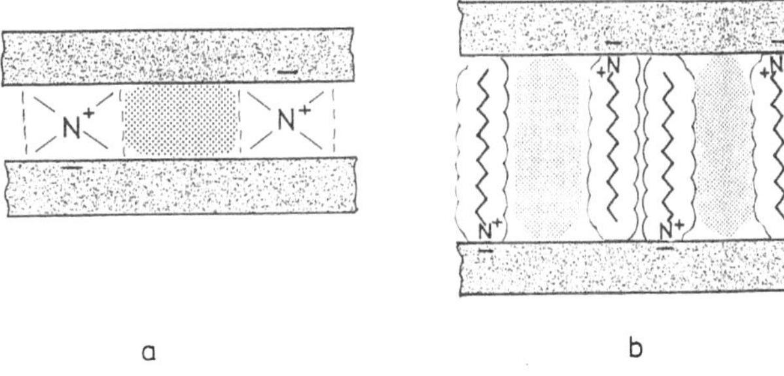


**Fig. 8** Montmorillonite modified with a) tetramethylammonium ions represents a more or less zeolitic system; modification with long chain alkylammonium ions b) provides a hydrophobic environment for the adsorptives penetrating into the interlayer space

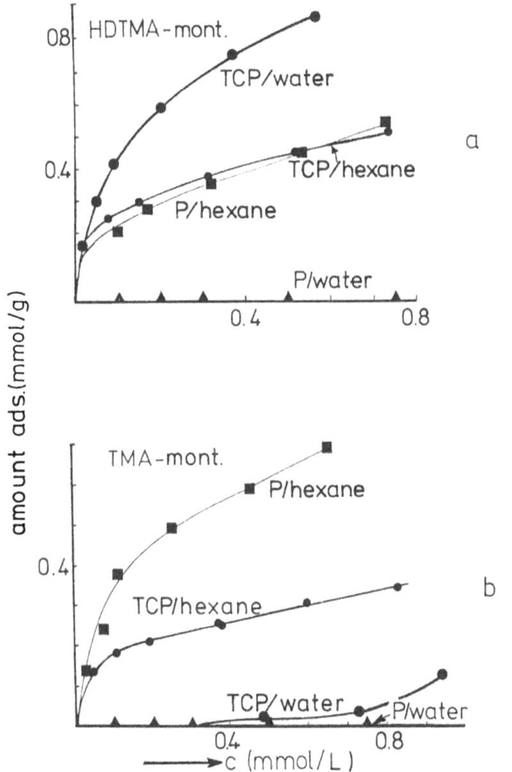


**Fig. 9** Adsorption of phenol (P) and 3,4,5-trichlorophenol (TCP) from solutions in hexane and water on tetramethylammonium (TMA) and hexadecyl trimethylammonium (HDTMA) montmorillonite [96]

TMA montmorillonite adsorbs aromatic hydrocarbons in the order benzene $\gg$ toluene $\geqslant$ p-xylene, ethyl benzene > o-xylene > dichlorobenzene [98]. The adsorption of benzene by TMA montmorillonite is not influenced by the presence of toluene but adsorption of toluene decreases when benzene is added [99].

The insecticide Lindan ($\gamma$-hexachloro cyclohexane) is not adsorbed by TMA montmorillonite. However, illites modified with TMA ions adsorb certain amounts of Lindan [98]. The saturated hydrocarbon overloaded with

chlorine atoms cannot penetrate into the interlayer space. Adsorption on the external surfaces of illite may be promoted by the high site density of TMA cations on illite.

## Polyhydroxoaluminum smectites

One way to prepare pillared smectites (and probably the simplest method for scaling up to industrial production) is introducing polyhydroxoaluminum cations by cation exchange. These cations are formed when NaOH is added to aluminum salt solutions in amounts NaOH/$Al^{3+}$ < 3 (typical 1—2.5) [100—105]. Usually, the polyhydroxoaluminum smectites are heated to 300—400°C to dehydrate them into polyoxoaluminum smectites. As adsorbents they may also be used in the uncalcinated form.

Zielke and Pinnavaia [106] compared the adsorption of several chlorinated phenols by pillared montmorillonite and Laponite (Laponites are synthetic hectorite-like materials). Pentachlorophenol, which is the strongest Brønsted acid, is best adsorbed by the polyoxoaluminum derivative and less by the polyhydroxo form. The adsorption capacity decreases strongly with increasing solution pH (from pH = 4.7 to 7.4), showing that this pollutant is adsorbed in undissociated form. The effectivity of the smectites (at pH = 4.7) is: polyoxoaluminum laponite > polyhydroxoaluminum laponite > polyoxoaluminum montmorillonite > polyoxochromium montmorillonite. Sodium laponite and montmorillonite show no tendency to adsorb the pollutant from aqueous solution.

The chief difference between laponite and montmorillonite is the high degree of disorder and delamination of laponite, which also persists in the pillared forms. The enhanced adsorption in comparison with montmorillonite results from the increased accessible surface area and may also be related to the more hydrophobic character of the bare silicate surface [8].

Nolan et al. [107] reported the adsorption of chloro dibenzo-p-dioxins on polyhydroxoaluminum montmorillonite. High distribution coefficients ($K_D$(ml/g) = amount $X$ adsorbed per g solid (mg/g)/amount $X$ in solution (mg/ml)) are obtained (octachloro dibenzodioxin on montmorillonite 2800 ml/g, on polyhydroxoaluminum montmorillonite 94000 ml/g). The distribution coefficient decreases sharply when the interlamellar polyhydroxoaluminum complexes are dehydrated to polyoxoaluminum pillars ($K_D$ = 1100 and 1800 ml/g for polyhydroxoaluminum montmorillonite heated to 170° and 550°C).

The adsorption properties of pillared clays can be improved by pre-adsorbing surfactants. Polyhydroxoaluminum montmorillonite covered with hexadecyl pyridinium ions is a better adsorbent for pollutants in industrial waste waters than hexadecyl pyridinium (HDPy) montmorillonite [108]. For benzopyrene $K_D$ is increased from 48000 ml/g (HDPy montmorillonite) to 95 800 ml/g (HDPy polyhydroxoaluminum montmorillonite) and for pentachlorophenol from 16 000 ml/g to 156 000 ml/g. No significant differences are noted in the sorption of 3,5-dichlorophenol on both adsorbents. The reduced adsorption of this pollutant may be related to its higher aqueous solubility.

The increased adsorption of certain pollutants on surfactant modified pillared montmorillonites results from the orientation of the surfactant cations different from that in the non-pillared smectites. As the polyhydroxoaluminium complexes are positively charged, the cations are assumed to be oriented with their positive head groups pointing away from the pillars (Srinivasan and Fogler, 1990). This makes the surface hydrophilic, whereas the space between the alkyl chains retains its hydrophobic character. It is also assumed that an electrostatic shielding against flocculation further increases the effective interface area.

Another way to modify pillared smectites is proposed by Michot and Pinnavaia [109]. Sodium montmorillonite is reacted with a solution containing the $[Al_{13}O_4(OH)_{24+x}(H_2O_{12-x}]^{(7-x)+}$ cation [110] and a technical alkyl pentaethylene oxide (Tergitol 15s-5). The reaction product is a pillared smectite loaded with the nonionic surfactant. The surfactant occupied micropores between the pillars are the adsorption sites for 3-chlorophenol, 3,5-dichlorophenol, 3,4,5-trichlorophenol and pentachlorophenol. The uptake of the pollutants increases with the number of chlorine atoms. The toxicant loaded clay can be recycled by calcination at 500 °C and re-adsorption of the surfactant.

## Conclusion

Smectite clay adsorbents (bentonites) offer a broad scale of applications and bind different kinds of compounds (Table 3). Percolation experiments are now widely used to test the adsorption behavior of absorbents for practical purposes. An enormous pool of data is produced. However, the data must be used very cautiously as they are often obtained under less defined conditions and far from equilibrium.

The interaction of organic materials with smectites often change the particle-particle interactions, the structure of the aggregates and mechanical properties. This can seriously affect the sealing properties of clay layers.

A further effect to be considered is the catalytic activity of the clay minerals which may cause degradation and rearrangement of the organic compounds. The behavior and fate of biocides are greatly influenced by such reactions. Catalytic hydrolysis is frequently observed [111—116]. The parent compound of many herbicides, s-triazine, is hydrolyzed to formamide [117]. The degree of protonation and hydrolysis of the derivatives is very different for the particular compounds [118].

The rate and mechanisms of degradation of parathion depend on the nature of the clay, its hydration status, and the saturating cation [119—121]. In particular cases (e.g., parathion) the rearrangement or degradation products are more toxic to mammals than the parent compound itself.

An instructive example of the influence of catalytic reactions on the mobility of organic compounds in soils is reported by Sawhney [122], who studied the oxidation and oligomerization of o- and m-methylphenol and -chlorophenol vapors in the presence of montmorillonite. As the polymerization of phenols is an oxidation process, a higher degree of polymerization and greater adsorption are observed in air than in nitrogen. As a consequence, the mobility of the phenols is reduced under aerobic conditions. In addition, the polymerized phenols also reduce the mobility of unreacted phenols by adsorption. Due to the greatly reduced polymerization in absence of oxygen an aerobic environment facilitates transport of the toxicants to ground water. Chemisorption and catalytic polymerization of halocarbons on montmorillonite reduces the mobility of halocarbons in soils [71].

**Table 3** Adsorption capacities of montmorillonite (exchanged with various cations from aqueous solutions)

| Species | Cation | Adsorption | | Ref |
|---|---|---|---|---|
| | | meq/g | mg/g | |
| $Cu^{2+}$ | $Na^+$, $Ca^{2+}$ | 1 | 32 | |
| $Cd^{2+}$ | $Na^+$, $Ca^{2+}$ | 1 | 56 | |
| $Pb^{2+}$ | $Na^+$, $Ca^{2+}$ | 1 | 104 | |
| lauryl ammonium | $Na^+$, $Ca^{2+}$ | 1 | 186 | |
| lauryl sulfate | $Na^+$ | 0.03[1] | 8[1] | |
| benzene | trimethylammonium | 1.3 | 101 | [98] |
| p-xylene | trimethylammonium | 0.14 | 15 | [98] |
| m-cresol | octadecyltrimethylammonium | 1.7 | 182 | [94] |
| p-nitrophenol | octadecyltrimethylammonium | 2.3 | 320 | [94] |
| 3,4,5-trichlorophenol | hexadecyltrimethylammonium | 0.8 | 159 | [96] |
| pentachlorophenol | hexadecylpyridinium | 0.12 | 32 | [97] |
| 3-monochlorophenol | pillared + Tergitol | 0.1 | 13 | [109] |

[1] apparent adsorption (from depletion of lauryl sulfate in solution at pH < 3; cf. p. 5).

# References

1. Lagaly G, Fahn R (1983) Ton und Ton-minerale. In: Ullmann's Encyklopädie der technischen Chemie. 23:311—326
2. Jasmund K, Lagaly G (1993) Ton-minerale und Tone — Struktur, Eigenschaften, Anwendungen und Einsatz in Industrie und Umwelt. Steinkopff Verlag Darmstadt
3. Brindley GW, Brown G (1980) Crystal structures of clay minerals and their X-ray identification. Min Soc London
4. Lagaly G (1992) From clay mineral crystals to colloidal clay mineral dispersions. In: Dobias B, Coagulation and flocculation. Marcel Dekker, New York, pp 427—494
5. Lagaly G (1993) Structural chemistry of silicates. In: Ullmann's Encyclopedia of Industrial Chemistry v. A23, pp 661—674
6. Lagaly G (1991) Erkennung und Identifizierung von Tonmineralen mit organischen Stoffen. In: Tributh H, Lagaly G (Hrsg) Ber Deutsch Ton- und Tonmineralgruppe (DTTG) Gießen 1991: S 86—13
7. Lagaly G (1981) Clay Min 15:1
8. Yariv S (1992) Internat Rev Phys Chem 11:345
9. 9a. Mortland MM (1970) Adv Agron 22:75
9b. Weber JB (1972) Interaction of organic pesticides with particulate matter in aquatic and soil systems. In: Advances in chemistry series, number 111: Am Chem Soc, pp 55—120
9c. White JL (1976) Clay-pesticide interactions. ACS Symposium Series, Number 29: Bound and conjugated pesticides residues. Kaufman DD, Still GG, Paulson GD, Suresh KB (eds) Am Chem Soc, pp 208—218
9d. Hance RJ (1980) Interactions between herbicide and soil. Academic Press, London
10. Ogawa M, Kuroda K, Kato C (1989) Chem Lett 1659 (keine Jahrgangs-nummer)
11. Ogawa M, Kato K, Kuroda K, Kato C (1990) Clays Clay Min 8:31
12. Ogawa M, Fujii K, Kuroda K, Kato C (1991) Mat Res Soc Symp Proc v 233: pp 89—94
13. Ogawa M, Hirata M, Kuroda K, Kato C (1992) Chem Lett: 365
14. Brindley GW, Moll WF (1965) Am Min 50:1355
15. Pfirrmann G, Lagaly G, Weiss A (1973) Clays Clay Min 21:239
16. Xu S, Harsh JB (1992) Clays Clay Min 40:567
17. Yariv S, Heller-Kallai L (1975) Clay Min 10:479
18. Mortland MM, Raman KV (1968) Clays Clay Min 16:393
19. Mingelgrin U, Tsvetkov F (1985) Clays Clay Min 33:62
20. Narine DR, Guy RD (1981) Clays Clay Min 29:205
21. Rupprecht H (1987) In: Müller BW (ed) "Controlled Drug Delivery" APV paperback No 17, WVG Stuttgart pp 197—225
22. Takahashi T, Yamaguchi M (1991) J Colloid Interface Sci 146:556
23. Samii AM, Lagaly G (1987) In: Schultz LG, van Olphen H, Mumpton FA (eds) Proc Intern Clay Conf, Denver (1985) The Clay Min Soc, Bloomington, Indiana, pp 343—351
24. Herrmann H, Lagaly G (1985) In: Konta J (ed) Proc 5th European Clay Conf Prague 1983, Univerzita K Praha, pp 269—277
25. Lagaly G (1984) Phil Trans R Soc London A 311:315—332
26. Lailach G, Brindley GW (1969) Clays Clay Min 17:95
27. Weber JB, Weed SB (1968) Soil Sci Soc Am Proc 32:485
28. Weed SB, Weber JB (1969) Soil Sci Soc Am Proc 33:379
29. Haque R, Lilley S, Coshow WR (1970) J Colloid Interf Sci 33:185
30. de Keizer A (1990) Progr Colloid Polymer Sci 83:118
31. Raupach M, Emerson WW, Slade PG (1979) J Colloid Interf Sci 69:398
32. Weber JG, Weed SB, Best JA (1969) Agricult Food Chem 17:1075
33. Cenens J, Schoonheyd RA (1988) Clays Clay Min 36:214
34. Schoonheyd RA, Heughebaert L (1992) Clay Min 27:91
35. Rytwo G, Serban C, Nir S, Margulies L (1991) Clays Clay Min 39:551
36. Yariv S, Nasser A, Bar-on P (1990) J Chem Soc Farad Trans 86:1593
37. Yariv S, Ghosh DK, Hepter LG (1991) J Chem Soc Farad Trans 87:1201
38. Margulies L, Rozen H, Nir S (1988) Clays Clay Min 36:270
39. Margulies L, Rozen H, Cohen E (1988) Clays Clay Min 36:159—164
40. Nir S (1986) Soil Sci Soc Am J 50:52
41. Hirsch D, Nir S, Banin A (1989) Soil Sci Soc Am J 53:716
42. Margulies L, Cohen E, Rozen H (1987) Pestic Sci 18:79
43. Brahimi B, Labbe P, Reverdy G (1992) Langmuir 8:1908
44. Weiss A (1963) Clays Clay Min 10:191
45. Favre H, Lagaly G (1991) Clay Min 26:19
46. Barrer RM, Kelsey KE (1961) Trans Farad Soc 57:625
47. Mackintosh EE, Lewis DG, Greenland DJ (1972) Clays Clay Min 20:125
48. McAteee JL (1962) Clays Clay Min 9:444
49. McAtee JL (1963) Clays Clay Min 10:153
50. Rheinländer T, Klumpp E, Rossbach M, Schwuger MJ (1992) Progr Colloid Polym Sci 89:190
51. White GN, Zelazny LW (1988) Clays Clay Min 36:141
52. Lagaly G (1989) Appl Clay Sci 4:105
53. Permien T (1993) Thesis, Univ Kiel
54. Wendelbo R, Rosenqvist IT (1987) In: Schultz LG, van Olphen H, Mumpton FA (eds) Proc internat clay conf 1985, The Clay Min Soc, Bloomington, Indiana pp 422—426
55. Peinemann N, Helmy AK (1992) Appl Clay Sci 6:419
56. Weiss A (1958) Kolloid Z 158:22
57. Bolt GH, Warkentin BP (1958) Kolloid Z Z Polymere 156:41
58. Chan DYC, Pashley RT, Quirk JP (1984) Clays Clay Min 32:131
59. Edwards DG, Quirk JP (1962) J Colloid Sci 17:872
60. Schott H (1967) Kolloid Z Z Polymere 219:42
61. Pusino A, Micera G, Gessa C (1991) Clays Clay Min 39:50
62. Gast RG, Klobe WD (1971) Clays Clay Min 19:311
63. Gast RG (1972) Soil Sci Soc Am Proc 36:14
64. Inoue A, Minato H (1979) Clays Clay Min 27:393
65. Bergseth H (1980) Acta Agriculturae Scand 30:460
66. van Bladel R, Halen H, Cloos P (1993) Clay Min 28:33
67. Brouwer E, Baeyens B, Maes A, Cremers A (1983) J Phys Chem 87:1213
68. Sposito G, LeVesque CS (1985) Soil Sci Soc Am J 49:1153
69. Stockmeyer M, Kruse K (1991) Clay Min 26:431
70. Weiss A (1989) Appl Clay Sci 4:193
71. Estes TJ, Vilker VL (1989) J Colloid Interface Sci 133:166
72. Hasenpatt R, Degen W, Kahr G (1989) Appl Clay Sci 4:179
73. Hasenpatt R (1988) Thesis ETH Zürich
74. Stul MS, van Leemput L (1982) Surface Technology 16:89; 101
75. Stul MS (1985) Clay Min 20:301
76. Hettinger WP (1991) Appl Clay Sci 5:445
77. Pinnavaia TJ, Tzou MS, Landau SD, Raythatha RH (1984) J Molecular Catalysis 27:195
78. Sawhney BL, Gent MPN (1990) Clays Clay Min 38:14
79. Fahn R (1973) Fette-Seifen-Anstrich-mittel 75:77
80. Fahn R, Fenderl K (1983) Clay Min 18:447
81. Barrer RM (1986) J Inclusion Phenomena 4:109

82. Barrer RM (1989) Pure Appl Chem 61:1903
83. Barrer RM, Hampton MG (1957) Trans Farad Soc 53:1462
84. Barrer RM, Perry SG (1961) J Chem Soc 842; 850 (keine Jahrgangs-nummer)
85. Barrer RM, Millington AD (1967) J Colloid Interface Sci 25:359
86. Lagaly G, Malberg R (1990) Colloids Surfaces 49:11
87. Stul MS, van Leemput L, Rutsaert M, Uytterhoeven JG (1983) J Colloid Interface Sci 92:222
88. Stul MS, van Leemput L, Leplat L, Uytterhoeven JB (1983) J Colloid Interface Sci 94:154
89. Kuhn K, Weiss A (1988) Schr Angew Geol, Karlsruhe 4:141
90. White D (1964) Clays Clay Min, Proc 12th Nat Conf, Atlanta 1963, pp 257—266
91. Bors J (1990) Radiochim Acta 51:139
92. Bors J, Gorny A (1992) Appl Clay Sci 7:245
93. Holz M, Sörensen M (1992) Ber Bunsenges Phys Chem 96:1441
94. Street GB, White D (1963) J Appl Chem 13:288
95. Stul MS, de Bock J (1985) Clays Clay Min 33:350
95a. Dékány I, Szántó F, Weiss A, Lagaly G (1986) Ber Bunsenges Phys Chem 90:422; 427
96. Mortland MM, Shaobai S, Boyd SA (1986) Clays Clay Min 34:581
97. Boyd SA, Shaobai S, Lee JF, Mortland M (1988) Clays Clay Min 36:125
98. Lee JF, Mortland MM, Boyd SA, Chiou CT (1989) J Chem Soc Farad Trans I 85:2953
99. Lee JF, Mortland MM, Chiou CT, Kile DE, Boyd SA (1990) Clays Clay Min 38:113
100. Seefeld V, Bertram R, Starke P, Geßner W (1988) Silikattechnik 39:239
101. Malla PB, Komareni S (1990) Clays Clay Min 38:363
102. Figueras F et al. (9 Namen!) (1990) Clays Clay Min 38:257
103. Serte J (1991) Clays Clay Min 39:167
104. Pa Ho Hsu (1992) Clays Clay Min 40:300
105. Molina R, Vieira-Coelho A, Poncelet G (1992) Clays Clay Min 40:480
106. Zielke RC, Pinnavaia TJ (1988) Clays Clay Min 36:403
107. Nolan T, Srinivasan KR, Fogler HS (1989) Clays Clay Min 37:487
108. Srinivasan KR, Fogler HS (1990) Clays Clay Min 38:277; 287
109. Michot LJ, Pinnavaia TJ (1991) Clays Clay Min 39:634
110. Botero JY, Axelos MAV, Tchoubar D, Cases JM, Fripiat JJ, Fiessinger F (1987) J Colloid Interface Sci 117:47
111. El-Amamy MM, Mill T (1984) Clays Clay Min 32:67
112. Fusi P, Ristori GG, Franci M (1982) Clays Clay Min 30:306
113. Fusi P, Franci M, Ristori GG (1989) Appl Clay Sci 4:235
114. Fusi P, Franci M, Ristori GG (1989) Appl Clay Sci 4:403
115. Ristori GG, Fusi P, Franci M (1987) Appl Clay Sci 2:145
116. Sánchez-Camazano M, Sánchez-Martin MJ (1991) Clays Clay Min 39:609
117. Nguyen TT (1986) Clays Clay Min 34:521
118. White JL (1976) In: Bailey SW (ed) Proc Internat Clay Conf Mexico City 1975, Publ Ltd, Wilmette, Illinois, pp 391—398
119. Mingelgrin U, Saltzman S (1979) Clays Clay Min 27:72
120. Yaron B (1978) Soil Sci 125:210
121. Gerstl Z, Yaron B (1981) Clays Clay Min 29:53
122. Sawhney BL (1985) Clays Clay Min 33:123
123. Martin RT et al. (9 Namen) (1991) Clays Clay Min 39:333
124. Müller-Vonmoos M, Kahr G (1983) Nagra Techn Ber 83—12 (Nagra CH-5401 Baden)
125. Müller-Vonmoos M, Kahr G, Madsen FT (1991) In: Tributh H, Lagaly G (eds) Identifizierung und Charakterisierung von Tonmineralen. Ber Dtsch Ton- und Tonmineralgruppe, Gießen, S 131—155

Progr Colloid & Polym Sci (1994) 95:73—90
© Steinkopff Verlag 1994

I. Dékány
M. Szekeres
T. Marosi
J. Balázs
E. Tombácz

# Interaction between ionic surfactants and soil colloids: adsorption, wetting and structural properties

Prof. Dr. I. Dékány (✉) ·
M. Szekeres · T. Marosi · J. Balázs ·
E. Tombácz
Department of Colloid Chemistry
Attila József University
Aradi Vŕtanúk tere 1
6720 Szeged, Hungary

**Abstract** The adsorption of ionic surfactants on different soil components such as silica, clay minerals, and humic acids was studied. The adsorption processes were controlled by flow microcalorimetry to determine the molar adsorption enthalpies of surfactant accumulation on clay and silicate surfaces. The evaluation of adsorption results for cationic surfactants has shown different mechanisms for solids having permanent (kaolinite, illite, montmorillonite) and pH-dependent surface charges (silica gels and powders). The adsorption mechanism for surfactants on silica surfaces with pH-dependent charges has been explained in terms of the development of charges on the surfaces and their interaction with surfactant cations and micelles. The surface hydrophobicity of clay-organocomplexes was characterized by batch microcalorimetry using pure liquids. The fractal dimensions of clays and their organocomplexes were determined by SAXS. The intercalation of organic compounds in the interlamellar space of layered silicates and humates was measured by x-ray diffraction using powder samples and suspensions. The structural properties of the surfactant coated clay suspension were characterized by rheology.

**Key words** Clay minerals — adsorption — ionic surfactants — heat of wetting — surface hydrophobicity — humic acids — rheology — SAXS experiments

## Introduction

Recently, difficulties arising from soil pollution have become conspicuous among the serious environmental protection problems pending. The surfactants are one group of organic pollutants which may come into the soils, and although the level of contamination is frequently low, their effect on interfacial and colloid structural properties of soils might be drastic. Considering such composite systems as soils in general, but especially from a interfacial and structural point of view, the colloidal fraction of soils is the most reactive. This fraction contains fine clay, oxide particles and humic substances, and their role in the interactions with dissolved materials and in the soil structure formation is fundamental.

The adsorption of ionic surfactants on several different kinds of natural solid materials (e.g., ores, minerals, clays) is of great practical importance. Therefore, a vast and ever-expanding literature is published on this subject. Because of the complexity of interactions involved in the accumulation of ionic surfactant at solid/liquid interface from dilute aqueous solution, no single and unifying model of adsorption process for a range of solid-surfactant combinations has yet emerged [1, 2].

Considering that the interaction of ionic surfactants with the colloidal components of soils is inherently influenced by their surface charge characteristics, components holding permanent and variable charges have to be distinguished. Phyllosilicates, e.g., kaolinite, illite, vermiculite or montmorillonite among swelling clays, are the major source of permanent negative charge, and oxides, e.g., alumina, silica, as well as organic humic substances have pH-dependent charges. The edges of clay particles also exhibit pH-dependent charge sites similarly to that on oxide surfaces which are due to ionization of broken $Si$-OH and $Al$-OH bonds.

Adsorption of ionic surfactants on oppositely charged surfaces, i.e., anionic surfactants at pH below the point of zero charge and cationic ones above that characteristic pH of oxides, and also cationic surfactants on clay lamellae, has been studied extensively. Besides the excellent reviews in the appropriate chapters of some books [1—3], some papers [4—10] among the previous publications from our department, and some recently published papers [11—15] have to be mentioned. Only a few publications on the interaction between cationic surfactants and predominantly negatively charged humic substances can be found [16—18]. Electrostatic interactions and hydrophobic bonding are important driving forces for adsorption. At lower concentration monolayer of surfactant molecules holding long alkyl chains can cover the oxide surface or the basal plates of clay particles, turning them hydrophobic. At higher concentration bilayers are developed or "hemimicelle" formation takes place due to the lateral interactions between alkyl chains [1, 2, 11, 13, 15, 19, 20—22].

Beyond the quantitative measurement of adsorption and the endeavor to describe the adsorption isotherm theroetically, innumerable instrumental methods are used to characterize the adsorption process, forming bonds, structure of interfacial layers and of whole dispersion, as well as the properties of organocomplex forming from surfactant and substrate. Frequently used spectroscopic and electrokinetic methods [1, 2, 11, 15, 19, 20] provide valuable data to interpret adsorption results and to reveal the mechanisms of adsorption. Microcalorimetry has recently been used [14, 23—26] to discuss the thermodynamics of adsorption and wetting properties of organocomplexes. The hydrophobicity of organocomplexes can be characterized by liquid mixture adsorption [5—8, 23—26]. X-ray diffraction method has proved to be excellent for lamellar systems [3, 4—8, 27—31] since the structure of organocomplexes, liquid mixture composition dependent swelling and the intercalation of solvent or other organic molecules can be studied by this method.

The present review is based on the research experience assembled at our department over decades in the field of ionic surfactant adsorption on several kinds of adsorbents, and it surveys our results related to some adsorbents being also important soil constituents, such as swelling and non-swelling clay minerals, silicas, and humic substances. Results of solution adsorption in parallel with flow microcalorimetry, immersional heat and rheological measurements, as well as x-ray diffraction and small-angle x-ray scattering of organocomplexes will be presented.

## Materials and methods

### Materials

*The following solids were used as adsorbents:*
Silicagel R (Reanal, Hungary, analytical grade), silica powder (precipitated from commercial sodium silicate solution by alcohol and then dried and sieved. The $Na^+$ content at the surface of the precipitated silica was 3.07 mmol per 100 g solid powder, determined potentiometrically.), silicates: Sepiolite (Spain), Zeolite 13X (Linde, FRG), kaolinite (Zettlitz, FRG), illite (Füzérradvány, Hungary), montmorillonite (Mád, Hungary), and vermiculite (South Africa). The specific surface areas of the adsorbents are summarized in Table 1. Their detailed description has been published [23—26].

*Surfactants:* n-alkylammonium-chloride, n-alkylpyridinium-chloride, sodium-dodecylsulphate (Fluka AG, FRG, purum).

*Humic substances:* Humic acids were obtained from brown coals (Tatabánya, Hungary, and Steinberg, FRG) by extraction with 0.1 mol dm$^{-3}$ NaOH. After the usual purification process, the sample from Tatabánya was fractionated with NaCl as described in [32]. The total acidity (method in [32]) and elemental composition of these humic acids are listed in Table 2. Humic acid solution was made by passing Na-humate solution through a column of hydrogen cation-exchange resin. The Na-humate solution used in the experiments was made by neutralization of the humic acid solution with NaOH.

**Table 1** Specific surface areas of the adsorbents used

| Adsorbent | $a_s^{BET}$, m$^2$/g | $a_s^L$, m$^2$/g |
|---|---|---|
| Quartz | 4.5 | — |
| Silicagel R | 557 | 580 |
| Precipitated silica | 78.9 | — |
| Sepiolite | 228 | — |
| Sepiolite acid treated | 463 | — |
| Zeolite 13X | 530 | — |
| Kaolinite | 28 | 80 |
| Illite | 44 | 101 |
| Montmorillonite | 76 | 760 |
| Vermiculite | 2.1 | 780 |
| HDP-Montmorillonite organocomplex | 38 | 778 |

$a_s^{BET}$ — specific surface area determined by BET method
$a_s^L$ — specific surface area determined by liquid adsorption method [4]

**Table 2** Total acidity and elemental composition of humic acids used in the experiments

| Humic acid from brown coal from | Total acidity mmol/g | Elemental composition % | | | | |
|---|---|---|---|---|---|---|
| | | C | O | N | H | Ash |
| Tatabánya | 5.974 | 54.0 | 34.9 | 3.1 | 4.3 | 3.7 |
| Steinberg | 6.700 | 56.1 | 30.6 | 3.5 | 4.2 | 5.6 |

The preparation of n-alkylammonium organocomplexes is described elsewhere [16]. The amounts of alkylammonium salts added to the humate solution were equivalent to the acidic groups (carboxyl and phenolic hydroxyl) of the humic acid sample.

## Methods

*Adsorption measurements:* The adsorption of various cationic surfactants was determined in aqueous suspensions and the specific excess amount was calculated using the relation $n_2^{\sigma(v)} = V^0 \Delta c/m$, where $V^0$ is the volume of the liquid phase, $m$ is the mass of the solid, $\Delta c$ is the difference between the initial and equilibrium concentrations ($\Delta c = c_0 - c_{eq}$). The equilibrium concentration of surfactants was determined by two-phase titration method [33].

*Microcalorimetric measurements:* The measurements were carried out with an LKB 2107 (Sweden) isothermal adsorption microcalorimeter, using two types of measuring techniques. The adsorption measurements were performed with the flow unit of the microcalorimeter at 25 ± 0.01 °C on porous silica gel, precipitated silica, and sepiolite clay minerals. The flow rate was 15—25 cm$^3$ surfactant solution/h. The heat effects at various concentrations were corrected by the heat of mixing [23—25]. The integral enthalpy of displacement $\sum_j \Delta_d h = f(c_{eq})$ in the process of adsorption was given as a function of the equilibrium surfactant concentration $c_{eq}$. Hydrophobilized kaolinite, illite, and montmorillonite were investigated by "batch" immersion in various wetting liquids. The molar enthalpy of immersional wetting, $\Delta_w H_i$ (J/mmol), was given as a function of the surface hydrophobicity [26].

*X-ray diffraction studies:* The x-ray diffraction measurements were performed with a Philips PW 1830 generator fitted with two measuring units: a powder diffractometer operating in a wide ($1°—90°$ in $2\,\Theta°$)-angle range and a compact Kratky camera in a small-angle range ($0°—5°$ in $2\,\Theta°$) (SAXS). The measurements were carried out at 40 kV and 35 mA with CuKα radiation ($\lambda = 1.54$ Å) using a Ni filter at 25 °C temperature. The sample support of the powder diffraction goniometer was used for powder tests and for the investigation of suspensions as well. In the latter case, the suspension was covered by Mylar foil to prevent evaporation of the dispersion liquids. The measurements were carried out with SAXS for solids at a layer thickness of 1 mm in a sample holder. A special sample holder was furnished with a Mylar foil window for measurements on clay mineral suspensions. In this sample holder, the measurements were performed in 10% suspension at a layer thickness of 1 mm.

The slit distance of the proportional counter detector was 100 μm. The measuring of control data and the processing of measurement data were carried out by computer in both techniques.

*Membrane osomometry:* The number average molar masses of humic acid and Na-humate (sample from Tatabánya) were determined at 5 and 500 mmol dm$^{-3}$ NaCl by membrane osmometry (Knauer membrane osmometer, Sartorius superfine membrane), using the extrapolation procedure described in [32].

## Results and discussion

Silica, silicate surfaces and cationic surfactants

*Precipitated silica, silica gel and quartz adsorbents with pH-dependent surface charge sites*

The adsorption of cationic surfactants from aqueous solutions on solid surfaces with dissociable surface sites depends greatly on the pH of the medium. On the one hand, it is determined by the pH-dependent surface charge state of the adsorbent. The point of zero charge (p.z.c.) of silica surfaces is usually at low pH, most frequently below 4 [34]. From the p.z.c. to pH about 6, the magnitude of the negative surface charge of silicas increases very slowly, especially at low electrolyte concentrations. Above pH 6 the builiding-up of the surface charge is similar to that for the other oxides, and at high pH values, i.e., above 9, they can obtain an enormously high surface charge density particularly in the case of precipitated silica samples [35]. On the other hand, the pH of the solution influences the electrolytic dissociation of the surfactant molecules. Pyridinium salts such as hexadecylpyridinium chloride (HDPCl) are fully dissociated over the whole pH range, but the primary amines such as dodecylammonium chloride (DDACl) are soluble only in strongly acidic media [ref. [2], pp 17—18].

Adsorption of HDP$^+$ on the precipitated silica adsorbent at pH 10.6 leads to monolayer coverage, whereas at pH 3 surface saturation is not observed up to an equilibrium concentration of 10 mmol/dm$^3$, as seen from the shape of the isotherms (Fig. 1a).

At pH 10.6, the surface of the substrate is negatively charged and sodium ions are adsorbed as counterions in the inner Helmholtz plane (IHP) of the electrical double layer (e.d.l.). Adsorption of HDP$^+$ gives a high-affinity isotherm, which reaches the plateau at equilibrium concentrations consistent with the critical micelle concentration (CMC) of HDPCl, 0.9 mmol/dm$^3$ [36]. The adsorbed amount is 0.7 mmol/g at saturation, and the surface area occupied by one molecule is about 0.2 nm$^2$. This means that the adsorbed amount is somewhat higher than would be expected for close-packed monolayer coverage (0.3 nm$^2$/molecule).

Measurements of the pH of the suspensions before and after adsorption (after standing for 24 h) shows that during the adsorption of HDP$^+$ there is no appreciable change in pH (the decrease is about 0.1 pH unit). It is a reasonable assumption that the adsorption of surfactant cations takes place via exchange between the sodium ions and the surfactant cations in the IHP of the e.d.l. of the substrate (see the schematic representation in Fig. 5a).

In acidic media (pH 3), the surface of the precipitated silica is nearly uncharged. The amount of HDP$^+$ adsorbed at very high equilibrium concentrations gives the area occupied by one molecule as 0.66 nm$^2$, which is twice the area demanded by one HDP$^+$ oriented perpendicularly to the surface. Because of the low pH (adjusted with HCl solutin) and the presence of sodium ions (7.5 × 10$^{-2}$ mol/dm$^3$ (originating from the silica preparation), the CMC of HDPCl is lowered. The CMC of HDPCl in 10$^{-2}$ mol/dm$^3$ NaCl solution is 10$^{-4}$ mol/dm$^3$ [11]. At high ionic strengths, the e.d.l. around the micelles has to be narrowed and the charge of the micelles is compensated almost fully within the compact layer. Thus, it can be supposed that adsorption of the micelles occurs on the uncharged surface of the precipitated silica in addition to the individual surfactant adsorption (Fig. 6b).

The adsorption of HDP$^+$ on Silicagel R and quartz (Fig. 1b) was investigated at neutral pH (6.5 for Silicagel R and 7.5 for quartz); thus, the surface of both substrates was negatively charged. In both cases, the experimental isotherm shows a plateau of monolayer coverage at equilibrium concentrations near the CMC. For Silicagel R, which has a porous structure, surfactant cations cannot be adsorbed inside the pores, but only on the external surfaces of the substrate. A singificant change in the pH of the bulk solution was detected during adsorption. With increasing adsorbed HDP$^+$ amount ΔpH rose from 0.6 to 2.0 pH units. The proton concentration in the bulk liquid increased due to their displacement from the IHP by adsorbed HDP$^+$ (Fig. 5a).

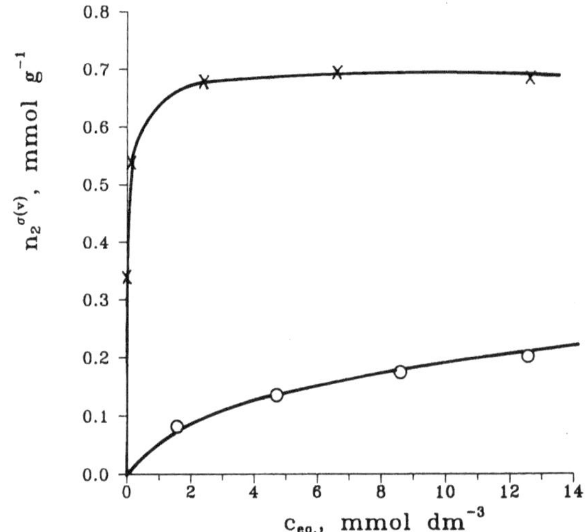

a

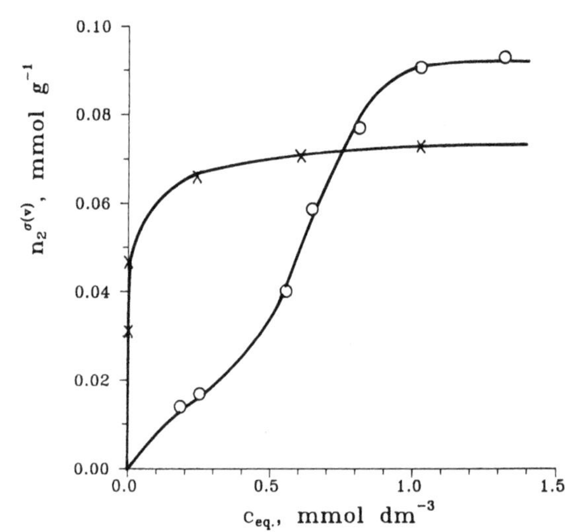

b

**Fig. 1** Adsorption isotherms of HDP$^+$ on a) precipitated silica at pH 10.6 (×) and pH 3 (○); b) Silicagel R at pH 6.5 (×), quartz at pH 7.5 (○)

The adsorption isotherm for the quartz adsorbent shows a bilayer character. The fitted value of the first plateau of the isotherm corresponds to the close-packed monolayer coverage, with an area of 0.3 nm$^2$ occupied by one HDP$^+$. It may be concluded from the amount adsorbed at the second plateau that further adsorption of the surfactant occurs in at least two additional layers, or rather via surface hemimicelle formation.

Figures 2a and 2b present the changes in the integral enthalpy of displacement ($\sum\Delta_d h$, J/g) as a function of the equilibrium concentration, and in the integral molar enthalpy of displacement ($\sum\Delta_d h/n_2^{\sigma(v)}$, kJ/mol) as a function of the relative surface coverage, in the process of HDP$^+$ adsorption on Silicagel R (at pH 6.5) and precipitated silica (at pH 10.6 and pH 3). The relative surface coverage is defined as the ratio of the amount

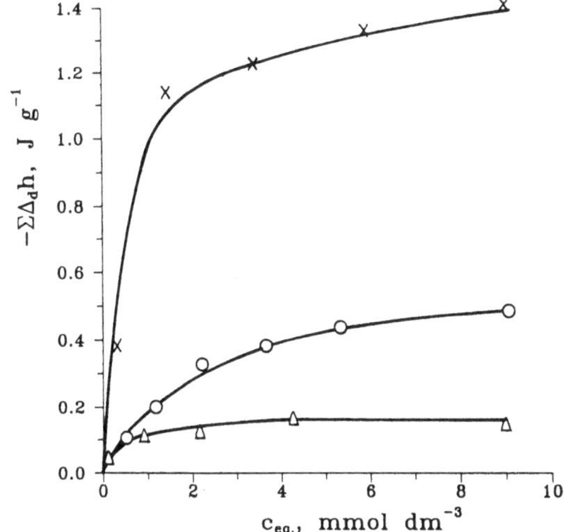

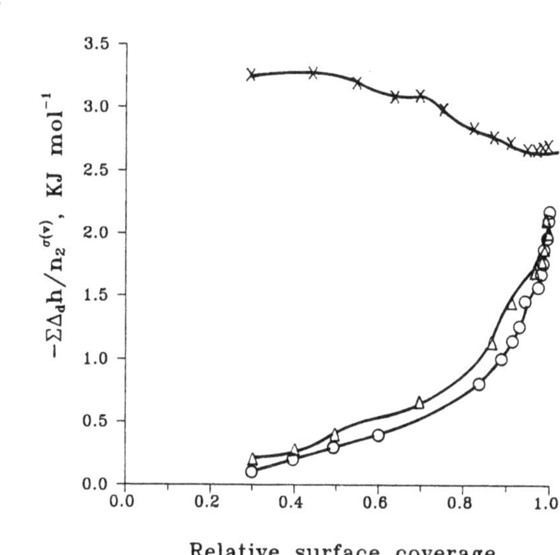

**Fig. 2** Enthalpy changes in the HDP$^+$-adsorption process: a) integral enthalpy function on precipitated silica at pH 10.6 ($\times$) and pH 3 ($\circ$), on Silicagel R at pH 6.5 ($\triangle$); b) integral molar enthalpy as a function of the relative surface coverage on precipitated silica at pH 10.6 ($\triangle$) and pH 3 ($\times$) and on Silicagel R at pH 6.5 ($\circ$)

adsorbed at a given equilibrium concentration to the saturation amount at the first plateau.

In those cases when the pH of the solution is higher than the pH of the p.z.c. of the substrate, i.e., at pH 6.5 for the Silicagel R and pH 10.6 for the precipitated silica, the solid surface is negatively charged and the adsorption process is governed by electrostatic interactions. Exchange of the counterions (H$^+$ for Silicagel R and Na$^+$ for precipitated silica) for surfactant cations in the IHP does not cause a significant change in the molar enthalpy at low relative surface coverage. The molar enthalpy increases sharply in the exothermal direction in the region of close-packed monolayer formation, due

to the lateral interactions between the hydrocarbon chains [21, 22].

Simultaneously, a change also takes place in the water structure [37]. Rearrangement of the water molecules escaping from the surroundings of the hydrocarbon chains of the surfactant molecules ("iceberg" water) is the driving force for the formation of the close-packed monolayer at the solid surface, just as for micelle formation in surfactant solutions.

For the precipitated silica at pH 3 (i.e., the surface is at or close to the state of zero charge), the situation is quite different from that for the charged surfaces. Adsorption of HDP$^+$ on the uncharged hydrophilic surface may be considered to be a result of dipole-dipole interactions between the oriented water molecules of the surface hydrate layer and the hydrate layer around the individual surfactant molecules [ref. [2], p. 56] and surfactant micelles (Figs. 6a and 6b). In this case there is no reason to suppose the existence of electrostatic interactions. The extent of the adsorbed amount increases monotonously, and at maximum adsorption, the surfactant layer is not a close-packed monolayer (0.66 nm$^2$/molecule). However, molar enthalpy of adsorption decreases monotonously up to the maximum relative surface coverage. The heat evolution for the uncharged hydrophilic surface displays quite a different character than that for the negatively charged surfaces, indicating a different mechanism of adsorption. The adsorption of the individual molecules together with their hydration shell seems to disturb the water structure in the hydrate layer of the solid surface and around the surfactant ions in such a way that some of the structured water molecules are discharged. This may be the reason for the large exothermic enthalpy effect at the first stages of the adsorption process. With increasing surface coverage, the integral enthalpy of displacement in proportion to the adsorbed amount gradually decreases (Fig. 2b).

The adsorption of DDA$^+$ on Silicagel R and precipitated silica powder (Fig. 3) was investigated at concentrations lower than 1 mmol/dm$^3$, i.e., well below its CMC (14.8 mmol/dm$^3$ [38]). The adsorption isotherms show constant adsorption above the equilibrium concentration of 0.6 mmol/dm$^3$. The amount of DDA$^+$ adsorbed is lower than that of HDP$^+$ by one order of magnitude for both substrates. DDA$^+$ is dissolved at low pH, and therefore the acidity of the systems was adjusted to pH 2 with HCl solution. In such an acidic medium, the surface of the adsorbents is weakly positively charged [15]. In spite of this, the adsorption of cationic surfactants occurs to a small degree. The occurrence of the second step of the isotherm for the precipitated silica can be explained in terms of the second layer or surface hemimicelle formation.

The integral enthalpy isotherms of displacement for the adsorption of DDA$^+$ on Silicagel R and precipitated silica adsorbents (Fig. 4) indicate that the mechanisms of

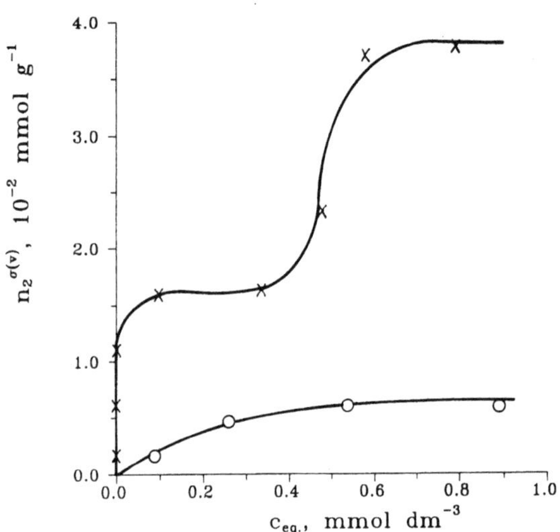
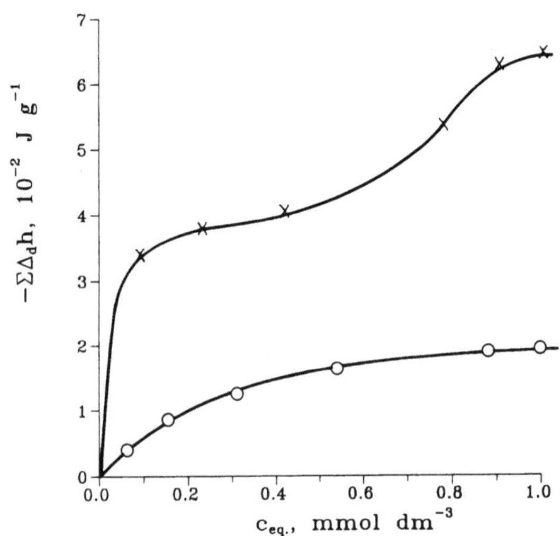

**a**

Fig. 3 Adsorption isotherms of DDA$^+$ on precipitated silica at pH 2 ($\times$) and Silicagel R at pH 2 ($\circ$)

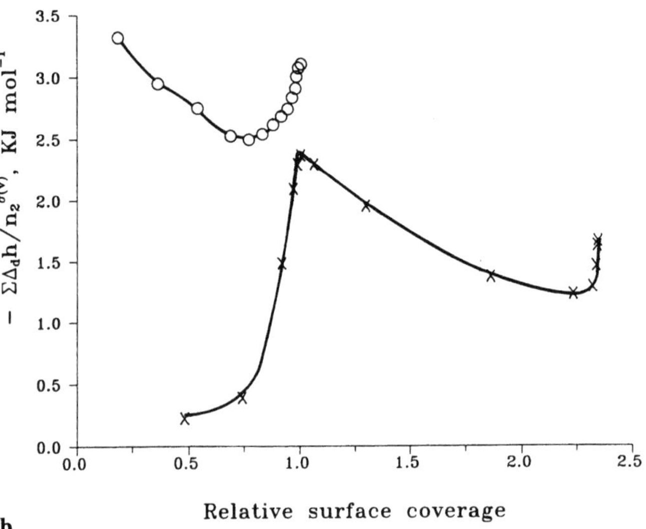

**b**

Fig. 4 Enthalpy changes in the DDA$^+$-adsorption process: a) integral enthalpy function on precipitated silica ($\times$) and Silicagel R ($\circ$), at pH 2; b) integral molar enthalpy as a function of the relative surface coverage on precipitated silica ($\times$) and Silicagel R ($\circ$), at pH 2

adsorption are similar to the mechanisms of adsorption of HDP$^+$ on the uncharged (pH 3) and negatively charged (pH 10.6) precipitated silicas, respectively. In the precipitated silica, the adsorption isotherm has a second step and comparison with the integral enthalpies of displacement suggests that the adsorption of whole micelles occurs due to laeral interactions in this region. For an interpretation, let us consider first the structure of the electrical double layer of Silicagel R and precipitated silica in the presence of the extremely high Cl$^-$ concentration originating from the pH adjustment. SilicagelR has a slightly positively charged surface at pH 2. The pH adjustment with HCl solution results in a small excess (about 0.5 mmol/dm$^3$) of Cl$^-$ in the bulk solution and it is very likely that these ions are adsorbed specifically in the IHP or nonspecifically in the OHP (outer Helmholtz plane), together with their hydration shell. This means that, due to their screening effect, the small positive surface charge is fully compensated within the compact layer of the e.d.l. and the surface behaves in the adsorption process as an uncharged hydrated surface (Figs. 6a and 6b). The precipitated silica adsorbent (slightly positively charged surface at pH 2) contained 33 mmol/dm$^3$ of Na$^+$ in suspension (0.25 g/20 cm$^3$ in the adsorption experiments) and 30—38 mmol/dm$^3$ of Cl$^-$ from the pH adjustment. In addition, the Cl$^-$ concentration is even higher because of the DDA$^+$ counterions. Under such conditions, it is quite reasonable to suppose that the small positive surface charge is overcompensated by Cl$^-$ in the IHP, and therefore the coions (Na$^+$ in the OHP) may be exchanged for DDA$^+$ during adsorption (Fig. 5b). Further, the high salt concentration causes a considerable lowering of the CMC.

*Chain silicates and zeolite*

The adsorption isotherms of HDP$^+$ on chain silicates such as sepiolite and sepiolite modified by acid treatment

(Fig. 7) exhibit a character similar to that for the precipitated silica adsorbent at pH 10.6 and 3 as regards the amount of surfactant adsorbed and the shape of the isoterms, but the integral molar enthalpies of displacement are considerably higher (Fig. 8a). It may be supposed that the reason for this difference is an additional effect due to the displacement of water from the micropores by the surfactant ions during the adsorption process.

The adsorption of HDP$^+$ on the surface of the original sepiolite gives a high-affinity adsorption isotherm. The amount of HDP$^+$ adsorbed in the region of ion exchange (i.e., for the immeasurably low equilibrium concentrations) is 0.2 mmol/g, which is in

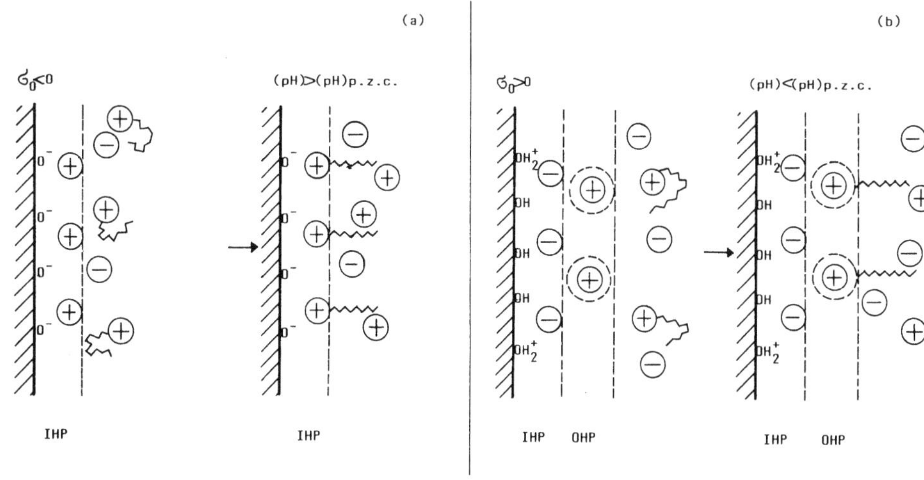

**Fig. 5** Schematic representation for the possible mechanism of interactions between cationic surfactant and charged oxide surface:
a) ion exchange in the Inner Helmholtz Plane;
b) ion exchange in the Outer Helmholtz Plane

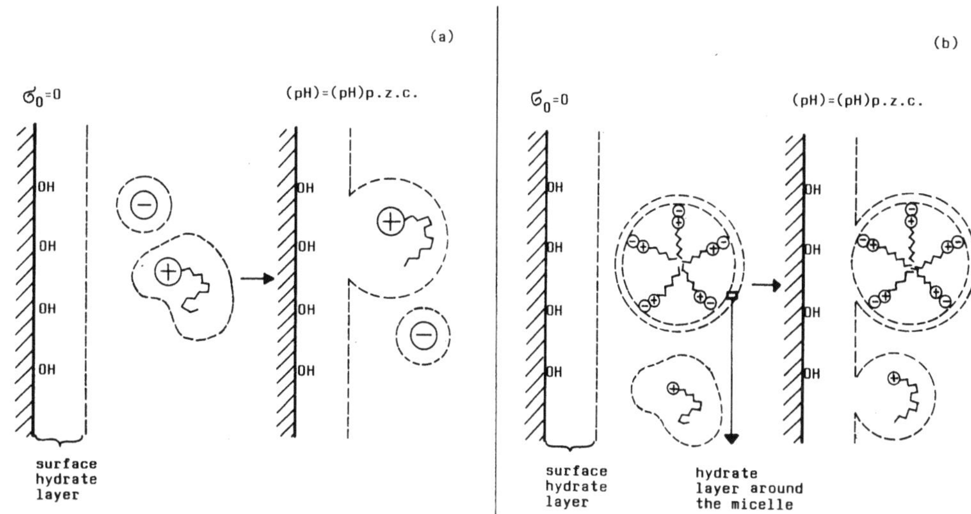

**Fig. 6** Schematic representation for the possible mechanism of adsorption of cationic surfactant on uncharged hydrophilic oxide surface via van der Waals interactions and H-bonding:
a) adsorption of individual surfactant ions;
b) adsorption of micelles

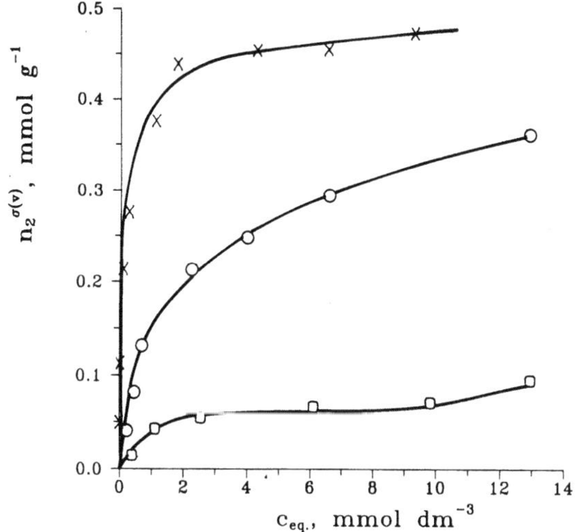

**Fig. 7** Adsorption isotherms of HDP$^+$ on sepiolite (×), acid treated sepiolite (○) and Zeolite 13X (□)

good agreement with the determined ion-exchange capacity of the sepiolite (0.26 mmol/g). At equilibrium concentrations near the CMC (0.9 mmol/dm$^3$), the isotherm turns to a plateau and at higher concentrations second layer formation is not observed. With increasing relative surface coverage, the integral molar enthalpy of displacement (Fig. 8b) increases in the exothermic direction. Thus, lateral interactions between the long hydrocarbon chains at the surface can be detected at a relatively loose packing of surfactant.

Treatment of the sepiolite with 2 mol/dm$^3$ HCl solution for 36 h leads to a decrease of the ion exchange capacity to 0.03 mmol/g. As a consequence, the adsorption of HDP$^+$ on the H-form of sepiolite does not display an ion exchange character, but resembles the adsorption of HDP$^+$ on the uncharged surface of precipitated silica at pH 3 (Fig. 1a). The specific surface area of the sepiolite increases from 228 m$^2$/g to 463 m$^2$/g on acid treatment.

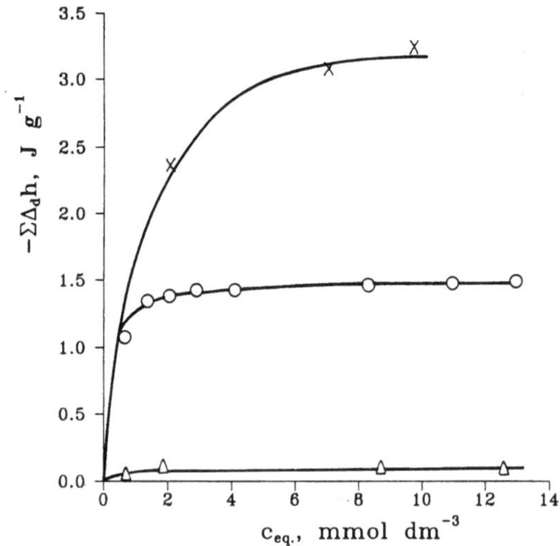

**a**

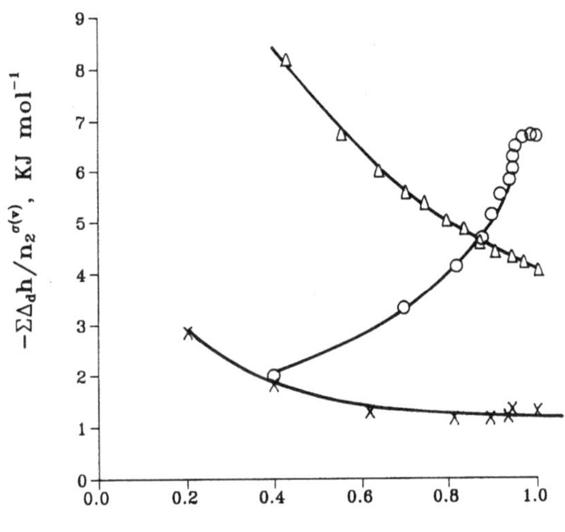

**b**

**Fig. 8** Enthalpy changes in the HDP$^+$-adsorption process: a) integral enthalpy function on sepiolite ($\times$), acid treated sepiolite ($\circ$) and Zeolite 13X ($\triangle$); b) integral molar enthalpy as a function of the relative surface coverage on sepiolite ($\circ$), acid treated sepiolite ($\triangle$) and Zeolite 13X ($\times$)

This increase is due to the partial degradationof the chain structure during displacement of the lattice metal ions (mainly $Al^{3+}$ and $Fe^{3+}$) by $H^+$ in the process of acid treatment. From the adsorption isotherm and the integral molar enthalpy isotherm of HDP$^+$, it can be seen that hydrated individual molecules and, at higher concentrations, hydrated micelles are adsorbed on the surface of the protonated sepiolite particles. Further evidence of such a mechanism is that the maximum amount of HDP$^+$ adsorbed is much smaller (2.19 nm$^2$/molecule) than required for close-packed monolayer coverage.

Figure 9 presents the changes int he adsorption properties of sepiolite as a result of treatment with 2 mol/dm$^3$

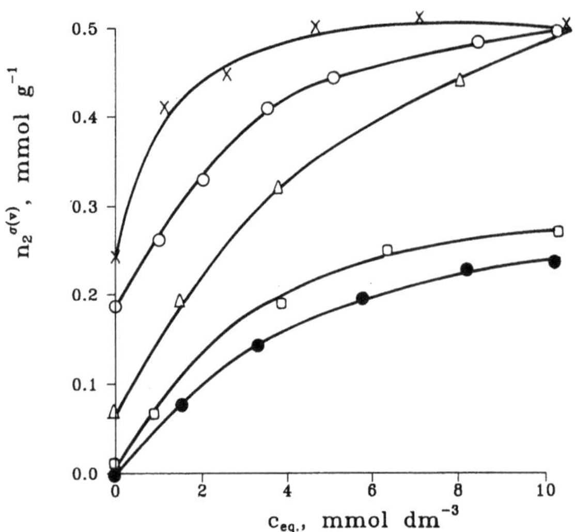

**Fig. 9** Effect of the acid treatment on the adsorption of HDP$^+$ on sepiolite: adsorption isotherms on sepiolite without acid treatment ($\times$), treated with 2 M HCl solution for 5 h ($\circ$), for 10 h ($\triangle$), for 36 h ($\square$) and for 72 h ($\bullet$)

HCl solution during 5, 10, 36, and 72 h. The adsorption of HDP$^+$ on the surface of the original sepiolite may be considered to be an apparently irreversible process. The integral enthalpy changes (Fig. 10) obtained by gradual dilution of the equilibrium surfactant solution in the flow cell after the adsorption process has been completed show that surfactant molecules adsorbed in the ion-exchange process are not desorbed. At the same time, adsorbed surfactant molecules and micelles are displaced from the surface of the H-form of sepiolite by water molecules. The changes in the integral enthalpy of adsorption and desorption are of the same magnitude, but opposite in sign.

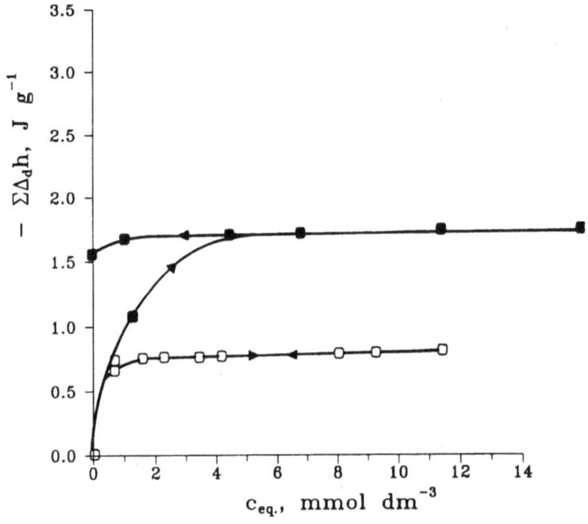

**Fig. 10** Integral enthalpy isotherms of adsorption ($\rightarrow$) and desorption ($\leftarrow$) of HDP$^+$ on sepiolite ($\blacksquare$) and 36-h acid-treated sepiolite ($\square$)

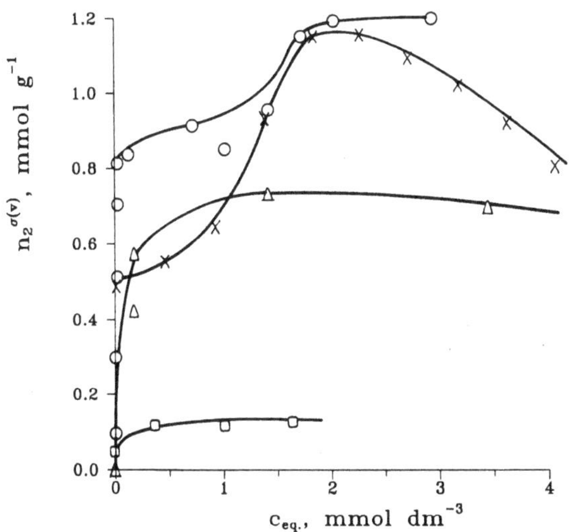

Fig. 11 Adsorption isotherms of HDP$^+$ on layer silicates montmorillonite (O), vermiculite (△), illite (×) and kaolinite (□)

Adsorption on Zeolite 13X (Figs. 7, 8a and 8b), which has a microporous structure, occurs only on the external surfaces and the adsorbed amount and the enthalpy of displacement are small compared with the corresponding data for adsorption on the two sepiolite adsorbents.

*Layer silicates*

Figure 11 depicts the adsorption isotherms of HDPCl on the layer silicates montmorillonite, vermiculite, illite and kaolinite. High-affinity isotherms were obtained in accordance with the CEC of the adsorbents [4]. The amounts of HDP$^+$ adsorbed on montmorillonite, vermiculite, and kaolinite in the process of ion exchange are 0.8, 0.5, and 0.05 mmol/g, respectively. Adsorption on montmorillonite and illite above the CEC occurs by second layer formation. In the adsorption of cationic surfactants on Na-kaolinite, Isfahany [12] concluded that the adsorption involves not only electrostatic interactions between the cationic head groups of the surfactant and the negatively charged surface sites, mainly on the basal plates, but also hydrophobic interactions of the surfactant molecules with the previously adsorbed molecules. The adsorption of HDP$^+$ on vermiculite was measured after allowing to stand for 48 h, and the exchange of cations by HDP$^+$ had not yet been completed.

*X-ray diffraction control of adsorption*
*of cationic surfactants on swelling layer silicates*

The quantities of surfactants adsorbed are well detectable by x-ray diffraction measurements, since the intercalation of surfactant molecules in montmorillonite results in a $d_L$ = 14.1 Å basal spacing if 20% of the total CEC is reached (Fig. 12) [5, 6]. Further adsorption of surfactants leads

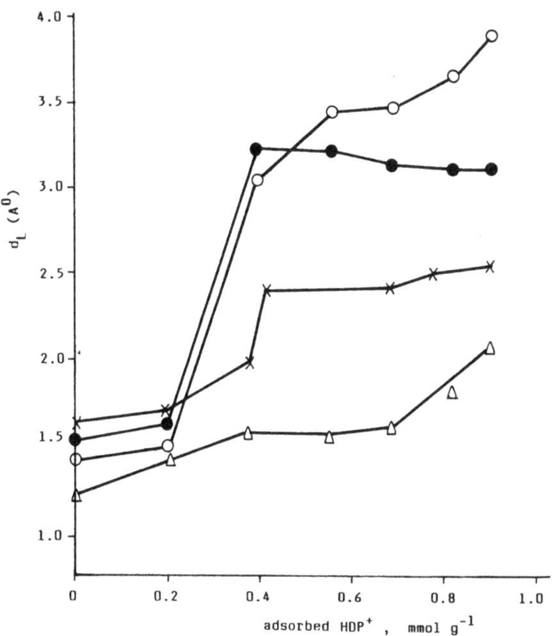

Fig. 12 Basal spacings of HDP$^+$ montmorillonites (△) dry sample, (×) in H$_2$O solution after adsorptionof HDP$^+$ cations, (●) in methanol, (O) in benzene

to a gradual increase in the basal spacing up to the formation of a bimolecular and a pseudo-three-layer organocomplex. The cationic surfactants in the suspension of a clay organocomplex can also be studied. In Fig. 12, $d_L$ is shown as a function of the quantity of surfactant adsorbed. It is worth studying the basal spacing in aqueous suspension because the $d_L$ values are higher than those in the dry material in spite of the increase in hydrophobicity of the surface. This means that a significant amount of water (with electrolyte counterions) can be found among the alkyl chains. If the organocomplexes are suspended in methanol (or in some other short-chain alcohol), the incorporation of the monolayer alkyl chain is sufficient for a significant swelling ($d_L$ = 30—35 Å). The swelling is even larger in benzene (or in other aromatic solvents) and reaches $d_L$ = 40 Å [5,6].

Adsorption of anionic surfactant on hydrophobic
(HDP$^+$-modified) surface

Adsorption isotherms of NaDS on montmorillonite and HDP-montmorillonite are shown in Fig. 13. Adsorption of the anionic surfactant via electrostatic interactions can take place only at the broken edges of the clay plates with pH-dependent surface charge sites under appropriate pH conditions. In our experiments the pH of the suspension was not adjusted, but it can be supposed that the edge sites are partly positively (*Al*-OH) and partly negatively (*Si*-OH) charged at neutral pH. The adsorption of DS$^-$ may

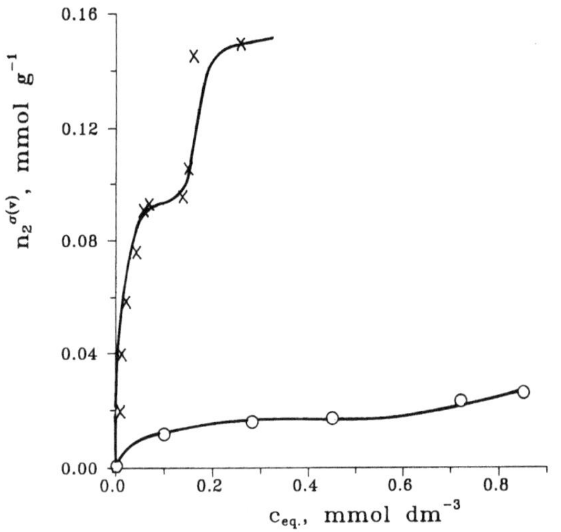

**Fig. 13** Adsorption of NaDS on montmorillonite (○) and HDP-montmorillonite (×)

occur at positively charged sites via electrostatic interactions. DS⁻ adsorption is considerably enhanced on the cationic surfactant-modified surface. The specific amount of DS⁻ adsorbed at the second plateau of the isotherm is 0.15 mmol/g, which is one order of magnitude lower than the amount of HDP⁺ adsorbed in the cation exchange process. This implies that the mechanism of DS⁻ ad-

sorption involves surface hemimicelle formation rather than incorporation of the hydrocarbon chains of DS⁻ into the layer of HDP⁺ fixed to the basal plates of clay particles.

## Results of SAXS experiments

By means of SAXS measurements, mainly the changes in certain structural properties (porosity and structure of gels) of clay minerals in the range 50—1000 Å can be studied. For the evaluation of structural changes, methods have been developed by Schmidt, Kriechbaum and Laggner [39, 40]. Here, however, we would like to illustrate only the fractal properties of clay minerals. Small-angle scattering of the clay material with layered structure is shown in Fig. 14. It can be established that in kaolinite $\log I = f(\log h)$, where the scattering vector $h = (2\pi/\lambda)\sin 2\theta$ follows the well-known Porod law $I(h) = k h^{-4}$ [39]. The fractal dimension estimated from the slope of the scattering function is $D_s = 2.02$. This relates to the fact that the surface is flat and the particles are compact. Actually, bentonite exhibits surface fractal properties, and synthetic hectorite, with high degree of dispersion, gives extremely high values. The scattering

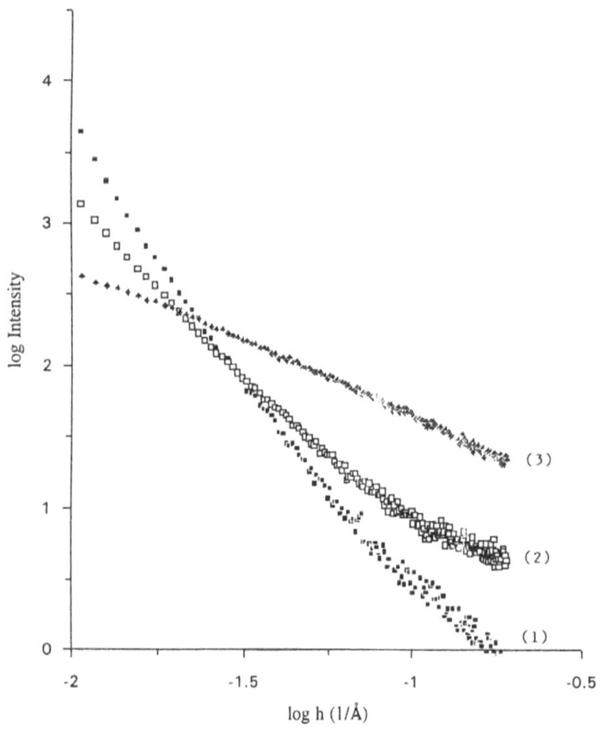

**Fig. 14** SAXS experiments on various layer silicates (1) kaolinite, (2) bentonite, (3) hectorite

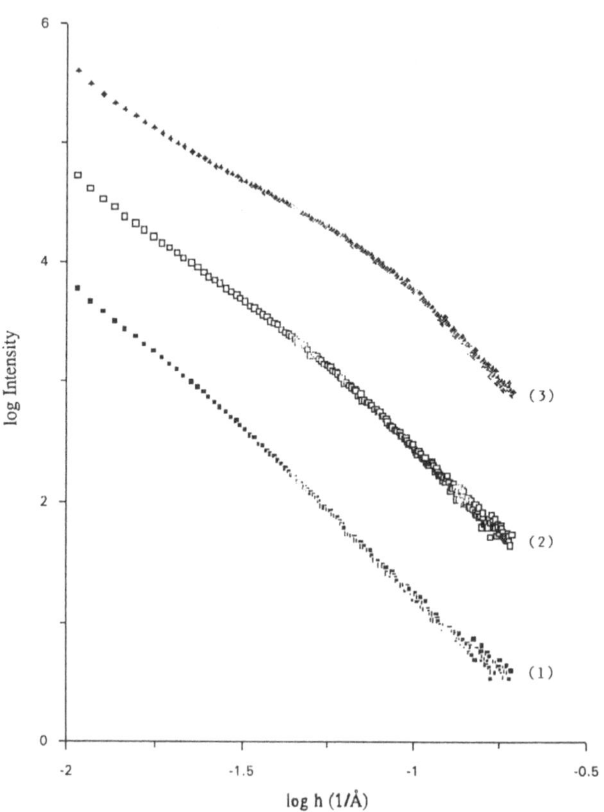

**Fig. 15** SAXS experiments on sepiolite minerals: (1) original sample, (2) treated 12 h with 2 M HCl, (3) treated 36 h with 2 M HCl. The intensities are moved by one (2) and two (3) units related to curve (1)

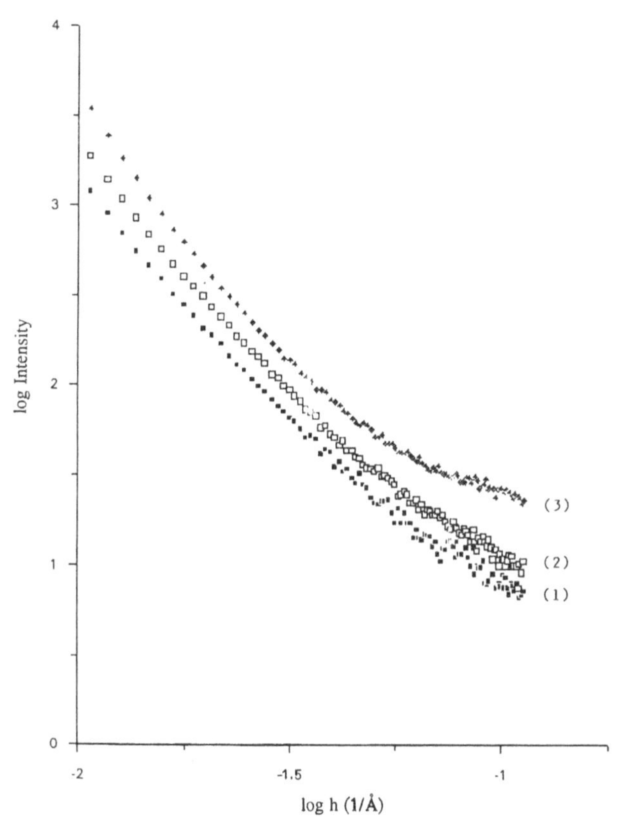

Fig. 16 SAXS experiments on (1) Na-montmorillonite, (2) 81 mequ. HDP$^+$/100 g montmorillonite, (3) 82 mequ. DMDH$^+$/100 g montmorillonite organoclay

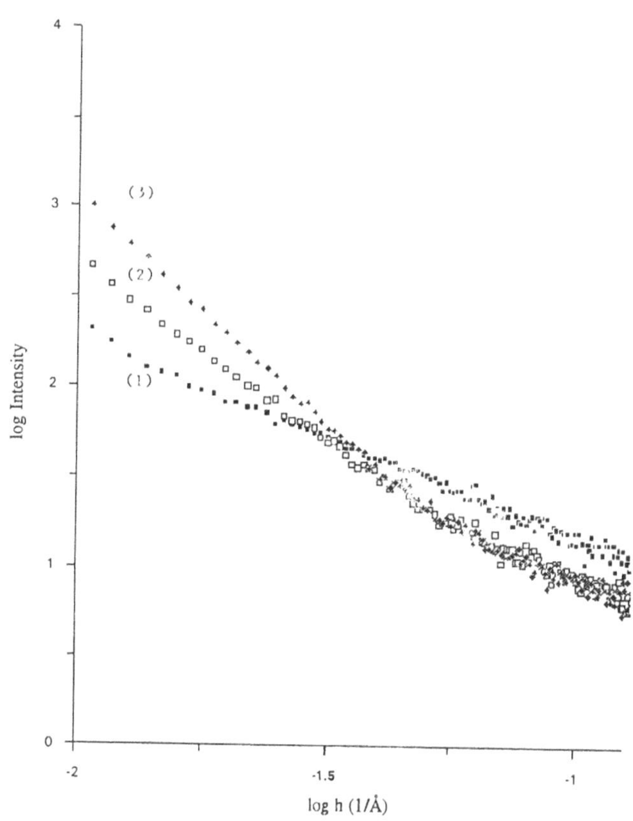

Fig. 17 SAXS experiments in montmorillonite suspension (1) Na-montmorillonite in water, (2) 30 mequ. HDP$^+$/100 g montmorillonite in H$_2$O, (3) 100 mequ. HDP$^+$/100 g montmorillonite in H$_2$O

function of the fibrous structure chain silicate sepolite is shown in Fig. 15. The double log diagram yields two linear sections for all three samples. The slope of the first section is $p_1$, from which the mass fractal dimension $D_m = |p_1| + 1$ can be established. The slope of the second is $p_2$, from which the surface fractal dimension $D_s = 6 - |p_2| + 1$ can be calculated (Table 3). In the original sepolite $D_m \to 3.0$, which is characteristic of smooth-surface particles. The acid treatment results in a decrease in $D_m$ which is caused by the destruction of the chain silicate structure. Hence, the acidic dissolution results in a decrease of the crystalline material and a porous xero gel — network comes into existence. Acid treatment causes practically no change in the surface fractal dimension. This means that the surface structure of the adsorbent does not change, but its adsorption capacity increases significantly (Table 3). Figure 16 shows the scattering curves of montmorillonite and the organocomplexes at maximal CEC. Montmorillonite exhibits surface fractal properties ($D_s = 2.40$). Its value is decreased by hydrophobization because of the aggregation of the surfactant caused by the adsorption of surfactant and drying ($D_s \to 2.0$). In aqueous suspension Na-montmorillonite forms a thixotropic gel and its mass fractal dimension is therefore $D_m = 2.17$.

If surfactants are added, the aggregation of lamellae occurs, the gel structure degrades due to coagulation, and its mass fractal properties cease to exist ($D_m \approx 3.0$) (Fig. 17).

## Immersional wetting properties of hydrophobic layer silicates

If the originally hydrophilic surface of clay minerals is partially covered by cationic surfactants, the heat of wetting is a function of the covered surface area [7, 26]. The non-swelling in aqueous medium kaolinite and illite modified by HDP$^+$ swollen by methanol and benzene can be seen in Fig. 18. The enthalpy of wetting ($\Delta_w H_i$) in polar methanol is exothermic and decreases exponentially as the covered surface area increases. For the interaction between benzene and the adsorbent, the enthalpy of wetting increase appears to be small. Because of intercalation of aromatic organic substances, the lamellae in the swelling montmorillonite organocomplexes move away from one another [8, 28—31]. Consequently, the wetting could also be an endothermic process. The exothermic or endothermic character of the heat effect depends on the wetting liquid and the degree of swelling. Figure 19 demonstrates that with a suitable surfactant

**Table 3** Fractal dimensions of various clay minerals and their organocomplexes with cationic surfactants

| Clay minerals | $|p_1|$ | $|p_2|$ | $D_m$ | $D_s$ | $a_s^{BET}$ (m²/g) |
|---|---|---|---|---|---|
| *layer silicates (dry powders)* | | | | | |
| Kaolinite | — | 2.98 | — | 2.02 | 28 |
| Bentonite | — | 2.20 | — | 2.80 | 52 |
| Hectorite (synt.) | 0.97 | — | 1.97 | — | 247 |
| *chain silicates (dry powders)* | | | | | |
| Sepiolite | 2.05 | 2.86 | 3.05 | 2.14 | 228 |
| Sepiolite (treated 12 h) | 1.95 | 2.85 | 2.95 | 2.15 | 452 |
| Sepiolite (treated 36 h) | 1.50 | 2.92 | 2.50 | 2.08 | 463 |
| *clay-surfactant complexes (dry powders)* | | | | | |
| Na-Montm. | — | 2.60 | — | 2.40 | 76 |
| HDP-Montm. | — | 2.61 | — | 2.39 | 38 |
| DMDH-Montm. | — | 2.88 | — | 2.12 | 29 |
| *clay-surfactant complexes (in aqueous suspension)* | | | | | |
| Na-Montm. | 1.17 | — | 2.17 | — | 778 |
| HDP-Montm. (0.3 mmol/g) | 2.02 | — | 3.02 | — | — |
| HDP-Montm. (1.0 mmol/g) | — | 2.35 | — | 2.65 | — |

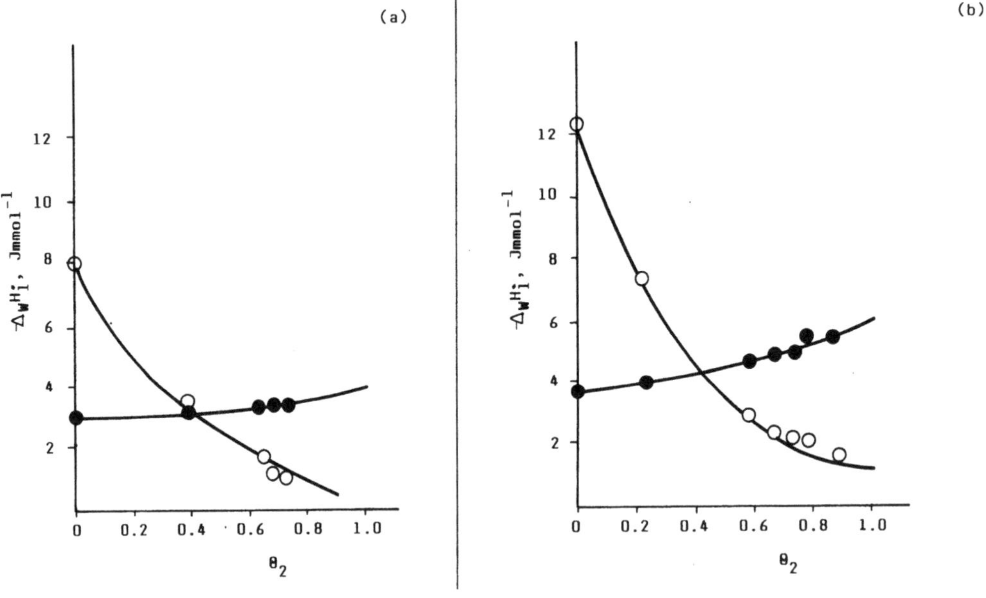


**Fig. 18** Heat of wetting in (○) methanol (1) and (●) benzene (2) on a) HDP-kaolinites; b) HDP-illites at different HDP⁺-coverage ($\theta_2$)

content and an alkyl chain length $n_c = 18$ during the intercalation, expansion of the lamellae takes place. Because of the increase in disorder in the chains, the swelling process is entropy driven [7, 26].

## Rheological properties of montmorillonite suspension with cationic and anionic surfactants

Since cationic and anionic surfactants are adsorbed on the silicate lamellae, the polarity of the surface of particles changes as well. The magnitude of the adhesion forces is influenced significantly by the increase in hydrophobic character of the surface in aqueous media. The extent of the interaction between the particles can be characterized by investigation of the rheological properties of the suspension. When a suspension of actual clay samples is tested in a rotation viscosimeter, the change in the shearing stress ($\tau$) can be given as a function of the speed gradient ($\dot{\gamma}$). The relation $\tau = \eta\,\dot{\gamma}$ is true for the Newtonian liquid, but the ideal plastic suspensions (Bingham model) have the Bingham yield value $\tau_B$ which yields information on interactions between the particles and on the stability of the house-of-cards structure.

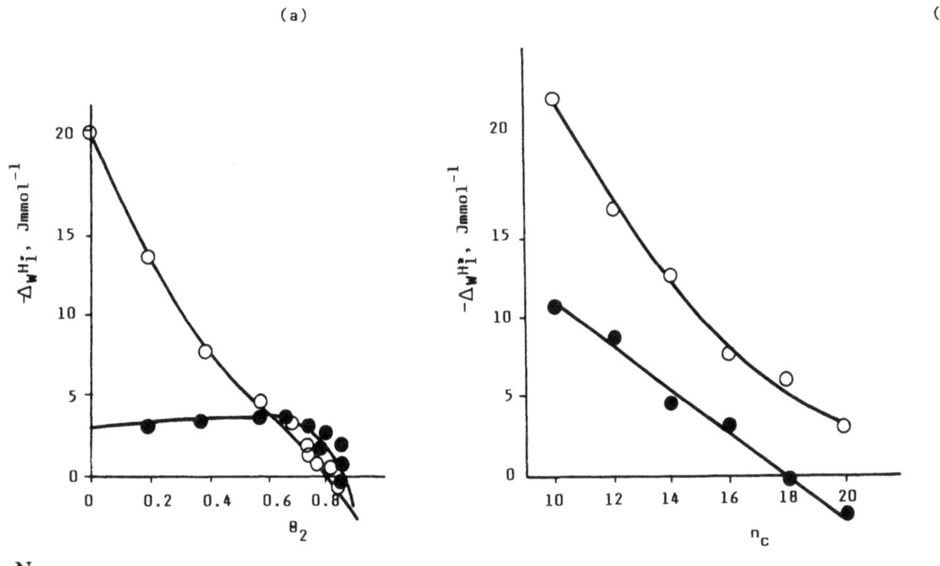

**Fig. 19** Heat of wetting in (○) methanol (1) and (●) benzene (2) on a) HDP-montmorillonite as a function of the HDP⁺-coverage; b) alkylammonium derivatives with different chain length ($n_c$)

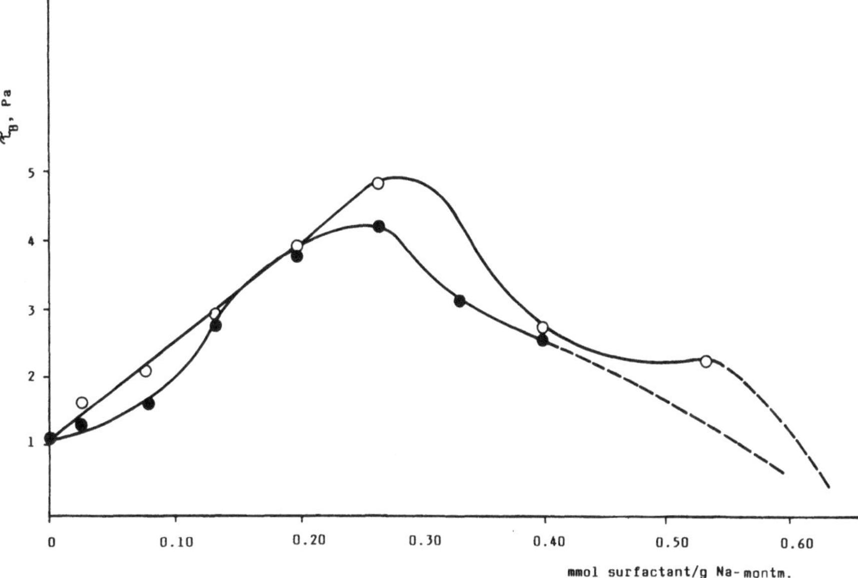

**Fig. 20** Rheological experiment in Na-montmorillonite suspension with (○) HDPCl and (●) HDTABr surfactant (suspension concentration: 7.5 g/100 cm³)

The effect of DTABr and HDPCl on Na-montmorillonite can be seen in Fig. 20. It emerges that the structure of the suspension is modified significantly by the binding of small amounts of surfactants. Increase in the hydrophobic character of the surface causes aggregation in aqueous media. The yield value $\tau_B$ then increases until a maximum is reached. If the surfactant quantity is further increased, the wetting improves again, because of the surfactant orientation and therefore the interaction between the lamellae aggregates decreases. Figure 21 demonstrates the surprisingly significant structure-changing effect of anionic surfactant (NaDS) when Ca-montmorillonite is used. It is well known that anionic surfactants adsorb only on the edges of the lamellae of montmorillonite. From this it could be expected that the house-of-cards structure will collapse. In contrast, Ca-montmorillonite, which swells very poorly in water,

undergoes disaggregation, and the structure-forming property of suspension increases considerably. In our opinion, the structure forming occurred at lower concentration because of the Na-montmorillonite formed. It dissolves poorly and forms a gel, which helps the structure of the suspension to solidify. The formation of $Ca(DS)_2$ compound is proved quite clearly by the x-ray diffraction picture. This can be seen in Fig. 22, which demonstrates the reflection of $Ca(DS)_2$ in the suspension.

## Humic substances aggregation state and interactions

Humic substances are the most important class of organic materials in the soils and surface waters, in terms of both quantity and ubiquity. One of the main characteristics

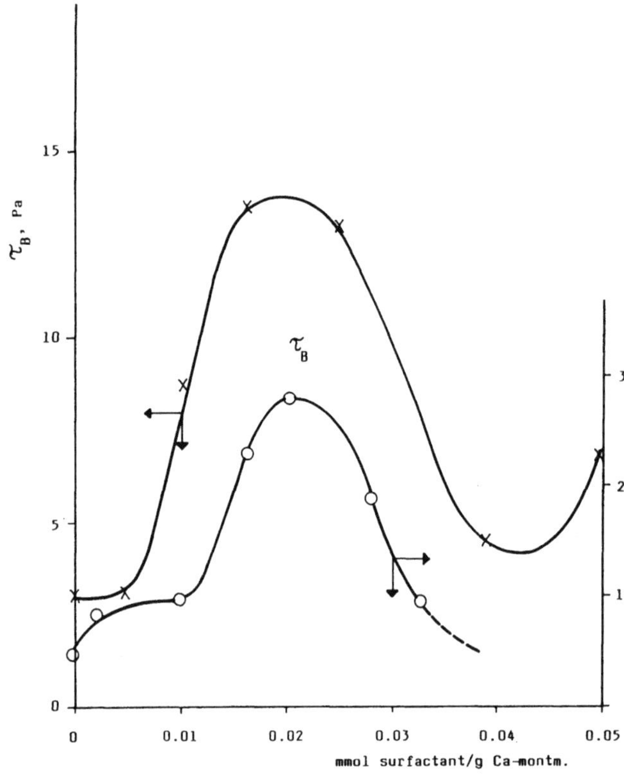

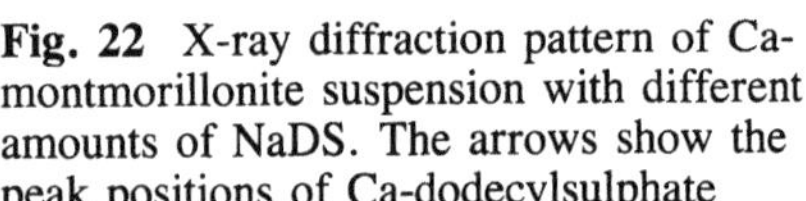

Fig. 21 Rheological experiment in Ca-montmorillonite suspension with (○) HDTABr and (×) NaDS (suspension concentration: 30 g/100 cm³)

of humic materials is their ability to interact with metal ions, oxides, hydroxides, minerals and organics, including toxic pollutants, to form water-soluble and water-insoluble associations. These interactions have been described as ion-exchange, adsorption, chelation, peptization and coagulation reactions [45], and it is likely that they affect many reactions which occur in soils and waters. These substances are polyanions of complex structure containing a carbon skeleton highly substituted with oxygen-containing functional groups, which can act as active sites in ion-exchange reactions or as ligands in complex formation with appropriate metal ions, either dissolved in aqueous medium or fixed in the surface layer of solids.

The colloidal state of humic substances is uniquely various. It is well known that their aggregation state depends not only on their individual molecular structure, but also on the solvation conditions under which they exist. The main external factors which influence the colloidal state of humic compounds are the pH, the ionic strength, the presence of di- or multivalent metal ions, cationic organic compounds, organic liquids and solid particles. In true solutions, humic materials, and particularly fulvic acids, are considered to be polyelectrolytes. However, they can separate from aqueous medium in response to a decrease in pH below a given level, to an increase in ionic strength, or to the addition of di- or multivalent metal ions or organic cations. Whether this phase separation should be called precipitation, coagulation or even micelle formation is open to discussion. Each term relates to a starting system (a real or macromolecular solution, or perhaps a subcolloidal dispersion)

Fig. 22 X-ray diffraction pattern of Ca-montmorillonite suspension with different amounts of NaDS. The arrows show the peak positions of Ca-dodecylsulphate

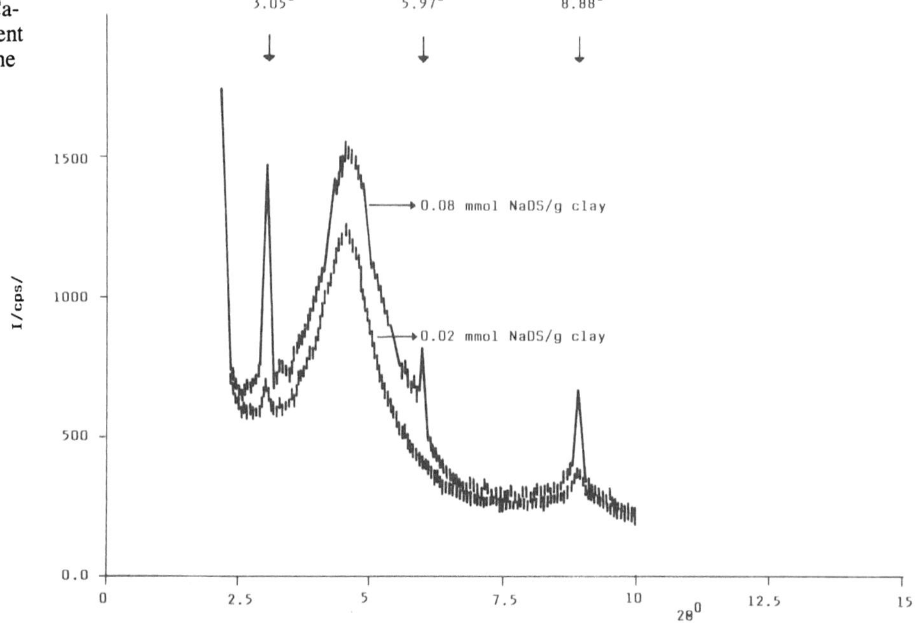

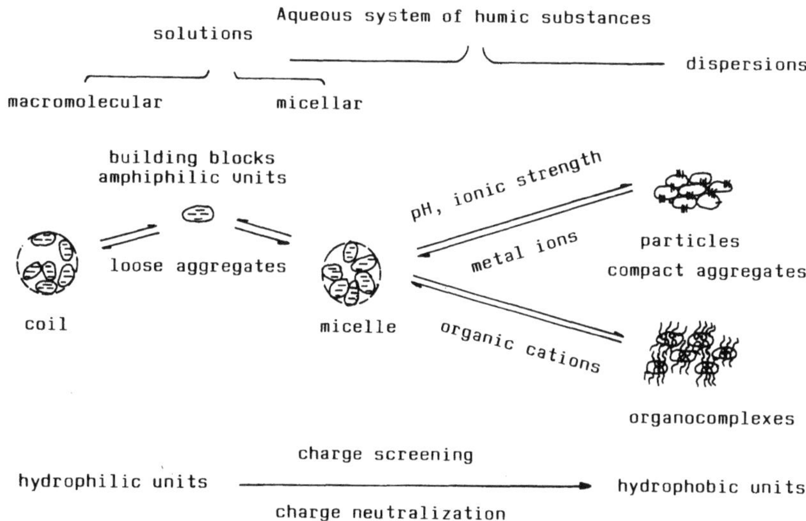


**Fig. 23** Schematic illustration of the variability of aqueous colloidal systems of humic substances

of given character and it can be defended with some success. Nevertheless, the systems formed are mostly colloidal or coarse dispersions of more or less hydrophobic particles.

The schematic illustration in Fig. 23 shows the different aqueous systems formed spontaneously from humic substances that have been reported in the literature [17]. Though humic substances still defy precise chemical description because of their inherent complexity and heterogeneity [46], they are considered to be composed of amphiphilic units (named building blocks in [46] or as molecules in [47]), which contain hydrophilic groups, predominantly negatively charged ones, linked chemically to an aromatic and/or aliphatic hydrophobic part. These amphiphilic units can associate to form loose aggregates, which can be regarded either as macromolecular coils or as micelles.

In dilute solution at appropriate pH, fulvic and humic acids and their alkali metal salts are real colloidal solutions, exhibiting the properties of both polyelectrolyte [45—47] and surfactant solutions [17, 48—50]. As may be seen in Fig. 23, each external factor which can induce the screening or neutralization of charges has a destabilizing effect on humic acid or humate solutions, and promotes the formation of more or less compact aggregates dispersed in the medium, i.e., the formation of colloidal or coarse dispersions of more or less hydrophobic aggregates.

*pH- and ionic strength-dependent aggregation state*

The pH- and ionic strength-dependent behavior of humic substances and their interactions with different metal ions have been investigated extensively for decades. Although models describing the ionization and complexation equilibria and charge distributions in aqueous systems areknown for both macromolecules and colloidal particles [51], aqueous humic systems are very difficult to

model. Apart from their imprecisely known molecules (size, shape and functional groups) and lack of uniformity, they always differ somewhat from the systems involved in the given models.

A typical example of their characteristic difference from polyelectrolytes is the reduced osmotic pressure vs. concentration functions (Fig. 24). While these functions for polyelectrolytes are characterized by a decreasing slope (reaching a quasi-ideal state) and the same ordinate intercept (indicating constant molar mass) with increasing neutral electrolyte concentration, these functions for humic acid and Na-humate solutions have practically constant slope (each system is quasi-ideal), but their ordinate intercepts change (characteristically with pH and NaCl (applied as neutral 1:1 electrolyte) concentration, indicating their strongly variable molar mass. The calculated molar masses of humic acid and Na-humate are shown in Table 4. It can be seen that both increasing salt concentration and decreasing pH induce a considerable increase in the measurable molar mass.

*Interaction with cationic surfactants*

The negatively charged functional groups of humates react not only with metal ions, but also with organic cations. Reactions between humate polyanions and surfactant cations seem to be similar to ion-pair formation between surfactant anions and cations. The formation of ion-pair complexes is a stoichiometric reaction in both cases, i.e., neutralization of equivalent amoaunts of negative (acidic functional groups on the humate carbon skeleton, and anionic head groups of the surfactant) and positive (cationic head groups of the surfactant) charges takes place. The bond formed is ionic in character. These ion-pair compounds are insoluble or poorly soluble in aqueous medium and can be characterized by the solubility product (precipitate formation analogy) or by the stability constant (complex analogy).

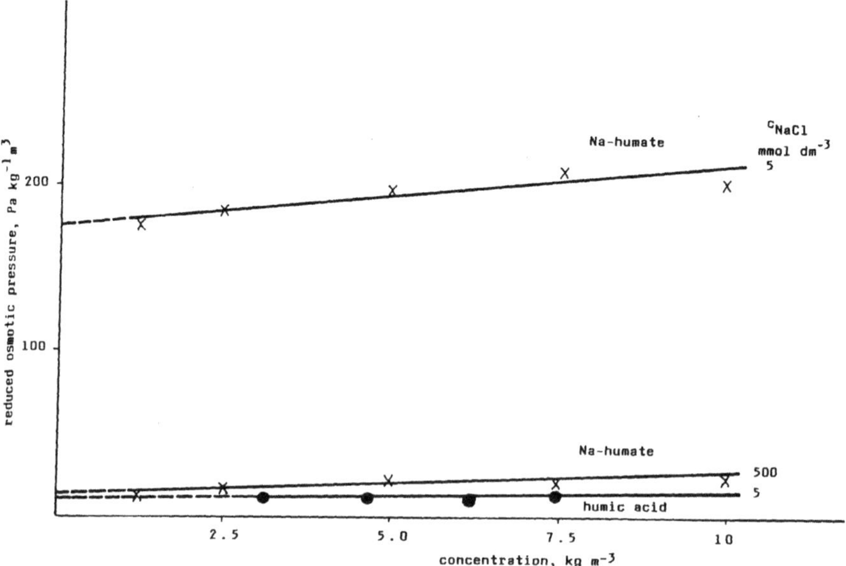

**Fig. 24** Dependence of reduced osmotic pressure of humic acid and Na-humate solutions on their concentration at 310 K



**Table 4** Experimental number average molar masses for humic acid and Na-humate, measured by membrane osmometry under different solution conditions

| Sample | pH | NaCl concentration (mmol dm$^{-3}$) | $M_n$ (g mol$^{-1}$) |
|---|---|---|---|
| Na-humate | 8 | 5 | 14,500 |
|  | 8 | 500 | 203,000 |
| Humic acid | 2.5 | 5 | 260,000 |

The dissolved hydrophilic units in a Na-humate solution can be precipitated by adding cationic surfactant solutions, in consequence of the neutralization of charges. The percentage of organic cations in the humate-organocomplex is high (about 50—60%) compared with the clay-organocomplexes (5—35%, [10]), since the amount of active sites in humates (i.e., total acidity) is about one order of magnitude higher than that in clays (i.e., cation exchange capacity). In this process, a coarse dispersion of aggregated hydrophobic units (humate-organocomplexes) forms in aqueous medium (schematic illustration in Fig. 23).

X-ray diffraction examination of the micro-structure of humate-organocomples has led to remarkable results; detailed investigations of different humic materials have been published [16]. While the solid humic acids show practically amorphous structure, n-alkylammonium humates have well-ordered, pseudolayered structures.

A series of x-ray diffraction patterns of humate organocomplexes containing alkyl chains with more than 10 carbon atoms are shown in Fig. 25. It seems that the alkylammonium humates, and especially those with longer alkyl chains, give relatively sharp reflections of high intensity, in contrast with the amorphous character of hydrophilic humic acid. The thickness of the parallel

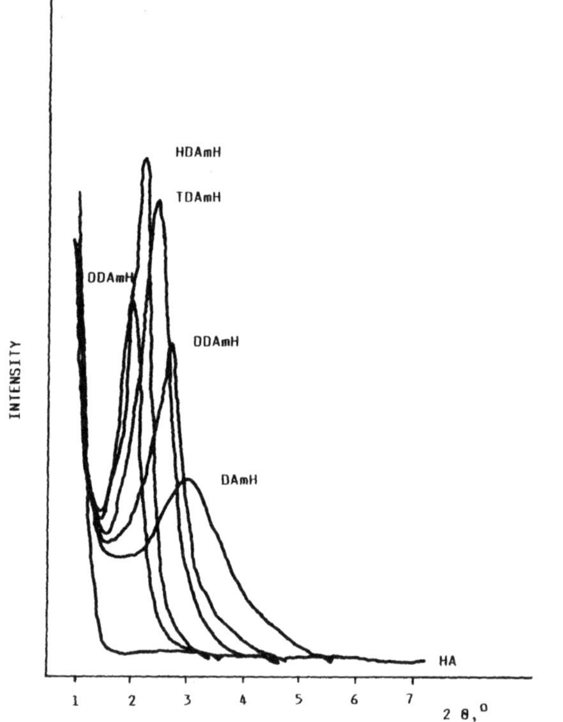



**Fig. 25** X-ray diffraction pattern of humic acid and the n-alkylammonium derivatives (HA: humic acid, DAmH: n-decyl ammonium humate, DDAmH: n-dodecylammonium humate, TDAmH: n-tetradecylammonium humate, HDAmH: n-hexadecylammonium humate, ODAmH: n-octadecylammonium humate)

lamellae increases from 2.95 nm to 4.46 nm with increasingly alkyl chain length. The evaluation of x-ray diffraction results reveals behavior similar to that of smectite-type clay organocomplexes. Data are summarized in Table 5. The results seem to support the layer structure

**Table 5** Results from x-ray diffraction powder patterns

| Sample | Amount of organic cation (%)*) | $2\Theta$ (degree) | $d_{001}$ (nm) exp. | calc. |
|---|---|---|---|---|
| Humic acid | 0 | amorphous | | |
| n-decyl-AmH | 51.42 | 3.00 | 2.95 | 2.89 |
| n-dodecyl-AmH | 55.48 | 2.71 | 3.26 | 3.27 |
| n-tetradecyl-AmH | 58.91 | 2.48 | 3.56 | 3.65 |
| n-hexadecyl-AmH | 61.85 | 2.21 | 4.00 | 4.02 |
| n-octadecyl-AmH | 64.40 | 1.98 | 4.46 | 4.40 |

*) It is equivalent to the total acidity of humic acid (6.70 mmol/g)

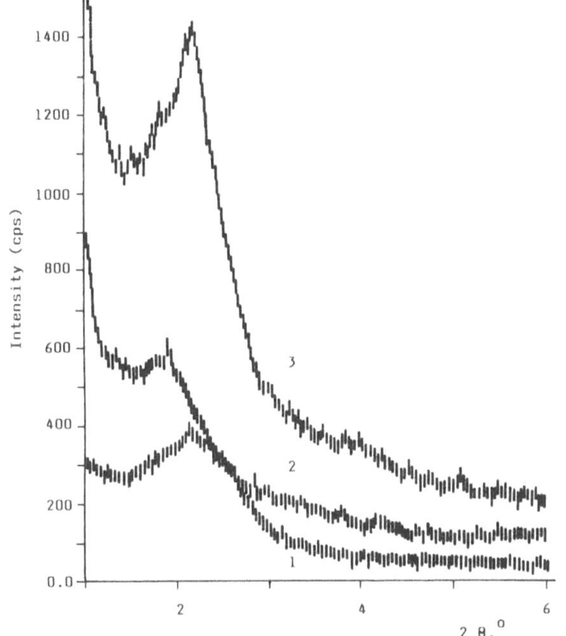

**Fig. 26** X-ray diffraction pattern of n-hexadecylammonium humate in solid phase (curve 1, $d_{001}$ = 4.07 nm) and wetted by toluene (curve 3, $d_{001}$ = 4.22 nm) and o-xylene (curve 2, $d_{001}$ = 4.93 nm)

of humic substances. The layers of this sample are about 0.45-nm-thick, and the bilayers of alkylammonium cations are bound to the carboxyl and phenolic hydroxyl groups with an inclination angle of 44.7°.

As mentioned above, the humate-organocomplexes are insoluble in water because of their hydrophobic character; however, these compounds can be wetted by organic liquids. Depending on the degree of solvation, from limited to unlimited swelling can take place when the humate-organocomplexes make contact with organic liquids. During limited swelling, the original pseudo-layered structure observed in the solid phase (Fig. 25) is still retained, as can be seen in Fig. 26. The increase in spacing between the humate layers, from 4.07 nm in the solid phase to 4.22 nm in toluene, and to 4.93 nm in o-xylene, indicates the penetration of aromatic molecules into the interlamellar space, i.e., the swelling of organocomplex due to the good solvation by these organic liquids. As shown in Fig. 26, the diffraction pattern becomes more diffuse when o-xylene is applied as wetting liquid. This indicates that n-hexadecylammonium humate can be partially dissolved in o-xylene. Complete dissolution is reached in an organic liquid or a liquid mixture ensuring better solvation. The solubility of humate-organocomplexes depends mainly on the length of the alkyl chain. Organocomplexes with shorter alkyl chains (e.g., n-butylammonium humate) can be dissolved in water, and those with longer chains (e.g., n-octadecyl-ammonium humate) in aromatic liquids (e.g. o-xylene).

**Acknowledgement** The authors wish to thank the Hungarian Science Foundation (OTKA 1442/1991, 1441/1991 and T 6077) for financial support.

# References

1. Hough DB, Rendall HM (1983) In: Parfitt GD, Rochester CH (eds) Adsorption from Solution at the Solid/Liquid Interface, Academic Press, London pp 247—319
2. Rosen MJ (1989) Surfactants and Interfacial Phenomena, John Wiley & Sons, New york
3. Theng BKG (1974) The Chemistry of Clay-Organic Reactions, Halsted Press, John Wiley & Sons
4. Szántó F, Dékány I, Patzkó Á, Várkonyi B (1986) Coll Surf 18:359
5. Dékány I, Szántó F, Weiss A, Lagaly G (1986) Ber Bunsenges Phys Chem 90:422—427
6. Dékány I, Szántó F, Weiss A, Lagaly G (1986) Ber Bunsenges Phys Chem 90:427—431
7. Dékány I, Szántó F, Nagy LG (1985) J Coll Interf Sci 103:321
8. Dékány I, Szántó F, Nagy LG, Schay G (1983) J Coll Interf Sci 93:151
9. Patzkó Á, Szántó F (1987) Coll Surf 25:173
10. Patzkó Á, Dékány I (1993) Coll Surf 71:299
11. Esumi K, Nagahama T, Meguro K (1991) Coll Surf 57:149
12. Isfahany TM (1990) Coll Surf 51:339
13. Gu T, Zhu BY, Rupprecht H (1992) Prog Coll Polym Sci 88:74

14. Noll LA, Gall BL (1991) Coll Surf 54:41
15. Backlund S, Sjöblom, Matijević E (1993) Coll Surf A 79:263
16. Tombácz E, Varga K, Szántó F (1988) Coll Polym Sci 266:734
17. Tombácz E, Regdon I (20—25 Sept, 1992) Humic substances as various colloidal systems, Proceedings of the 6th International Meeting of the IHSS, Monopoly (Bari) Italy, in press
18. Pfirrmann GL (1968) Thesis, Heidelberg
19. Somasundaran P, Healy TW, Fuerstenau DW (1964) J Phys Chem 68:3562
20. Fuerstenau DW (1956) J Am Chem Soc 60:981
21. Kern HE, Piechocki A, Brauer U, Findenegg GH (1978) Prog Coll Polym Sci 65:118
22. Kern HE, Findenegg GH (1980) J Coll Interf Sci 75:346
23. Dékány I, Zszednai Á, Király Z, László K, Nagy LG (1986) Coll Surf 19:47
24. Dékány I, Zsednai Á, Király Z, László K, Nagy LG (1987) Coll Surf 123:41
25. Dékány I, Ábrahám I, Nagy LG, László K (1987) Coll Surf 23:57
26. Dékány I, Szántó F, Nagy LG (1986) J Coll Interf Sci 109:376
27. Weiss A, Mehler A, Hofmann U (1956) Z Naturforsch 11b:431
28. Lagaly G, Weiss A (1971) Kolloid Z Z Polym 243:48
29. Lagaly G, Weiss A (1971) Kolloid Z Z Polym 248:979
30. Lagaly G, Stange H, Weiss A (1972) Kolloid Z Z Polym 250:675
31. Lagaly G (1982) Ullmanns Encyklopädie 21:366
32. Tombácz E, Meleg E (1990) Organic Geochemistry 15:357
33. Reid VW, Longman GF, Heinrich E (1968) Tenside 5:90
34. James RO, Parks GA (1982) In: Matijević E (ed) Surface and Colloid Science, Vol 12, Characterization of Aqueous Colloids by their Electrical Double-Layer and Intrinsic Surface Chemical Properties, Plenum Publishing Corporation
35. Yates DE, Healy TW (1976) J Coll Interf Sci 55:9
36. Hartley GS (1938) J Chem Soc 1968
37. Tanford C (1973) Formation of Micelles and Biological Membranes, The Hydrophobic Effect, John Wiley & Sons
38. Raltson AW, Eggenberger DN, Broome FK (1949) J Am Chem Soc 71:2145
39. Schmidt PW (1989) In: Anvir A (ed) The Fractal Approach to Heterogeneous Chemistry, John Wiley, New York, London
40. Kriechbaum M, Degovics G, Tritthardt I, Laggner P (1989) Prog Colloid Polym Sci 79:101
41. Nagy B, Bradley WF (1955) The Amer Mineral 40:885
42. Hofmann U, Endell K, Wilm D (1933) Z Kristallog 86:340
43. van Olphen H (1965) J Coll Interf Sci 20:822
44. Grim RE (1953) Clay Mineralogy, McGraw-Hill Book Comp, New York
45. Schnitzer M (1986) In: Huang PM, Schnitzer M (eds) Binding of Humic Substances by Soil Mineral Colloids Ch 4, Interactions of Soil Minerals with Natural Organics and Microbes, SSSA Special Publication Number 17, Soil Science Society of America, Inc Madison Wisconsin USA
46. Theng BKG (1979) Formation and Properties of Clay-polymer Complexes Ch 12, Elsevier, Amsterdam, pp 283—326
47. Wershaw RL (1989) In: Everett RC, Leenheer DM, Thorn KA (eds) Humic Substances in the Suwannee River, Georgia: Interactions, Properties and Proposed Structures, US Geological Survay Open File Report 87—557, p 354
48. Tombácz E, Sipos S, Szántó F (1981) Agrokémia Talajtan 30:365
49. Shinozuka N, Lee Ch (1991) Marine Chemistry 33:229
50. Guetzlhoff TF, Rice JA (1992) Does humic acid form a micelle?, International Conference on Organic Substances in Soils and Sediments, Lanchester, UK
51. Buffle J (1988) Complexation Reactions in Aquatic Systems: an Analytical Approach, Ellis Horwood Limited Publisher, Chichester

Progr Colloid & Polym Sci (1994) 95:91—97
© Steinkopff Verlag 1994

J. Lyklema

# Adsorption of ionic surfactants on clay minerals and new insights in hydrophobic interactions

J. Lyklema
Wageningen Agricultural University,
Department of Physical
and Colloid Chemistry,
Deijenplein 6, 6708 HB Wageningen,
The Netherlands

**Abstract** The adsorption of dodecylpyridinium chloride and dodecyltrimethylammonium bromide on Na-kaolinite is investigated by studying the effect of electrolyte and temperature on the adsorption, by means of electrophoresis and microcalorimetry. A sharp distinction between adsorption in the first and second layer is observed. The point where these two regions meet is retrieved in all experiments. Second layer adsorption is driven by hydrophobic bonding, for which the enthalpy is now established as a function of temperature. It changes from endothermic at low temperature to exothermic at higher temperature. A new molecular interpretation for hydrophobic bonding is proposed, based on quasichemical lattice statistics with orientation-dependent interactions. All experimental features can at least semiquantitatively be explained by this new model.

**Key words** Adsorption — surfactants — clay minerals — hydrophobic bonding

## Introduction

The present paper deals with the physico-chemical background of surfactant adsorption, where "background" is the interpretation in terms of basic physicochemical laws. The purpose is threefold. In the first place there is, of course, the desire to understand the phenomena better than just in a descriptive way. Second, a better understanding of the underlying principles assists us in predicting adsorption behavior under less familiar conditions. Last, but not least, from surfactant adsorption inferences may be drawn on physico-chemical laws.

Interpretations can be offered on two levels, phenomenological or in terms of a model. The former usually rests on thermodynamics and generally helps to establish generally valid conclusions. Moreover, by establishing the signs of the adsorption enthalpy $\Delta_{ads}H$ and entropy $\Delta_{ads}S$, basic information on the driving forces may be extracted. Having established these, models can be developed to describe the adsorbate in terms of distributions and interactions of molecules.

Adsorption of ionic surfactants is particularly challenging in these respects because, apart from the enormous relevance for practice, the phenomenon invariably involves the sum effect of a number of interactions, including

i) electrostatic interaction between the head group and the adsorbent;

ii) non-electrostatic, or "chemical" interaction between the head group and the surface;

iii) hydrophobic (and other) tail-tail interactions;

iv) hydrophobic and other "chemical" tail-surface interactions.

Thermodynamic considerations may contribute to unraveling these various contributions. In this approach, one should always keep in mind that surfactant adsorption is an example of adsorption from solution, i.e., it is virtually an exchange process: ions may replace other ions (one of these may be a surfactant ion) and uncharged molecular moieties may replace water.

In the present work, the emphasis of the experiments will be on the adsorption of cationic surfactants on kaolinite. Apart from the arguments given in the introduction this choice was motivated by the theme of the meeting: it is likely that this system functions as a model for the adsorption of pesticides in soils. Moreover, we want to learn how far thermodynamics can take us with "difficult" systems. Because of our intention to analyze the date thermodynamically, adsorption isotherms under a variety of conditions (temperature, salt concentration) will be supplemented by directly measured enthalpies.

Part of this work is based on experiments by T. Mehrian Isfahany [1].

# Materials and methods

*Sodium kaolinite* was a well-crystallized (98%) sample of Sigma Company, repeatedly washed with 1 M NaCl. The cation exchange capacity was 57 $\mu$mole g$^{-1}$ according to the silver-thiourea method [2], but 30 $\mu$mole g$^{-1}$ according to the ammonium acetate procedure [3]. The BET (N$_2$) area was 12 m$^2$ g$^{-1}$. The aspect ratio (plate area/edge area) of the particles is about 80:20%.

*Dodecylpyridinium chloride* (DPC or DP$^+$ Cl$^-$) has been synthesized from 1-chloroalkane and pyridine, and purified according to Colichman [4]. Concentration determinations were done by complexation with bromothymol blue [1, 5].

*Dodecyltrimethylammonium bromide* (DTAB or DTA$^+$Br$^-$) was obtained from Aldrich Chemie and used without further purification.

*Adsorption isotherms* were obtained by depletion.

*Adsorption enthalpies* were measured in a Thermal Activator Monitor (TAM), an isothermal number 2277 microcalorimeter from LKB, Sweden. It contains a 25 ml stainless steel titration cell, fitted into a single detector measuring cylinder. The cell, with a reference ampoule, was especially designed for mixing liquids and adsorption from solution. For more details, and the execution of the measurements, see [6]. Basically, the heat evolved is measured by adding the surfactant solution to the kaolinite dispersion, where the heat of dilution is subtracted as the blank. In this way a plot $q_{ads}(\Gamma)$ of the heat evolved as a function of the amount adsorbed $\Gamma$ is obtained.

The slope of this curve

$$\Delta_{ads}H_m = \partial q_{ads}/\partial\Gamma \tag{1}$$

gives the partial molar adsorption enthalpy. This quantity was reproducible within 0.7 kJ mol$^{-1}$.

*Electrophoretic mobilities* were measured in a Malvern-Mark II Zetasizer.

For further experimental details, see [1].

# Adsorption isotherms — Influence of electrolyte

Figure 1 gives adsorption isotherms of DPC on Na-kaolinite. Figures 1a and b give the same information, but b relays information that is hardly visible in a, viz. a different electrolyte effect between the ranges of low and high adsorption, henceforth distinguished as region I and region II, respectively. There is a common intersection point (c.i.p) at $\Gamma \approx 38$ $\mu$mole g$^{-1}$ where there is no salt effect. Trends like in Fig. 1b are also found for DTAB (c.i.p. about 35 $\mu$mole g$^{-1}$) and for DPC at 60° (c.i.p. $\approx 37$ $\mu$mole g$^{-1}$).

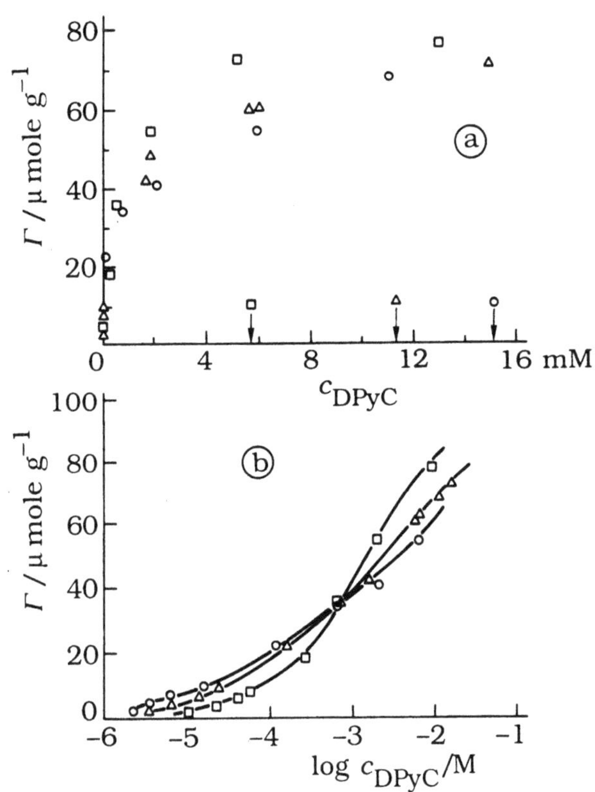


**Fig. 1** Adsorption of DPC on Na-kaolinite Temperature 20°C, pH = 5, NaCl-concentration: ○ 5 mM, △ 20 mM, □ 100 mM; Top: linear plot with the critical micelle concentration indicated; Bottom: semi-logarithmic scale

The common intersection point demarcates two regions with a different adsorption mechanism. At low $\gamma$, NaCl competes with DPC, at high $\Gamma$ it promotes the adsorption of DPC.

Before offering a model interpretation, the two regimes can be thermodynamically characterized through surfactant adsorption that is accompanied by expulsion and uptake of NaCl, respectively. This follows immediately from Gibbs' law, written as

$$d\gamma = -S_a dT - \Gamma_{DPC} d\mu_{DPC} - \Gamma_{NaCl} d\mu_{NaCl} \, , \tag{2}$$

where $\gamma$ is the interfacial tension, $T$ the absolute temperature, and $\mu$ the chemical potential of the substance mentioned. Cross-differentiation at fixed temperature yields

$$\left(\frac{\partial \Gamma_{DPC}}{\partial \mu_{NaCl}}\right)_{T,\mu_{DPC}} = \left(\frac{\partial \Gamma_{NaCl}}{\partial \mu_{DPC}}\right)_{T,\mu_{NaCl}} \tag{3}$$

As $c_{NaCl} \gg c_{DPC}$, $d\mu_{NaCl} = 2RT d \ln c_{NaCl}$ and $(d\mu_{DPC})_{\mu_{NaCl}} = d\mu_{DP^+} = RT d\ln c_{DPC}$, hence

$$\frac{1}{2}\left(\frac{\partial \Gamma_{DPC}}{\partial \ln c_{NaCl}}\right)_{T,c_{DPC}} = \left(\frac{\partial \Gamma_{NaCl}}{\partial \ln c_{DPC}}\right)_{T,c_{NaCl}} \tag{4}$$

Below the c.i.p. (region I) adsorption of DPC is accompanied by expulsion of electrolyte, above it (region II) electrolyte is co-adsorbed.

Mechanistically speaking, the obvious interpretation of region I is that a $Na^+$ counterion is exchanged against a $DP^+$ ion. At pH 5 the plates and the edges are both negatively charged with $Na^+$-ions as the dominant counterions. The surface charge (most of it on the plates) is the result of isomorphic substitution and constant. Representing a negative charge on (or behind) the surface as $R^-$, with $Na^+$ as the counterion, the exchange process obeys

$$R^-Na^+ + DP^+Cl^- \leftrightharpoons R^-DP^+ + NaCl \, , \tag{5}$$

thus explaining the release of NaCl.

With this reasoning, one cannot prove whether, in addition to the purely electrostatic exchange (5), some specific adsorption of DPC may also take place. Additional measurements are required for that. One of these is determinations of $\Delta_{ads}H$, which is about zero for process (5) if $Na^+$ and $DP^+$-adsorb at (about) the same potential. However, for energetically driven specific adsorption $\Delta_{ads}H < 0$. Adsorption by hydrophobic bonding, for which $\Delta_{ads}H$ might be slightly endothermic, is unlikely because the plate surface is hydrophilic.

A quantitative argument that over region I exchange (5) dominates is that the amount of surfactant cation required to compensate the surface charge (i.e. the value of $\Gamma$ at the c.i.p.) is within less than 8% independent of the

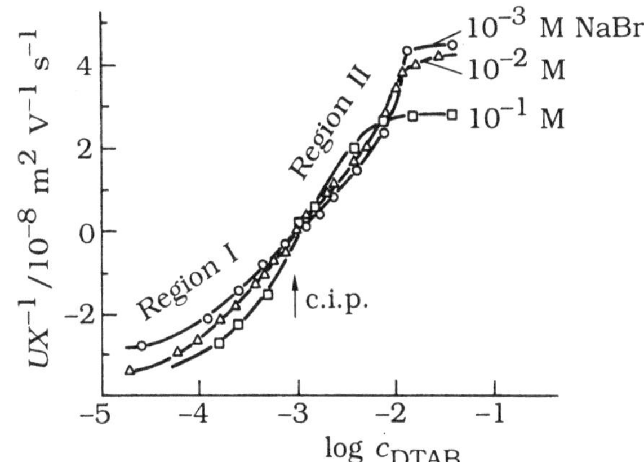


**Fig. 2** Electrophoretic mobility of Na-kaolite in the presence of dodecyltrimethylammonium bromide and NaBr pH = 5, $T = 20\,°C$

nature of the cation ($DP^+$ or $DTA^+$) and of the temperature (20°, 60°). Moreover, this value (35—38 µmole $g^{-1}$) is not too far from the CEC value measured by the ammonium acetate method [3], equal to ~30 µmole $g^{-1}$. Apparently, these two methods measure about the same surface charge, whereas the thiourea method [2] which gives 57 µmol $g^{-1}$, measures a larger exchanging area. Perhaps the former mainly "sees" the plates, the latter plates and edges.

At the c.i.p. the charges on the particle (dominated by the plates) must be virtually compensated. Figure 2 demonstrates this to be the case: the isoelectric point coincides within experimental error with the c.i.p. Very similar results are obtained with DPC.

All of this suggests that at the c.i.p. the first layer is completed. The surfactants are not adsorbed as hemimicelles or admicelles [7—9].

In region II superequivalent adsorption takes place (Fig. 2). A second layer is formed due to a non-electrostatic interaction that is strong enough to overcome the formation of an electrical double layer with the cationic head groups directed to the solution. Addition of electrolyte screens the ensuing electrostatic repulsion and therefore promotes the adsorption (Fig. 1). Chloride ions are electrostatically attracted by the cationic head groups, the accompanying $Na^+$ ions remain in the diffuse part of the double layer. Hence, NaCl (or NaBr for DTAB) is co-adsorbed, in line with (4).

The most likely driving force for the formation of the second layer is hydrophobic bonding.

## Adsorption isotherms — Influence of temperature

It is typical for hydrophobic bonding that, at least at low temperature, the adsorption increases with $T$. At high $T$ a decrease is observed. Figure 3 shows that indeed a max-

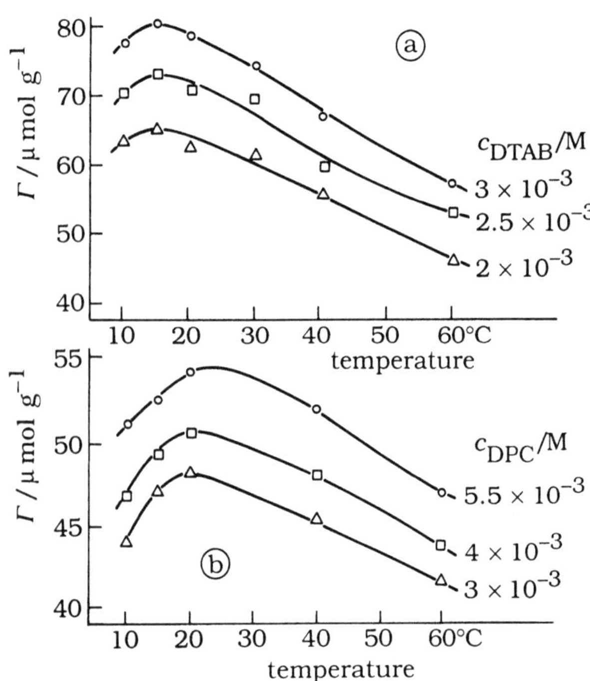



**Fig. 3** Temperature dependence of the adsorption of DTAB (top) and DPC (bottom) on Na kaolinite Region II, pH = 50 $c_{\mathrm{NaCl}} = 10^{-3}$ M

imum is found. Adsorptions increasing with $T$ point to entropically driven processes. This can be inferred from (2) by differentiation with respect to $T$:

$$\left(\frac{\partial \Gamma_{\mathrm{DPC}}}{\partial T}\right)_{\mu_{\mathrm{NaCl}},\mu_{\mathrm{DPC}}} = \left(\frac{\partial S_a}{\partial \mu_{\mathrm{DPC}}}\right)_{T,\mu_{\mathrm{NaCl}}} \sim \frac{1}{RT}\left(\frac{\partial S_a}{\partial \ln c_{\mathrm{DPC}}}\right)_{T,c_{\mathrm{NaCl}}} \tag{6}$$

The l.h.s. of this equation can be converted to $(\partial \Gamma_{\mathrm{DCP}}/\partial T)_{c_{\mathrm{NaCl}},c_{\mathrm{DPC}}}$. When this derivative is positive, the interfacial excess entropy rises with increasing surfactant concentration, that is: with increasing adsorption.

In passing, it may be noted that in region I the temperature dependence of $\Gamma_{\mathrm{DPC}}$ is zero.

From the temperature dependence of the adsorption the isosteric enthalpy of adsorption can also be obtained, using

$$\left(\frac{\partial \ln c_{\mathrm{DPC}}}{\partial (1/T)}\right)_{\mathrm{pH},c_{salt},\Gamma_{\mathrm{DPC}}} = \frac{\Delta_{\mathrm{ads}}H(\mathrm{isost.})}{R} \tag{7}$$

for region II. It is plotted in Fig. 4 as curve 1. In view of the analogy between the second layer formation and micelle formation, corresponding plots for micellization are included. Curve 2 is taken from ref. [10], and is obtained from the temperature coefficient of the critical micelle concentration using

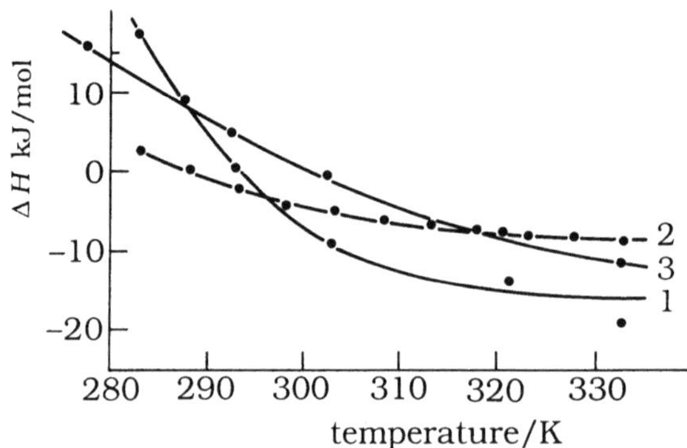



**Fig. 4** Comparison between the isosteric adsorption enthalpy (region II) of DPC on Na-kaolinite, pH = 50, $c_{\mathrm{NaCl}} = 10^{-3}$ M, curve 1, with the enthalpy of micellization according to different sources (curves 2 and 3) further discussion in the text

$$\Delta_{\mathrm{mic}}G^0 \approx RT\ln\mathrm{CMC} \tag{8}$$

and the Gibbs-Duhem relation. These data apply to zero salt concentrations. Curve 3 are own microcalorimetric measurements [11].

Considering the very different sources and methods, the analogy between the three curves of Fig. 4 is gratifying. First, this points to the great similarity with micelle formation and to hydrophobic bonding as the driving force. This last process is known to be entropically driven [12, 13], but the enthalpic part is under dispute. One of the reasons is that it is small and therefore difficult to measure at various temperatures. We confirm that $\Delta H$ is small, between 0 and 0.7 $kT$ per segment, depending on $T$, and has a sign that changes with temperature. At low $T$ hydrohobic bonding is endothermic, i.e., it would oppose the process which therefore must be entropically driven. However, at high $T$ the phenomenon is partly enthalpically, partly entropically driven.

## Direct measurement of the adsorption enthalpy

In a parallel paper, we report on the details of the method and experimental results [6]. Plots of the cumulative heat of adsorption as a function of $\Gamma$ consist of two perfectly straight lines, with a break at the c.i.p. Using (1) it follows that in regions I and II $\Delta_{\mathrm{ads}}H$ is independent of coverage, though different between the first and second layer. Constancy of $\Delta_{\mathrm{ads}}H$ implies that each molecule adsorbs in the same way, with the same surfactant-adsorbent and surfactant-surfactant interaction energy. There is no indication for the formation of hemimicelles at a certain $\Gamma$, that rearrange into other aggregate structures at higher coverages.

**Table 1** Partial molar enthalpy of adsorption of DPC on Na-kaolinite region II

| Temperature (°C) | $c_{NaCl}$ (mM) | $\Delta_{ads}H_m$(cal) (kJ mole$^{-1}$) | $\Delta_{ads}H_m$(isost) (kJ mole$^{-1}$) |
|---|---|---|---|
| 6 | 5 | 13.2 | 12.5 ± 15 |
| 20 | 5 | 1.7 | 0.0 |
| 60 | 5 | −30.0 | −19.0 ± 2 |
| 20 | 100 | 0.7 | — |
| 60 | 100 | −10.6 | — |

In view of our emphasis on hydrophobic bonding as the driving force for the second layer formation, we represent the results for DPC in Table 1, with, for comparison, the isosteric heats included (Fig. 4, curve 1). The already observed trends are confirmed. The calorimetric enthalpies are low, especially at room temperature where they change sign. Most other microcalorimetric enthalpy measurements for surfactant adsorption, reported in the literature [14—18] (for both regions) are close to zero. This may be one of the reasons for confusion about the sign of $\Delta_{ads}H$ and, hence, about the mechanism. The electrolyte effect is small (also for $\Delta_{ads}H$ in region I [6] except at 60° for region II. For this phenomenon no interpretation can be offered.

## Hydrophobic bonding revisited

The overall conclusion is that the formation of the second layer is analogous to micelle formation, i.e., to a large extent determined by hydrophobic bonding. This process is characterized by the following features

i) Binding passes through a maximum as a function of temperature.
ii) the enthalpy changes sign from endothermic to exothermic at a temperature close (or identical) to the temperature where the binding has a maximum.
iii) The entropy increases upon binding, but $T\Delta S$ decreases with temperature.
iv) Because of enthalpy-entropy compensation, the Gibbs energy of the process, $\Delta G$, is not very temperature-dependent.

These trends are now more or less established, also in other types of measurements. For instance, the solubility of small apolar solutes in water has a minimum as a function of $T$, and enthalpy-entropy compensation has also been found by others [19-21].

At issue is now how to interpret these observations in terms of water structure.

The current interpretation of the hydrophobic effect is via "iceberg" formation, i.e., the tendency of water molecules around apolar molecules to cluster into molecular aggregates of low entropy. This is supposed to happen because introduction of the foreign molecule requires the breaking of hydrogen bonds; the water tries to minimize the incurred energy loss by making internal bridges, though at the loss of configurational entropy. A variety of molecular interpretations has been offered, most of them including pictures for the "iceberg," "ice-like strukture," "clathrate-like hydrate" or however they may be called [12, 13, 20, 22—28].

Recently, N.A.M. Besseling of our Department proposed an alternative interpretation, that fits very well into known peculiarities of water, and can account for the trends i)—iv) above without invoking icebergs or similar structures. Anticipating more detailed publications, the basic principles are now reviewed.

The first point is that water has a relatively open structure. Otherwise stated, fluid water contains numerous vacancies that are spontaneously present. Traditionally, this open, about four-co-ordinated structure is interpreted as caused by hydrogen bonds in tetrahedral geometry. However, if each water molecule were co-ordinated by eight others, it could also be involved in four hydrogen bridges. Such a structure does occur at very high pressures in ice VII, but not at ambient pressure and temperature. The inference is that, in addition to the strong attraction that water molecules experience under orientations leading to hydrogen bonds, there must be repulsions between them in other orientations. Considering the charge distribution over a water molecule this is likely (repulsion between contacting lone pairs).

Dissolution of a small apolar molecule into liquid water is not very different from introducing a vacancy, the only difference is that in the former case an additional exothermic enthalpy is observed due to Van der Waals attraction between the apolar and the water molecules. However, this contribution is relatively small and independent of temperature. Because of this the minimum in the solubility of apolar molecules as a function of temperature has the same root as the maximum in the isobaric density. Differences between the temperatures where these extrema are found, are due to the differences in enthalpy. More precisely, introduction of an apolar molecule takes place in addition to vacancies.

To elaborate these ideas, Besseling considered aqueous solutions of apolar molecules to be three-component systems, containing water ($w$), the solute ($s$), and vacancies ($v$), distributed over a body-centered cubic lattice. Such a lattice allows for eight coordinations, but at normal $p$ and $T$ only about four of them are realized; the others refer to vacancies. Orientation dependence is accounted for in the quasi-chemical approximation, assigning a pair energy $u_{Hb} = -20.1$ kJ mol$^{-1}$ to a hydrogen bond contact and $u_{nHb} = +1.75$ kJ mole$^{-1}$ for non-hydrogen bond contacts. Of course, interaction energies with vacancies are zero and for apolar solutes, an orientation-independent energy can be introduced. These values have been obtained by fitting the density and pressure at water-vapor coexistence, i.e., they should be physically realistic.

Basically, the energy and entropy of the system are given by

$$U = \sum_{\alpha \leqslant \beta} n_{\alpha\beta} u_{\alpha\beta} \tag{9}$$

and

$$S = S(\text{id.}) + S^E$$

$$= -k \sum_{A} n_A \ln \frac{n_A}{\omega_A} - k \sum_{\alpha \leqslant \beta} n_{\alpha\beta} \ln \frac{n_{\alpha\beta}}{n_{\alpha\beta}(\infty)} , \tag{10}$$

respectively. Here, $n_{\alpha\beta}$ is the number of contacts between face $\alpha$ of one molecule and face $\beta$ of the other with which it is in contact, and $u_{\alpha\beta}$ is the corresponding energy (see above), $n_{\alpha\beta}(\infty)$ is the value that $n_{\alpha\beta}$ would assume if all contacts were random, i.e., if there would not be orientation correlations. Further, $k$ is Boltzmann's constant, $n_A$ is the number of molecules of type $A$ and $\omega_A$ is the number of distinguishable orientations that $A$ can have.

From (9) and (10) the Helmholtz energy is obtained and from that all relevant mechanical and thermodynamic quantities, including the pressure, density and the chemical potential of each component, the properties of water near surfaces, etc.

By way of example, Fig. 5 gives the thermodynamic functions for the hydration of vacancies. At given $T\Delta_{\text{hydr}}S_m$, $\Delta_{\text{hydr}}H_m$ can be adjusted to obtain the corresponding parameters for small apolar molecules. The inverse functions relate to dehydration, as happens for instance in hydrophobic bonding. The theory can be extended to dissolved chain molecules like tails of surfactants, but even in the present picture it is seen that the switch from endothermal $\Delta H_m$ for hydrophobic bonding at low temperature to exothermal $\Delta H_m$ at higher temperatures is well predicted. However, the endothermic nature of $\Delta H_m$ at low temperature is not due to additional hydrogen bond formation, as in traditional iceberg models, but to an increase of the number of repulsive water-water interactions, induced by the dissolution of vacancies or foreign apolar molecules. This last conclu-

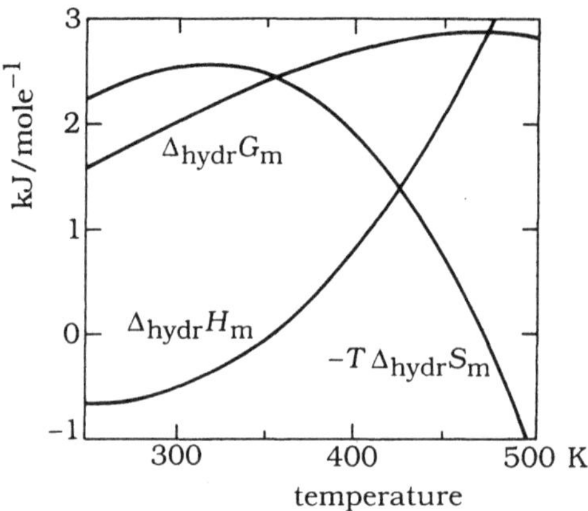


**Fig. 5** Thermodynamic functions for the hydration of apolar molecules, quasichemical approximation; lattice with orientation-dependent interactions after Besseling (1993)

sion simply follows from counting $n_{\text{Hb}}$ and $n_{\text{nHb}}$. It is further noted that $\Delta H$ is relatively small, as observed. In agreement with observation, $T\Delta S_m$ is positive but decreasing with $T$ above room temperature. At very high temperatures $T\Delta S_m$ even becomes negative; then $\Delta G_m$ passes through a maximum. The occurrence of such a maximum was recently put forward by Privalov and Gill [28].

Finally, it is noted that $\Delta H_m$ and $T\Delta S_m$ depend rather strongly on $T$, whereas $\Delta G_m$ is much less dependent on it. In other words, the new model also accounts for the entropy-enthalpy compensation.

Hence, it is concluded that the new model can at least semi-quantitatively account for all the main features for hydrophobic bonding. In forthcoming papers, we hope to return to all of this in more detail.

## Conclusion

By combining a number of experimental techniques the role of hydrophobic bonding in surfactant adsorption is better defined. The results can be interpreted with a new molecular theory.

## References

1. Mehrian Isfahany T (1992) Thermodynamics of the Adsorption of Organic Cations on Kaolinite, PhD Thesis, Agricult Univ Wageningen, NL
2. Chabra R, Pleysier J, Cremers A (1975) Proc Int Clay Conf 439
3. Schofield RK (1949) J Soil Sci 1:1
4. Colichman EL (1950) J Am Chem Soc 81:1834
5. Mehrian T, de Keizer A, Klyklema J (1991) Langmuir 7:3094
6. Mehrian T, de Keizer A, Kortweg AJ, Lyklema J (1993) Colloids Surf A73: 133
7. Fuerstenau DW, Wakamatsu T (1968) Adv Chem Ser 79:161; (1973) Trans AIME 254:123
8. Somasundaran P, Goddard ED (1979) In: Modern Aspects of Electrochemistry 13:207

9. Harwell JH, Hoskins JC, Schechter RS, Wade WH (1985) Langmuir 1:251
10. Adderson JE, Taylor H (1964) J Colloid Interface Sci 19:495
11. Mehrian T, de Keizer A, Korteweg AJ, Lyklema J (1993) Colloids Surf A71:255
12. Némethy G, Scheraga HA (1962) J Chem Phys 36:3401
13. Tanford C (1980) The Hydrophobic Effect: Formation of Micelles and Biological Membranes, Wiley, New York
14. Rouquerol J, Partyka S (1981) J Chem Tech Biotechnol 31:584
15. Van Os NM, Haandrikman G (1987) Langmuir 3:1051
16. Partyka S, Keh E, Lindheimer A, Groszek A (1989) Colloids Surf 37:309
17. Partyka S, Rudzinski W, Brun B (1989) Langmuir 5:297
18. Denoyel R, Rouquerol J (1991) J Colloid Interface Sci 143:555
19. Shinoda K, Fujihira M (1968) Bull Chem Soc Japan 41:2612
20. Shinoda K (1977) J Phys Chem 81:1300
21. Patterson D, Barbe M (1976) J Phys Chem 80:2435
22. Pratt LR (1985) Ann Rev Phys Chem 36:433
23. Jorgensen WL, Gao J, Ravimohan C (1985) J Phys Chem 89:3470
24. Wallqvist A (1991) J Phys Chem 95:8921
25. Scheraga HA (1966) J Chem Phys 45:3296
26. Evans DF, Ninham BW (1986) J Phys Chem 90:226
27. Privalov PL, Gill SJ (1988) Adv Protein Chem 39:191
28. Privalov PL, Gill SJ (1989) Pure Appl Chem 61:1097

Progr Colloid & Polym Sci (1994) 95:98—107
© Steinkopff Verlag 1994

P. F. Low

# The clay/water interface
# and its role in the environment

P. F. Low
Agronomy Department,
Purdue University,
West Lafayette,
Indiana 47907, USA

**Abstract** Data are presented to show that the surfaces of a common clay mineral, montmorillonite, modify the properties of the nearby water to a depth of at least 3.5 nm and that the value of every water property depends exponentially on $t$, the thickness of the films of adsorbed water, but is independent of the character of the surfaces. The viscosity and yield point are among the properties of the water that are modified. A modification of either of these properties has a commensurate effect on the flow of water between adjacent surfaces. Data are also presented to show that the surface-induced modification of the water is responsible for the swelling of the clay mineral and affects its ability to adsorb solutes. The swelling of clay reduces the permeability of the soil and the adsorption of solutes by the clay reduces their mobility. Thus, clay-water interaction has a significant impact on the convective and diffusive transport of pollutants through the soil.

**Key words** montmorillonite — clay — water — clay/water interface — clay/water interaction— pollution

## Introduction

Soils have been the repository of much of the world's refuse. Toxic substances from this refuse may be transported by convection or diffusion to underground reservoirs, lakes, and streams. In this way, our supplies of drinking water are jeopardized. Neither convection nor diffusion will occur appreciably in the soil in the absence of water. Soil water is in close contact with the surfaces of soil particles, especially the surfaces of clay minerals. If these surfaces modify the water, its role as a medium of transport for toxic substances will be affected. It is important, therefore, that we know how such surfaces interact with the water adjacent to them. The present paper reports some of the author's observations on clay/water interaction and indicates how this interaction affects the convective and diffusive transport of ions and neutral molecules.

## The interaction of water with clay minerals

Clay particles occur abundantly in the soil. They are mostly colloidal aluminosilicates. To learn how their extensive surfaces react with water, we have studied the following properties of water associated with montmorillonite and other clay minerals: threshold gradient [1], thermal expansibility [2, 3], isothermal compressibility [4], frequency of O—H stretching [5, 6], molar absorptivity [7], freezing point depression [8, 9] specific volume [10], specific heat capacity [11], heat of compression [12], viscosity [13], and free energy, enthalpy and entropy [6, 14, 15]. Not all of these properties will be discussed here. Instead, we will discuss only a few of them to illustrate the kind of results obtained.

The volume, $V$, of a clay/water system can be regarded as being the sum of two components, one due to unperturbed clay (i.e., clay that retains the properties it has in the pure state) and another due to water that is perturbed by the surfaces of the clay particles. Thus,

$$V = m_c v_c^0 + m_w \phi_v , \tag{1}$$

where $v_c^0$ is the specific volume of the pure clay, $\phi_v$ is the apparent specific volume of the water, and $m_c$ and $m_w$ are the repsective masses of clay and water. Note that all effects of clay/water interaction are included in $\phi_v$. Differentiation of Eq. (1) with respect to the temperature, $T$, yields

$$(\partial V/\partial T)_p = m_c (\partial v_c^0/\partial T)_p + m_w (\partial \phi_v/\partial T)_p \tag{2}$$

if the pressure, $P$, is held constant. Hereafter, we will call $(\partial \phi_v/\partial T)_p$ the apparent specific expansibility of the water.

According to one of Maxwell's relations,

$$(\partial S/\partial P)_T = -(\partial V/\partial T)_p , \tag{3}$$

where $S$ is the entropy of the system. For any reversible, isothermal process that occurs in the system:

$$T (\partial S/\partial P)_T = (\partial Q/\partial P)_T , \tag{4}$$

where $Q$ is the heat released in the process. By combining Eqs. (3) and (4), we obtain

$$(\partial Q/\partial P)_T/T = -(\partial V/\partial T)_P . \tag{5}$$

We used a Calvet microcalorimeter to measure $Q$ after each of several, successive, small increments of $P$ were applied to a sytem consisting of Na-montmorillonite and water. Between increments of $P$, the system was allowed to come to thermal equilibrium at the initial value of $T$. Hence, the process was essentially isothermal. It was also found to be reversible. We were able, therefore, to use Eq. (5) to obtain the value of $(\partial V/\partial T)_p$ from the slope of the graph of cumulative $Q$ vs. cumulative $P$. We also obtained the value of $(\partial V/\partial T)_p$ directly by confining the system in a dilatometer and measuring its change of $V$ with $T$ at atmospheric pressure. Then, we used these values of $(\partial V/\partial T)_p$ and estimated values of $(\partial v_c^0/\partial T)_p$ in Eq. (2) to calculate the corresponding values of $(\partial \phi_v/\partial T)_p$. These procedures were repeated with systems having different values of $m_w/m_c$. The resulting data, taken from the work of Clementz and Low (2), are presented in Fig. 1.

Observe from Fig. 1 that the data obtained calorimetrically were in excellent agreement with those obtained dilatometrically. The respective equations for the curves in Fig. 1 are

$$(\partial \phi_v \partial T)_p = 2.58 \times 10^{-4} \exp[0.6/(m_w/m_c)] \tag{6a}$$

and

$$(\partial \phi_v/\partial T)_p = 2.58 \times 10^{-4} \exp[0.5/(m_w/m_c)] . \tag{6b}$$

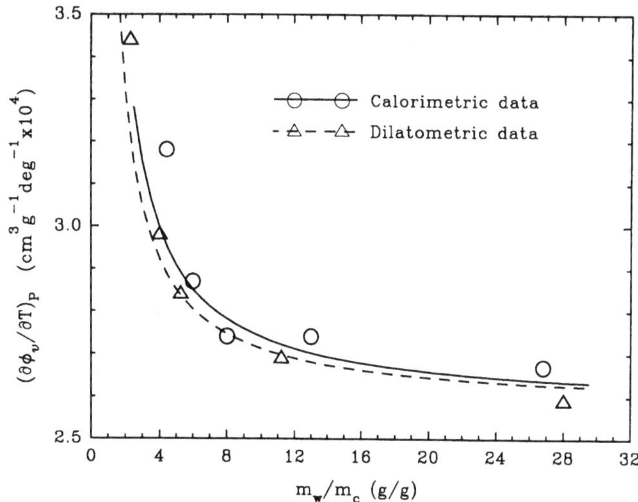


**Fig. 1** Relation between $(\partial \phi_v/\partial T)_p$ and $m_w/m_c$ for mixtures of Na-montmorillonite and water

The factor preceding the exponential equals $(\partial v_w^0/\partial T)_p$, the specific expansibility of pure bulk water. It is clear, therefore, that the apparent specific expansibility of the water in a system of Na-saturated montmorillonite and water increases exponentially with decreasing water content.

A common exponential equation applies to every property of water in the clay/water systems that we have examined thus far. A general form of this common equation is

$$J_i/J_i^0 = \exp \beta_i (m_c/m_w) , \tag{7}$$

where $J$ and $J^0$ are the values of the property, $i$, in the clay/water system and pure bulk water, respectively, and $\beta$ is a characteristic constant. Now, the average thickness, $t$, of the water films on the surfaces of the clay particles is given by

$$t = m_w/m_c \rho_w S , \tag{8}$$

where $\rho_w$ is the density of the water and $S$ is the specific surface area of the clay. If

$$\beta_i = k_i S , \tag{9}$$

which would be the case if the water interacted non-specifically with the surfaces of the clay particles, we can combine Eqs. (7), (8) and (9) to obtain

$$J_i/J_i^0 = \exp k_i/\rho_w t . \tag{10}$$

Thus, if the variation of $\rho_w$ with $t$ is insignificant, i.e., if $\rho_w$ is essentially constant, $J$ becomes a function of $t$ only. That this equation is obeyed when $J = (\partial \phi_v/\partial T)_p$ is shown in Fig. 2. The relevant data for this figure were taken from a paper by Sun et al. [16].

**Fig. 2** The effect of $t$ on $(\partial\phi_v/\partial T)_p$ and $\varepsilon$ in mixtures of Na-montmorillonite and water

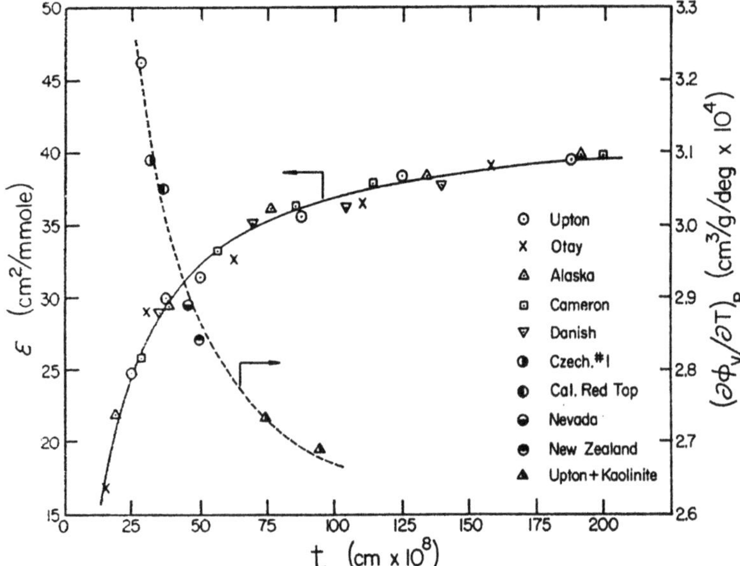


The dashed curve in the figure is described by Eq. (10) with $J^0 = 2.5 \times 10^{-4}$ cm$^3$ g$^{-1}$ deg$^{-1}$, $k = 7.22 \times 10^{-8}$ g cm$^{-2}$, and $\varphi_w = 1.0$ g cm$^{-3}$.

Beer's law describes the absorption of radiant energy by any fluid medium. It may be written

$$A = \varepsilon c I \tag{11}$$

where $A$ is the absorbance at a given frequency, $\varepsilon$ is the molar absorptivity at this frequency, $c$ is the concentration of the absorbing dipole, and $I$ is the length of the path over which absorption occurs. According to spectroscopic theory, $\varepsilon$ is related to the dipole moment of the absorbing dipole [17]. Changes in $\varepsilon$ are determined primarily by charge transfer [18]. With respect to water and other hydrogen-bonded substances, $\varepsilon$ has been found to depend on the strength of the hydrogen bonds [19]. We decided, therefore, to mix several Na-saturated smectites (i.e., expandable clay minerals) in different proportions with a 3% solution of D$_2$O in H$_2$O, determine $A$ for each mixture at the frequency of O—D stretching by infrared spectroscopy, use Eqs. (8) and (11) to calculate the corresponding values of $t$ and $\varepsilon$, and then plot $\varepsilon$ against $t$. The results, taken from a paper by Mulla and Low [7], are shown in Fig. 2. The different symbols in this figure represent different smectites with different values of $\sigma$, the surface charge density. The solid curve is represented by Eq. (10) with $J^0 = 42.3$ cm$^2$ mmole$^{-1}$, $k = -13.68 \times 10^{-8}$ g cm$^{-2}$ and $\rho_w = 1.0$ g cm$^{-3}$. It is obvious that the value of $\varepsilon$ deviates increasingly from its value in pure bulk water as the water content of the system decreases, i.e., as the water films surrounding the particles become thinner. This implies that intermolecular hydrogen bonds in the water are stretched and become weaker as the sur-

face of the clay particle is approached. Moreover, it is obvious that $\varepsilon$ depends only on the thickness of the water films and not on the density of charge or other characteristics of the surface. From the latter observation, we conclude that the water interacts non-specifically with the clay surface.

Now let us return to Figs. 1 and 2. Since $\phi_v$ includes all volume changes due to clay/water interaction, it may be argued that deviations of $(\partial\phi_v/\partial T)_p$ from $(\partial v_w^0/\partial T)_p$ are not due only to perturbation of the water by the clay particles, but also to the perturbation of the clay particles by the water. This argument can be tested in a simple way. We can write Eq. (10) for any two properties of the water, designated by the subscripts 1 and 2, respectively, and then combine the two equations to get

$$\ln (J_1/J_1^0) = \frac{k_1}{k_2} \ln (J_2/J_2^0)$$

or

$$(J_1/J_1^0) = (J_2/J_2^0)^{k_1/k_2} . \tag{12}$$

Equation (10) applies to both $(\partial\phi_v/\partial T)_p$ and $\varepsilon$ and so Eq. (12) must apply to them also. That it does is illustrated in Fig. 3. Now, the measured values of $\varepsilon$ are characteristic of the water alone because montmorillonite is transparent to infrared radiation at the frequency of O-D stretching. It necessarily follows that $(\partial\phi_v/\partial T)_p$ is also characteristic of the water only. Otherwise Eq. (12) would not hold for these two properties. The same conclusion would follow for any other property of the interparticle water that is described by Eq. (10). As mentioned earlier, Eq. (10) describes every property of the water in the clay/water

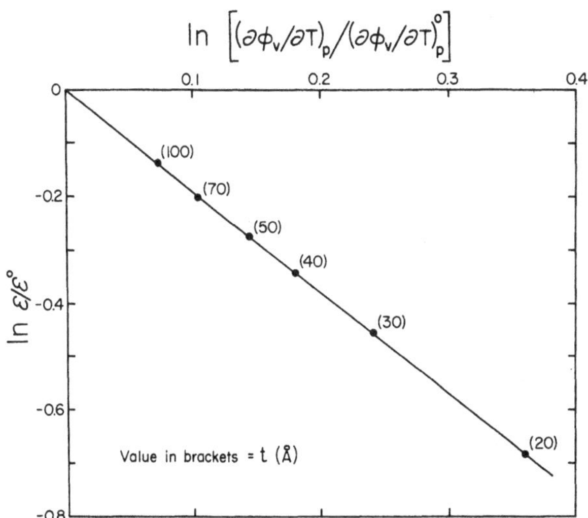



**Fig. 3** Relation between $\varepsilon/\varepsilon^0$ and $(\partial\phi_v/\partial T)_p/(\partial\phi_v^0/\partial T)$ in mixtures of Na-montmorillonite and water

systems we have investigated thus far. A ramification of the general applicability of Eq. (10) is that, if the respective values of $k$ are known, Eq. (12) can be used to calculate the value of any property of the interparticle water in a given system if the value of any other property of the interparticle water in that system is known.

Values of $J$ in Eq. (10) are average values for the water in films having the thickness $t$. These values do not provide information about the distribution of the given property with respect to the surface of the particle. Therefore, we are left with the problem of determining the depth to which the clay surface perturbs or modifies the structure (and, hence, the properties) of the adjacent water. This problem can be solved as follows. If sufficient water is present in a clay/water system, water that is perturbed by the clay particles coexists with water that retains the properties characteristic of bulk water. It is reasonable to assume that the perturbed water exists in films (hydration shells) next to and parallel to the planar surfaces of the particles and that the unperturbed or bulk water exists in films beyond but contiguous with the hydration shells. Both kinds of water contribute to the value of $J$. Hence, we can write

$$J = f^* J^* + f^0 J^0 ,\tag{13}$$

where $f^*$ and $f^0$ are the fractions of perturbed water and bulk water, respectively, in the system and $J^*$ is the *average* value of the given property for the perturbed water. Now

$$f^* + f^0 = 1\tag{14}$$

$$t = t^* + t^0\tag{15}$$

and

$$f^* = \frac{m_w^*}{m_w} = \frac{t^*}{t} ,\tag{16}$$

where $t^*$ is the average thickness of the films of perturbed water next to the particle surfaces, $t^0$ is the average thickness of the films of unperturbed water, and $m_w^*$ is the mass of perturbed water. Note that the second equality in Eq. (16) holds only if the density of the water is invariant with respect to the surfaces of the clay particles. Although this is not strictly true (10), the actual variation in the density of the water is small enough to be ignored for the present purpose. Combination of Eqs. (13), (14) and (16) yields

$$J = (J^* - J^0)\frac{t^*}{t} + J^0 .\tag{17}$$

From Eq. (17) it is evident that $J$ will be a linear function of $1/t$ as long as $J^*$ and $t^*$ remain constant. These two parameters will remain constant until the hydration shells come into contact and begin to overlap, i.e., until all the bulk water is eliminated and $t^* = t$. Then they will begin to vary. It follows, therefore, that the thickness of the hydration shells can be determined from the value of $1/t$ at which the graph of $J$ vs $1/t$ departs from linearity.

Shown in Fig. 4 are plots of $J$ vs $1/t$ when $J = (\partial\phi_v/\partial T)_p$ and when $J = \varepsilon$. Observe that both plots depart from linearity at a value of $1/t$ corresponding to $t = t^* \cong 35 \times 10^{-8}$ cm (3.5 nm). This is not necessarily the maximum thickness of the water perturbed by the particle surfaces. The value of $1/t$ at which curvature of the

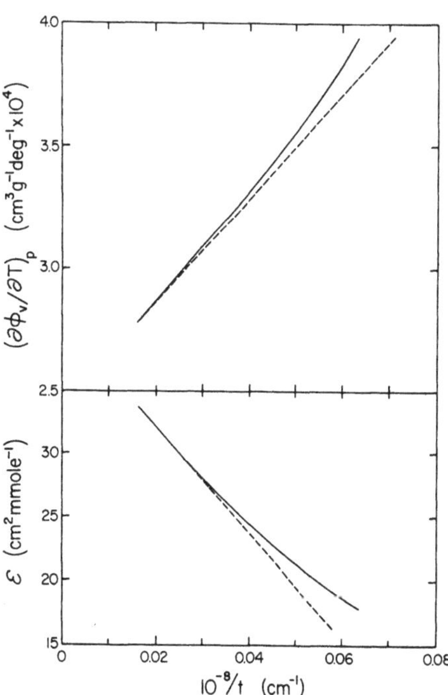



**Fig. 4** Dependence of $(\partial\phi_v/\partial T)_p$ and $\varepsilon$ on $1/t$ in mixtures of Na-montmorillonite and water

graph of $J$ vs $1/t$ becomes apparent depends on how sensitive the given property is to perturbation of the structure of the water and on how sensitive the measuring device is to changes in this property. Consequently, we conclude that water is perturbed by the surfaces of clay particles to a depth of *at least* 3.5 nm. In this regard, it is important to note that even extremely small changes in a water property can have important consequences. For example, consider the thermodynamic equation developed by Low and Anderson [20]

$$\bar{g}_w - g_w^0 = -\bar{v}_w \Pi \, , \tag{18}$$

where $\bar{g}_w$ and $\bar{v}_w$ are the partial specific free energy and partial specific volume of the water in a clay/water system, $g_w^0$ is the specific free energy of pure bulk water, and $\Pi$ is the swelling pressure of the clay. This equation shows that a difference between $\bar{g}_w$ and $g_w^0$ of only 0.1 J/g produces a P of 0.1 MPa ($\sim 1.0$ atm). A $\Pi$ of 0.1 MPa is quite significant in nature.

In this section of the paper we have established the following concepts:

1) interaction between clay and water perturbs the water and alters its properties;
2) interaction between clay and water does not depend on specific characteristics of the clay surface;
3) interaction between clay and water extends to a distance of at least 3.5 nm from the clay surface;
4) the properties of the water in a clay/water system are described by a common exponential equation with parameters that depend specifically on the property.

## The swelling of clay

When the clay in a soil absorbs water, it swells into the neighboring pores and reduces their radii. According to Poiseuille's law, the rate of fluid flow through a circular pore is proportional to the fourth power of the radius of the pore. Consequently, the swelling of clay in a soil reduces the permeability of the soil to water and, thereby, the convective transport of toxic substances dissolved in the water. It is clear, therefore, that clay swelling and environmental pollution are related.

The clay mineral, montmorillonite, is found in most soils. It swells because water enters the region between its superimposed layers and forces them apart. Other expandable clay minerals swell for the same reason. The pressure that must be applied to keep the layers from moving apart is the swelling pressure, $\Pi$. From a thermodynamic viewpoint, water migrates from an external phase surrounding the particles to the interlayer phase because its partial specific free energy is lower there than in the external phase; and $\Pi$ is the pressure that must be applied to the clay to make the partial specific free energy of the water the same in the two phases. In keeping with

Eq. (18), if the external phase is pure bulk water, $\Pi$ depends on the difference between $\bar{g}_w$ and $g_w^0$. In the osmotic concept of swelling as described by electric double layer theory [21], $\bar{g}_w$ is supposed to be lowered in the interlayer phase because of the high concentration of exchangeable cations there. But is should be noted that interaction of the water with the surfaces of the layers can also lower $\bar{g}_w$ and enhance $\Pi$.

Experiments of Low [22] with expandable clays established the following empirical relation between $\Pi$ (expressed in atmospheres) and $m_c/m_w$

$$\Pi + 1 = B \exp a m_c/m_w \, , \tag{19}$$

where $B$ and $a$ are constants. If $\Pi$ is expressed in units other than atmospheres, the lefthand-side of this equation becomes $(\Pi + P_a)/P_a$ or $P/P_a$, where $P$ is the absolute pressure acting on the clay and $P_a$ is the pressure of the atmosphere. Equation (19) can be combined with Eq. (8) to give

$$\Pi + 1 = B \exp(k_a'/\rho_w t) \, , \tag{20}$$

provided that $a = k_a'S$. Then, $\Pi$ would be independent of the specific characteristics of the clay surface and would be a function of $t$ alone. Figure 5, which was taken from a paper by Low [23], shows that this is so. The solid curve in Fig. 5 obeys Eq. (20).

By solving Eqs. (10) and (20) simultaneously to eliminate $\rho_w t$, we find that

$$\ln(\Pi + 1) = (k_a'/k_i) \ln(J_1/J_i^0) + \ln B$$

or

$$(\Pi + 1) = B (J_1/J_1^0)^{k_a'/k_i} \, . \tag{21}$$

The validity of Eq. (21) is demonstrated in Fig. 6 for the case of $J_i/J_i^0 = \varepsilon/\varepsilon^0$. This figure was published earlier by Low [24].

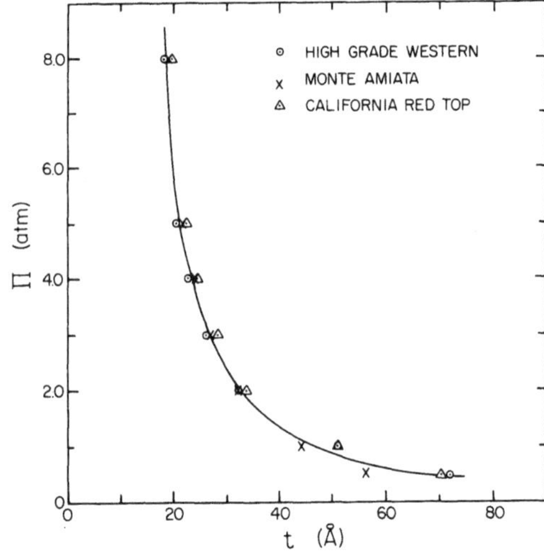


**Fig. 5** Relation between $\Pi$ and $t$ for three different Na-montmorillonites

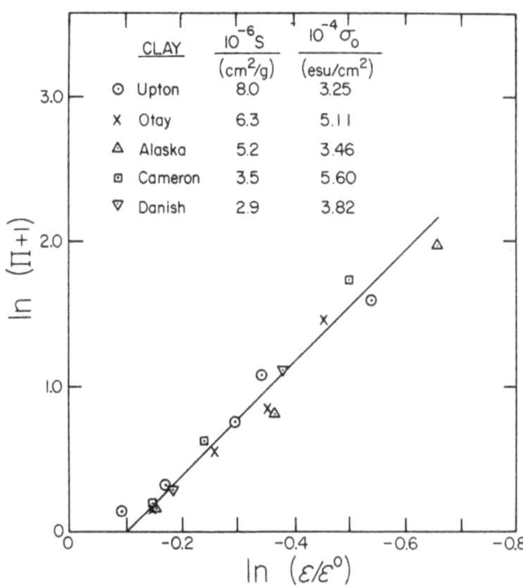

**Fig. 6** Relation between $\Pi + 1$ and $\varepsilon/\varepsilon^0$ for mixtures of Na-montmorillonite and water

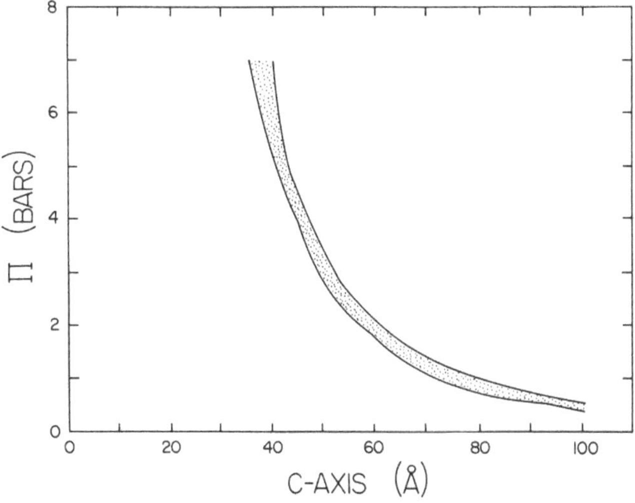

**Fig. 7** The limits (solid lines) encompassing the curves of $\Pi$ versus $c$-axis spacing ($\lambda + 9.3$ Å) for eight different Na-montmorillonites

Recall that $\varepsilon$ depends on intermolecular bonding in the water. By virtue of the relation between $\Pi + 1$ and $\varepsilon/\varepsilon^0$ in Fig. 6, we are led to the conclusion that $\Pi$ is governed by intermolecular bonding as influenced by clay/water interaction.

Viani et al. [25] used x-ray diffraction to measure $\lambda$, the interlayer distance, at several values of $\Pi$ for eight expandable clays having different values of $\sigma$ and $Na^+$ as the exchangeable cation. Figure 7 was taken from their work. The curves of $\Pi$ vs $\lambda$ for all eight clays fell within the stippled area between the solid lines. Hence, the rela-

tion between $\Pi$ and $\lambda$ was essentially the same for every clay. This relation is described by the following empirical equation

$$\Pi + 1 = b \exp(k_a/\lambda) , \tag{22}$$

where $b$ and $k_a$ are constants. Observe that Eqs. (19), (20) and (22) are the same equation expressed in different interrelated variables.

The surface charge density, $\sigma$, is an inherent electrical property of a clay that arises from the substitution of a cation of lower valence for a cation of higher valence within the crystal structure of the clay. Its value is obtained experimentally by dividing the cation exchange capacity of the clay by the surface area of the clay. Any electrostatic theory of swelling, including the electric double-layer theory, would require that the interlayer distance, $\lambda$, depend on $\sigma$ at any value of $\Pi$. Such a theory would also require that $\lambda$ depend on $\varepsilon$, the dielectric constant of the interlayer solution, at any value of $\Pi$. Figure 8, which was taken from the paper of Viani et al. [25], shows that $\lambda$ is independent of $\sigma$; and Fig. 9, which was constructed from the results of Zhang [26], shows that $\lambda$ is independent of $\varepsilon$. In both cases, $\lambda$ was measured by x-ray diffraction. The necessary conclusion is that the swelling of clay is not primarily electrostatic in nature. As indicated in Fig. 6, it is due to non-specific interaction between the particle surfaces and the water. This interaction decreases $\bar{g}_w$ and increases $\Pi$ in keeping with Eq. (18). Evidently, the electric charge of the clay layers makes a negligible contribution to $\Pi$ because this charge is effectively screened by exchangeable cations condensed on the clay surface. In other words, almost all the exchangeable cations are in the Stern layer.

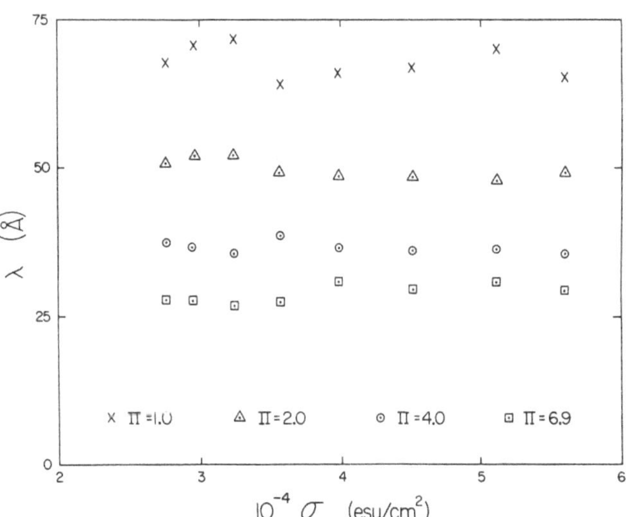

**Fig. 8** Relation between $\lambda$ and $\sigma$ for eight different Na-montmorillonites

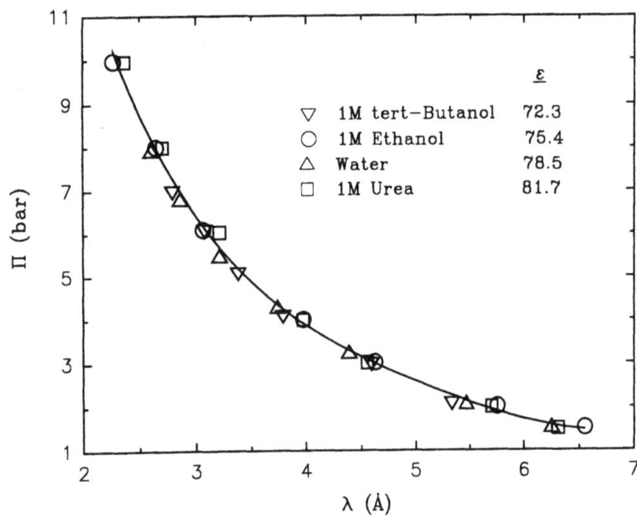


**Fig. 9** Relation between $\Pi$ and $\lambda$ for Na-montmorillonite equilibrated with aqueous solutions having different values of $\underline{\varepsilon}$

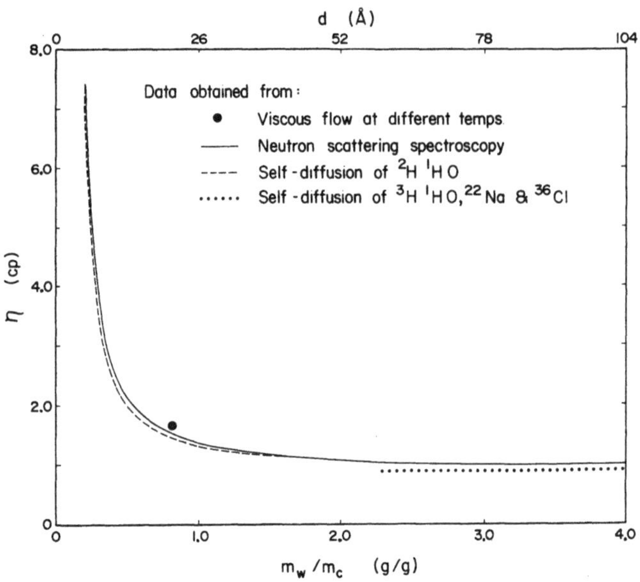


**Fig. 10** Relation between $\eta$ and $d$ (or $m_w/m_c$) in mixtures of Na-montmorillonite and water

On the basis of the results discussed in this section of the paper, we propose the following concepts:

1) the swelling of expandable clays is not due to electrostatic interaction between the clay layers;
2) the swelling of expandable clays is due to interaction between the clay layers and the interlayer water;
3) the swelling of expandable clays depends only on the distance between the clay layers and not on the characteristics of those layers;
4) a common exponential relation between $\Pi$ and $\lambda$ describes the swelling of all expandable clays.

## Effect of clay/water interaction on the transport of solutes

By using data from experiments on neutron scattering by clay/water systems and from experiments on the viscous flow and self-diffusion of water in such systems, Low [13] was able to determine the average coefficient of viscosity, $\eta$, of the water in the system as a function of $m_w/m_c$. His results are shown in Fig. 10. The coincident curves in this figure obey Eq. (10) with $J/J^0$ replaced by $\eta/\eta^0$ and $t$ replaced by $d$, where $d = 2t$. According to Darcy's law,

$$Q = Ak\rho gi/\eta = AKi \tag{23}$$

in which $Q$ is the volume of fluid flowing through a porous medium in unit time across the area $A$ under the hydraulic gradient $i$, $k$ is the intrinsic permeability of the porous medium, $\rho$ is the density of the fluid, $g$ is the acceleration of gravity, and $K$ is the hydraulic conductivity.

It is evident, therefore, that an increase in $\eta$ will reduce convective flow. From Fig. 10, we see that clay/water interaction increases $\eta$ substantially at low values of $m_w/m_c$ for any given value of $k$. The value of $k$ for compacted clays is very low [13]. Thus, in keeping with Eq. (23), clay-rich soils or clay barriers impede the escape of toxic wastes from disposal sites.

An example of how clay/water interaction influences convective flow in natural sediments is presented in Fig. 11. This figure was taken from the paper of Young et al. [27]. It shows the effect of temperature on the observed and calculated values of $Q_T/Q_{20}$, the ratio of the flow rate at $T = T$ to that at $T = 20°C$, for water flow through a core of argillaceous shale that was removed from the earth at a depth of $\sim 1500$ meters. The calculated value of $Q_T/T_{20}$ was obtained by assigning the water in the core the handbook values of $\eta$ at $T = T$ and $T = 20°C$. In considering Fig. 11, it is important to note that the difference between the observed and calculated curves cannot be attributed to either thermally induced changes in the swelling of the clay or to thermal expansion of the mineral grains. Zhang et al. [28] found that the swelling of clay is quite insensitive to changes in temperature, and calculations of Young et al. [27] showed that thermal expansion of the mineral grains could be neglected if reasonable values for the coefficient of thermal expansion of the grains were employed. The necessary conclusion is, therefore, that clay/water interaction modified the effect of temperature on the viscous flow of water in the shale.

Not only is the viscous flow of water modified in the pores of fine-grained media, but the water in these pores

Progr Colloid Polym Sci (1994) 95:98–107
© Steinkopff Verlag 1994



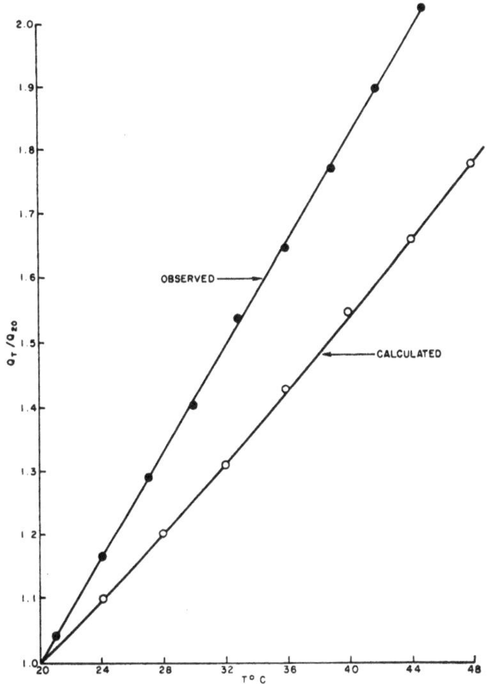

**Fig. 11** Effect of $T$ on $Q_T/Q_{20}$ in a natural shale

can have sufficient structural integrity to resist shear until the hydraulic gradient exceeds a critical value called the threshold hydraulic gradient. The existence of such a gradient was demonstrated by Miller and Low [1]. One of their graphs is reproduced in Fig. 12. Under the condition that the prevailing hydraulic gradient in a porous medium is below the threshold gradient, the transport of toxic substances through the medium would have to be by diffusion instead of convection. Diffusive transport is much slower than convective transport.

As any ionic or molecular constituent moves through the interparticle water along a gradient in electric potential or concentration in a clay/water system, it does so in a series of hops over energy barriers between equilibrium positions, i.e., positions of minimum energy. At the crest of an energy barrier, the moving constituent is supposed to be in an activated state and the height of the barrier in units of energy is the activation energy. The activation energy is the energy required by the constituent to push aside water molecules ahead of it to form a hole plus the energy required to break bonds with neighboring water molecules so that it can move into the hole thus formed. Consequently, the activation energy for ionic conductance or molecular diffusion should depend on the integrity of the structure of the surrounding water; and the integrity of this water should be indicated by the magnitude of any of its structure-sensitive properties. With these thoughts in mind, Low [29] compared measured values of the specific volume of the water with measured values of the activation energy for exchangeable ion conductance in montmorillonite/water systems having $Li^+$, $Na^+$ and $K^+$ as the exchangeable cations. His results are reproduced in Fig. 13. Observe how well the specific volume of the water correlates with the activation energy for exchangeable ion conductance. The values of both of these quantities differ from the corresponding values in bulk solutions. We conclude, therefore, that the modified water in a clay/water system affects ionic migration.

**Fig. 12** Relation between flow rate and hydraulic gradient in a mixture of Na-montmorillonite and water

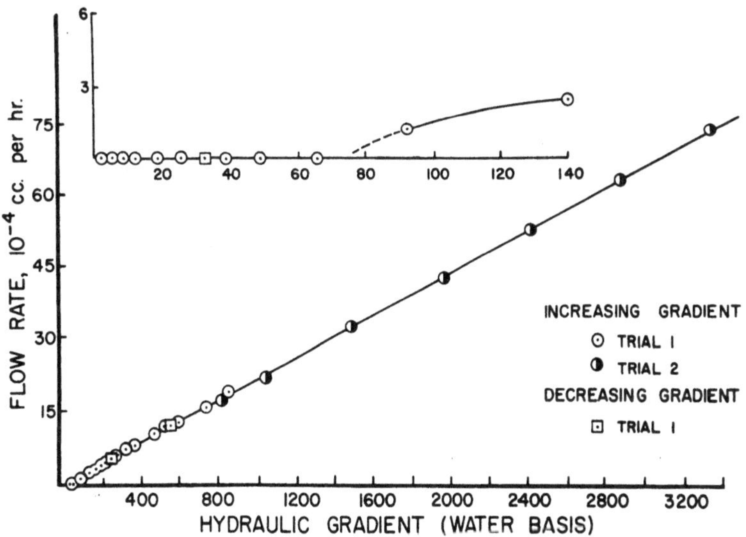

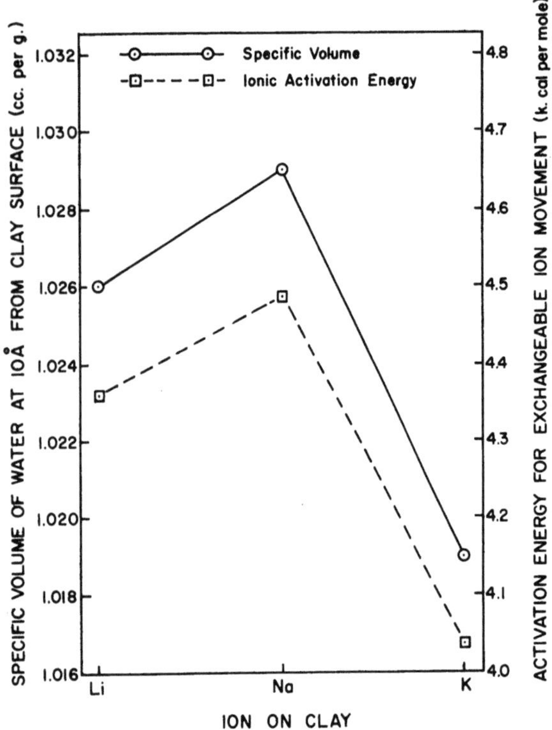

Fig. 13 Relation between the activation energy for exchangeable ion movement and the specific volume of adsorbed water in Li-, Na- and K-montmorillonite

In the theory of rate processes [30], the diffusion coefficient, $D$, of any constituent, $i$, is given by

$$D_i = B_i \exp(-E_i/RT) \tag{24}$$

where $E$ is the activation energy, $R$ is the molar gas constant, $B$ is a constant that depends on the entropy of activation, and $T$ is the absolute temperature. If we apply this equation to two constituents, a neutral organic molecule and water, diffusing simultaneously through the same system and divide one equation by the other, we obtain

$$\frac{D_0}{D_w} = \frac{B_0}{B_w} \exp(E_w - E_0)/RT , \tag{25}$$

where the subscripts $o$ and $w$ denote the organic molecule and water, respectively. In recent experiments in our laboratory, T. X. Wang used radioactive isotopes to measure the simultaneous diffusion of 1,4-dioxane and water in clay/water systems ($m_w/m_c = 2.5$) and aqueous solutions containing different concentrations of non-radioactive 1,4-dioxane. Hence, Eq. (25) applies to his experiments. What he observed is reported in Fig. 14. Evidently, $E_w - E_0$ changed differently with the concentration of 1,4-dioxane in the clay/water system than it did in aqueous solution. In this regard, it is important to note

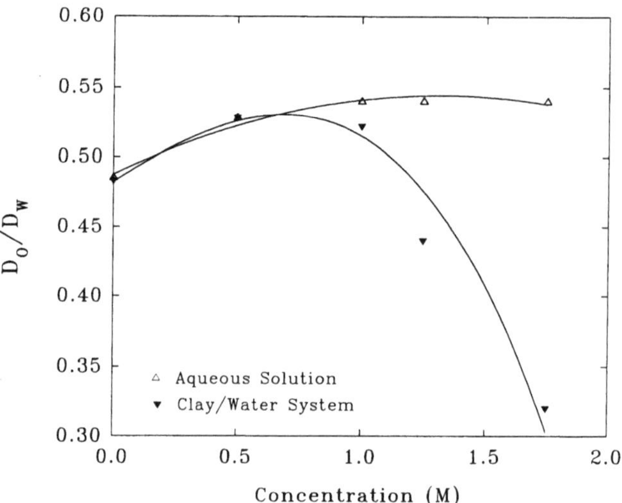


Fig. 14 Relation between $D_0/D_w$ and the concentration of 1,4-dioxane in mixtures of Na-montmorillonite and water

that both the 1,4-dioxane and water were diffusing simultaneously through each medium and so the geometry of the diffusion path was the same for both in each medium. This is especially noteworthy with respect to the clay/water system. We conclude, therefore, that the 1,4-dioxane interacted differently with the water molecules in the interparticle solution than with the water molecules in the bulk solution. In other words, the interparticle water had a special effect on the diffusion of 1,4-dioxane.

From thermodynamic theory, we know that

$$\Delta G = \Delta H - T \Delta S , \tag{26}$$

where $\Delta G$, $\Delta H$, and $\Delta S$ are the changes in free energy, enthalpy, and entropy of a system, respectively, that occur as the result of any reaction or process. If the process is the spontaneous, reversible adsorption of a solute on a solid adsorbent at constant $P$ and $T$, $\Delta G < 0$ and $\Delta H = -Q_a$, where $Q_a$ is the heat of adsorption. Hence, Eq. (26) becomes

$$\Delta G = -Q_a - T \Delta S < 0 . \tag{27}$$

Zhang et al. [31] used a Calvet calorimeter to measure $Q_a$ as a function of $T_2^{(n)} S$, the apparent adsorption, for the adsorption from aqueous solution of several different organic compounds on Na-montmorillonite. The apparent adsorption is the difference between the amount of organic compound actually present in the system per unit mass of montmorillonite and that which would be present if the concentration of the compound in the equilibrium solution persisted up to the montmorillonite/water interface. Their results are reproduced in Fig. 15. Observe from this figure that $Q_a$ was negative for some of the

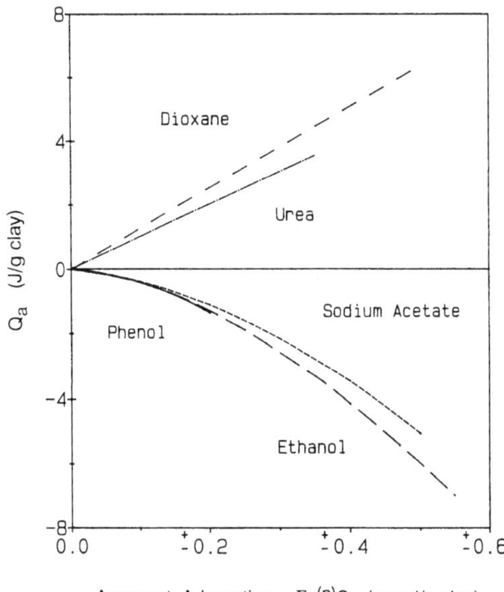



**Fig. 15** Relation between $Q_a$ and $T\Gamma_2^{(n)}S$ for the adsorption of five different organic compounds on Na-montmorillonite (negative values on the abscissa apply to the acetate ion only)

organic compounds, namely, phenol, ethanol, and sodium acetate. For negative values of $Q_a$, Eq. (27) requires that $\Delta S$ be positive, i.e., that the degree of randomness of the system increases with adsorption. But the localization of the organic compound in a region near the montmorillonite/solution interface should decrease the entropy of the system. We conclude, therefore, that the observed increase in entropy was due to a randomization of the water molecules in the interfacial region under the influence of the adsorbed organic compounds. In other words, the adsorption of the compounds was accompanied, and affected, by structural changes in the interfacial water. This conclusion is important as regards environmental pollution because adsorption affects the rate of migration of pollutants in the soil.

In summary, we have shown that interaction between the surfaces of clay particles and water modifies the structure of the water and that this modification: 1) affects the swelling of the clay and, thereby, the permeability of the system, 2) affects the viscosity and yield point of the interparticle water, and 3) affects adsorption of solutes by the clay. Hence, it is clear that clay/water interaction will affect the convective and diffusive transport of solutes through a soil, especially one that is rich in clay.

## References

1. Miller RJ, Low PF (1963) Soil Sci Soc Am Proc 27:605—609
2. Clementz DM, Low PF (1976) In: Kerker M (ed) Colloid and Interface Sci Vol 3, Academic Press, New York, pp 485—502
3. Ruiz HA, Low PF (1976) In: Kerker M (ed) Colloid and Interface Sci Vol 3, Academic Press, New York, pp 503—515
4. Oliphant JL, Low PF (1983) J Colloid Interface Sci 95:45—50
5. Sallé de Chou J, Low PF, Roth CB (1980) Clays Clay Minerals 28:111—118
6. Fu MH, Zhang ZZ, Low PF (1990) Clays Clay Minerals 38:485—492
7. Mulla DJ, Low PF (1983) J Colloid Interface Sci 95:51—60
8. Kolaian JH, Low PF (1963) Soil Sci 95:376—384
9. Low PF, Anderson DM, Hoekstra P (1968) Water Resources Res 4:379—394
10. Anderson DM, Low PF (1958) Soil Sci Soc Am Proc 22:99—103
11. Oster JD, Low PF (1964) Soil Sci Soc Am Proc 28:605—609
12. Kay BD, Low PF (1975) Clays Clay Minerals 23:266—271
13. Low PF (1976) Soil Sci Soc Am J 40:500—505
14. Oliphant JL, Low PF (1982) J Colloid Interface Sci 89:366—373
15. Zhang ZZ, Low PF (1989) J Colloid Interface Sci 133:461—472
16. Sun Y, Lin HH, Low PF (1986) J Colloid Interface Sci 112:556—564
17. Steele D (1971) Theory of vibrational spectroscopy, Saunders, Philidelphia
18. Tsubomora H (1956) J Chem Phys 24:927
19. Pimentel GC, McClellan AL (1960) The hydrogen bond. Freeman and Co., San Francisco
20. Low PF, Anderson DM (1958) Soil Sci 86:251—253
21. Verwey EJW, Overbeek JThG (1948) Theory of the stability of lyophobic colloids, Elsevier, Amsterdam
22. Low PF (1980) Soil Sci Soc Am J 44:667—676
23. Low PF (1987) Langmuir 3:18—25
24. Low PF (1987) In: Schultz LG, van Olphen H, Mumpton FA (eds) The Clay Minerals Soc, Bloomington, IN, pp 247—256
25. Viani BE, Low PF, Roth CB (1983) J Colloid Interface Sci 96:229—244
26. Zhang F (1992) Ph D Thesis, Purdue Univ
27. Young A, Low PF, McLatchie AS (1964) J Geophysical Res 69:4237-4245
28. Zhang H, Zhang ZZ, Low PF, Roth CB (1993) Clay Minerals 28:25—31
29. Low PF (1962) Clays Clay Minerals 9:219—228
30. Glasstone S, Laidler KJ, Eyring H (1941) The theory of rate processes, McGraw-Hill, New York
31. Zhang ZZ, Low PF, Cushman JH, Roth CB (1990) Soil Sci Soc Am J 54:59—66

*Progr Colloid & Polym Sci* (1994) 95:108—112
© Steinkopff Verlag 1994

N. Kallay
S. Žalac
J. Ćulin
U. Bieger
A. Pohlmeier
H. D. Narres

# Thermochemistry and adsorption equilibria at hematite — water interface

Prof. N. Kallay (✉) · S. Žalac ·
J. Ćulin
Laboratory of Physical Chemistry
Faculty of Science
University of Zagreb
P.O. Box 163
41001 Zagreb, Croatia

U. Bieger · A. Pohlmeier ·
H. D. Narres
Institut für Angewandte Physikalische
Chemie
Forschungszentrum Jülich GmbH
Postfach 1913
52425 Jülich, FRG

**Abstract** Enthalpy of charging of hematite surface was determined by calorimetry and also by measuring the dependence of the point of zero charge (p.z.c.) on temperature. The calorimetric experiment was designed in such a way that the point of zero charge was in the middle between the initial and final pH. In that case, the reaction heat corresponds to the difference in standard reaction enthalpies of surface deprotonation andprotonation reactions, without electrostatic contribution. The following obtained calorimetric value: $\Delta_b H^0 - \Delta_a H^0 = 33.3$ kJ mol$^{-1}$ agrees well with the result of p.z.c. measurements; $\Delta_b H^0 - \Delta_a H^0 = 33.2$ kJ mol$^{-1}$. In addition to the thermochemistry results, the adsorption of dodecylsulfate and salicylate ions on the hematite was examined.

**Key words** Adsorption enthalpy — adsorption equilibrium — hematite — calorimetry

## Introduction

Transport or organic substances, such as pesticides or synthetic compost, through soil is related to the adsorption — desorption equilibria in this heterogeneous system. Physicochemical investigations concern model systems, i.e., well-defined pure oxides, representing the soil, and some model compounds such as phosphates, oxalates, salicylates, humic acids, representing the pollutants. Adsorption equilibrium can be represented either by adsorption isotherms at different pH values, or by appropriate equilibrium parameters resulting from a theoretical model. For the latter purpose the surface complexation model was found to be suitable [1—4]. The treatment involves consideration of equilibria in the bulk of the solution and also the postulation of reactions taking place at the interface. The complications, commonly involved in the interpretation, are due to electrostatic forces affecting the energy state of the adsorbed organic ions. One possibility is to take the electrostatic term into account by assuming a certain structure of electrical inter-

facial layer and another, experimental approach is to determine the adsorption isotherm at the pH corresponding to the isoelectric point (i.e.p.) and consequently, to avoid any assumption regarding the electrical interfacial layer structure [3, 4].

More recently, in addition to the adsorption equilibria studies, calorimetry was employed in examining of the metal oxide aqueous interface. The results deal either with relatively simple situations, such as protonation and deprotonation reactions at the surface or with the heats of the adsorption of organic molecules onto surfaces. In both cases, one encounters two problems. The first one is due to the fact that, upon addition of a reactant, several reactions take place in the calorimeter. In order to evaluate enthalpy of a specific surface reaction, one has to be able to distinguish between different contributions to the heat and also to determine the extents of all reactions taking place in the system. The next problem is related to the electrostatic contribution to the enthalpy. Recently, an experimental design was developed which enabled the proper interpretation of the calorimetric data so that one is able to obtain the "chemical" contribution to the difference of the reaction enthalpies of deprotonation and protonation, i.e., the same quantity which one gets from the temperature dependence of the point of zero charge. The method was demonstrated on the example of titania [5].

Since hematite represents some aspects of the soil behavior, the aim of this study is to apply the proposed method to this colloid, and to obtain corresponding thermodynamic parameters. In addition, some results of the adsorption of organic substances on the hematite will be discussed.

## Theoretical

According to surface complexation model [1, 2] protonation and deprotonation of hydrated metal oxide surface concerns amphotheric surface sites (MOH) as follows

$$\text{MOH} + \text{H}^+ \rightarrow \text{MOH}_2^+ \, ; \quad K_a, \, \Delta_a H, \, \Delta \xi_a \tag{1}$$

$$\text{MOH} \rightarrow \text{MO}^- + \text{H}^+ ; \quad K_b, \, \Delta_b H, \, \Delta \xi_b \, , \tag{2}$$

where $M$ denotes metal at the solid surface, while $\Delta \xi_r$ and $\Delta_r H$ are reaction extent and reaction enthalpy, respectively. Surface equilibrium constants $K$ are given by

$$K_a = K_a^0 e^{F\Psi_0/RT} = \frac{\Gamma_{\text{MOH}_2^+}}{a_{\text{H}^+} \Gamma_{\text{MOH}}} \tag{3}$$

$$K_b = K_b^0 e^{-F\Psi_0/RT} = \frac{\Gamma_{\text{MO}^-} a_{\text{H}^+}}{\Gamma_{\text{MOH}}} \tag{4}$$

where $K^0$ denotes standard ("intrinsic") equilibrium constant, $\Gamma$ is the surface concentration (amount of sites divided by the surface area), $\Psi_0$ is the electrostatic potential at the 0-plane in which charged surface groups are located, and $F$, $R$ and $T$ have their usual meaning. At the point of zero charge ($\text{pH}_{\text{pzc}}$), one can take

$$\Gamma_{\text{MO}^-} = \Gamma_{\text{MOH}_2^+} \quad \text{and} \quad \Psi_0 = 0 \, , \tag{5}$$

so that Eqs. (3)—(5) yield

$$\text{pH}_{\text{pzc}} = 0.5 \lg \frac{K_a^0}{K_b^0} \, , \tag{6}$$

and

$$-RT \ln K^0 = \Delta_r G^0 = \Delta_r H^0 - T \Delta_r S^0 \, . \tag{7}$$

The temperature dependence of the point of zero charge is given by

$$\text{pH}_{\text{pzc}} = \frac{\Delta_b H^0 - \Delta_a H^0}{2RT \ln 10} - \frac{\Delta_b S^0 - \Delta_a S^0}{2R \ln 10} \, . \tag{8}$$

Thus, from these measurements, one cannot obtain the singular reaction enthalpy but only the difference, which does not include the electrostatic contribution. This "enthalpy difference" may be obtained from the slope in presentation $\text{pH}_{\text{pzc}}$ vs $1/T$.

Calorimetric determination of the enthalpies of deprotonation and protonation, reactions (1) and (2), involves addition of an acid or a base to suspension at certain (initial) pH. In such a case, neutralization also takes place

$$\text{OH}^- + \text{H}^+ \rightarrow \text{H}_2\text{O} \, ; \quad K_n^0, \, \Delta_n H, \, \Delta \xi_n \, , \tag{9}$$

with the equilibrium constant related to the "ionic product" of water

$$K_n^0 = 1/K_w^0 = (a_{\text{H}^+} a_{\text{OH}^-})^{-1} \, . \tag{10}$$

The measured heat $Q$ is a sum of products of extents of reactions (1), (2) and (9) and corresponding reaction enthalpies since the extents of association of counterions are negligible in the vicinity of p.z.c. at low ionic strength:

$$\begin{aligned} Q &= \sum \Delta_r H \Delta \xi_r \\ &= \Delta_a H \Delta \xi_a + \Delta_b H \Delta \xi_b + \Delta_n H \Delta \xi_n \, . \end{aligned} \tag{11}$$

It is possible to obtain the value of the difference $\Delta_b H^0 - \Delta_a H^0$ by performing the calorimetry experiment at low ionic strength in such a manner that the difference between $\text{pH}_{\text{pzc}}$ and initial pH of suspension is equal to the difference between the final pH and $\text{pH}_{\text{pzc}}$. In that case, the p.z.c. value lies in the middle between the initial and final pH. Due to the compensation of the electrostatic effects, this experimental design produces the difference of standard reaction enthalpies, i.e., only chemical difference corresponds to the one obtained from the temperature dependence of p.z.c. (5).

By measuring the initial and final pH values of the suspension ($\text{pH}_1$ and $\text{pH}_2$, respectively), one obtains the changes in the amounts of $\text{H}^+$ and $\text{OH}^-$ ions in the liquid bulk which are related to the amounts of added acid or base ($\Delta n_{\text{H}^+(\text{add})}$, $\Delta n_{\text{OH}^-(\text{add})}$) and also to the extents ($\Delta \xi$) of protonation (a), deprotonation (b), and neutralization ($n$) reactions.

$$\begin{aligned} \Delta n_{\text{H}^+} &= c^0 \left( \frac{V_2 \, 10^{-\text{pH}_2}}{y^2} - \frac{V_1 \, 10^{-\text{pH}_1}}{y_1} \right) \\ &= \Delta n_{\text{H}^+(\text{add})} - \Delta \xi_a + \Delta \xi_b - \Delta \xi_n \end{aligned} \tag{12}$$

$$\begin{aligned} \Delta n_{\text{OH}^-} - K_w^0 c^0 \left( \frac{V_2 \, 10^{\text{pH}_2}}{y^2} - \frac{V_1 \, 10^{\text{pH}_1}}{y_1} \right) \\ = \Delta n_{\text{OH}^-(\text{add})} - \Delta \xi_n \, , \end{aligned} \tag{13}$$

where $V_1$ and $V_2$ denote initial and final volumes of the liquid phase, respectively. The above relationships include activity coefficients ($y$) and the standard value of concentration ($c^0 = 1$ mol dm$^{-3}$) according to

$$a_i = y_i \frac{c_i}{c^0} \tag{14}$$

Equations (3), (4), (12), and (13) yield the ratio of the extents of surface reactions (1) and (2) [5]

$$\frac{\Delta\xi_a}{\Delta\xi_b} = -\frac{K_a^0}{K_b^0} \frac{a_{H+(2)}\, a_{H+(1)}}{e^{F\Psi_2/RT}\, e^{F\Psi_1/RT}} . \tag{15}$$

Surface potential in the vicinity of p.z.c. can be approximated by using the Nernstian approach [2]

$$\Psi_0 = \frac{RT\ln 10}{F}\, (\text{pH}_{pzc} - \text{pH})a , \tag{16}$$

where the coefficient $a$ takes into account the deviation from the ideal slope. From Eqs. (15) and (16), one obtains

$$\frac{\Delta\xi_a}{\Delta\xi_b} = -10^{(1-a)(2\text{pH}_{pzc}-\text{pH}_2-\text{pH}_1)} . \tag{17}$$

According to (17)

$$\Delta\xi_b = -\Delta\xi_a , \tag{18}$$

if the system obeys completely the Nernst equation, e.g., if $a = 1$. However, Eq. (18) still holds if the experiment is designed so that initial and final pH values are related by

$$\text{pH}_1 - \text{pH}_{pzc} = \text{pH}_{pzc} - \text{pH}_2 . \tag{19}$$

In the second case the extents are equal in magnitude but opposite in sign (18), regardless of the possible disobedience of the Nernstian behavior. According to Eq. (13) the extent of neutralization should be calculated from initial and final pH values by

$$\Delta\xi_n = \Delta n_{OH-(add)} - K_w^0 c^0 \left( \frac{V_2\, 10^{\text{pH}_2}}{y_2} - \frac{V_1\, 10^{\text{pH}_1}}{y_1} \right) . \tag{20}$$

Extents of surface reactions, according to (12) and (18), are

$$\Delta\xi_b = -\Delta\xi_a$$
$$= 0.5 \left[ \Delta n_{H+(add)} - \Delta\xi_n - c^0 \right.$$
$$\left. \cdot \left( \frac{V_2\, 10^{-\text{pH}_2}}{y_2} - \frac{V_1\, 10^{-\text{pH}_1}}{y_1} \right) \right] . \tag{21}$$

In the calorimetry experiment requirements of Eq. (19) should be respected so that reaction extents $\Delta\xi_n$, $\Delta\xi_a$, and $\Delta\xi_b$ may be calculated by Eqs. (20) and (21), respectively. The next step is to evaluate the difference in reaction enthalpies by the relationship

$$\Delta_b H^0 - \Delta_a H^0 = \frac{Q - \Delta_n H\, \Delta\xi_n}{\Delta\xi_b} , \tag{22}$$

which comes from Eq. (11).

As stated in (22), the experiment in which initial and final pH values are related to p.z.c. by (19) yields the difference in standard enthalpies describing "chemical" interactions only and does not incorporate the electrostatic effects, just as in the case of the value obtained from p.z.c. dependence on temperature. The compensation of the electrostatic effect is based on the assumption of "symmetrical" surface potential with respect to p.z.c. and does not necessarily assume Nernstian behavior. It is sufficient that surface equilibria results, e.g., half pH unit below p.z.c., positive surface potential equal in magnitude to that one half pH unit above p.z.c. If such a behavior is found within a certain temperature range, then the derivative of potential with respect to temperature is again the same in magnitude but opposite in sign. Since the electrostatic contribution to the total enthalpy exhibits the same behavior as a derivative of the potential, the compensation of electrostatic contribution to the enthalpies takes place.

## Experimental

### Chemicals

The chemicals used in this study were of the analytical purity grade. Hematite powder was a product of Alfa Division, Danvers, Mass. U.S.A. Standard NBS buffers [6] were: i) aqueous solution of $KH_2PO_4$ ($m = 0.05$ mol kg$^{-1}$) and $Na_2HPO_4$ ($m = 0.025$ mol kg$^{-1}$), and ii) the solution prepared from potassium hydrogen phtalate ($m = 0.025$ mol kg$^{-1}$). The pH values of these buffers are tabulated for the wide temperature range.

### Point of zero charge

The effect of temperature on the p.z.c. was determined by measuring pH of a concentrated suspension as described earlier [7]. The ionic strength was low ($10^{-4}$ mol dm$^{-3}$ NaNO$_3$). This method was found to be suitable due to its accuracy and simplicity. The accuracy of p.z.c. is directly related to the accuracy of the pH measurement, while the simplicity in this case is due to the fact that one just measures the electromotive force between glass and reference electrode for two buffers and the suspension. In the measurements, the Jenway 3030 pH-meter was used together with Jenway glass electrode and the Iskra (Kranj, Slovenia) calomel electrode with a salt bridge ($10^{-4}$ mol dm$^{-3}$ NaNO$_3$).

### Calorimetry

Calorimetric measurements were performed by isoperibol reaction calorimeter constructed by Simeon et al.

[8]. The suspension was prepared by mixing (under ultrasound) 16 grams of hematite in 80 cm$^3$ of aqueous solutionof HNO$_3$. After equilibration the pH was measured as 4.92. The suspension (70 $^3$) was transferred into the calorimetric vessel and thermostated at 25 °C. In the course of the experiment 0.41 cm$^3$ of NaOH solution ($c$ = 0.5 mol dm$^{-3}$) was added. By means of separate potentiometric titration, the amount of base was chosen so that final pH was 7.28. Since the p.z.c. at this temperature was found to be at pH = 6.10, the condition of Eq. (19) was satisfied.

## Results and discussion

### Calorimetry

Figure 1 displays the dependence of electromotive force on temperature for the hematite suspension as well as for two buffers having pH approximately 4 and 7. The pH values of the suspension were calculated by using the tabulated values for two standard buffers [4]. These pH values were taken as points of zero charge [7]. The temperature dependence of p.z.c. of hematite is presented in Fig. 2. Linear regression according to Eq. (8) resulted in

$$\Delta_b H^0 - \Delta_a H^0 = 33.2 \text{ kJ mol}^{-1}$$

and

$$\Delta_b S^0 - \Delta_a S^0 = 122.5 \text{ kJ K}^{-1} \text{ mol}^{-1} .$$

Point of zero charge of $a$-Fe$_2$O$_3$ suspension at 25 °C was found to be at pH = 6.1.

In the calorimetry experiment, the measured heat was —8.02 J. The extent of neutralization, calculated by Eq. (20), was found to be $\Delta\xi_n = 2.044 \cdot 10^{-4}$ mol. Extents of surface deprotonation and protonation reactions were

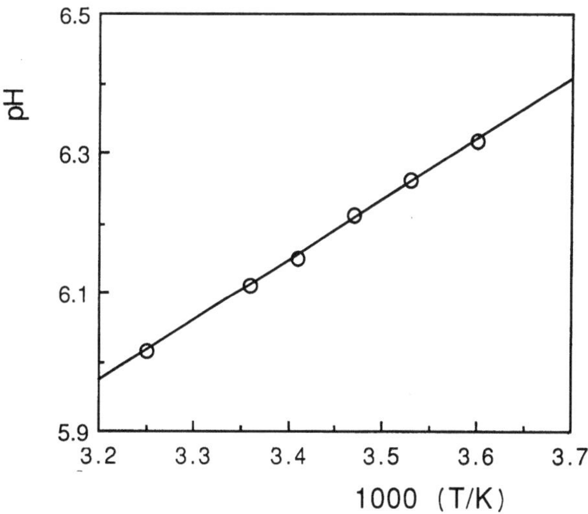



**Fig. 2** The dependence of pH$_{pzc}$ of α-Fe$_2$O$_3$ suspension on temperature

$\Delta\xi_b = -\Delta\xi_a = 1.018 \cdot 10^{-4}$ mol from Eq. (21). The difference in standard reaction enthalpies ($\Delta_b H^0 - \Delta_a H^0$) was calculated by Eq. (22), taking the literature value [9] for $\Delta_n H^0 = -55.84$ kJ mol$^{-1}$, as

$$\Delta_b H^0 - \Delta_a H^0 = 33.3 \text{ kJ mol}^{-1} .$$

The proposed approach to the design of the calorimetry experiments, concerning charging of metal oxide surfaces, resulted in the difference of standard enthalpies of deprotonation and protonation reactions (2) and (1). As already demonstrated for anatase [5], the calorimetry result for hematite agrees with the corresponding measurements of the p.z.c. dependence on temperature. The obtained value also agrees with other published data [10, 11].

The method described here enables the proper assignment of the enthalpy to a certain surface reaction. It also makes possible to avoid electrostatic contribution to the enthalpy. The next step should deal with different initial and final conditions to enable the determination of singular values of $\Delta_a H^0$ and $\Delta_b H^0$ and also to examine the electrostatic effect on the enthalpy.

### Adsorption

Adsorption isotherms of dodecylsulfate (Ds$^-$) and salicylate (Sal$^-$) ions, are presented in Fig. 3. They were fitted with Langmuir isotherm yielding:

$$\Gamma_{max} = 4.3 \times 10^{-6} \text{ mol m}^{-2}, \quad a = 0.4 \text{ nm}^2, \quad \lg K = 2.8$$

for Ds$^-$

$$\Gamma_{max} = 1.9 \times 10^{-6} \text{ mol m}^{-2}, \quad a = 0.9 \text{ nm}^2, \quad \lg K = 2.8$$

for Sal$^-$

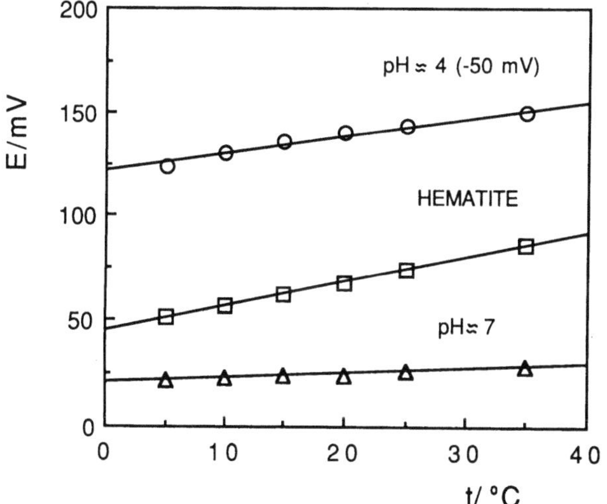



**Fig. 1** The dependence of electromotive force on temperature for α-Fe$_2$O$_3$ suspension and two buffers of approximnate pH values 4 and 7, respectively. Note that 50 mV were subtracted for the buffer of pH 4, in order to enable presentation of data by using one scale

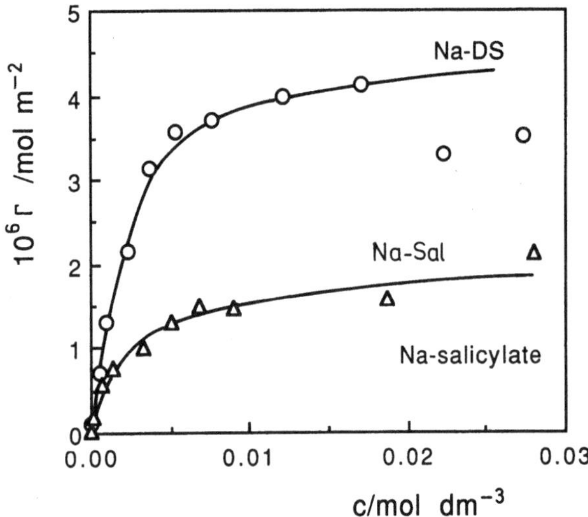


**Fig. 3** Adsorption isotherms of organic ions on hematite. The dependence of surface concentration ($\Gamma$) on the equilibrium concentration of dodecylsulfate ions at pH = 4.6 (o), and of salicylate ions at pH = 4.2 ($\triangle$)

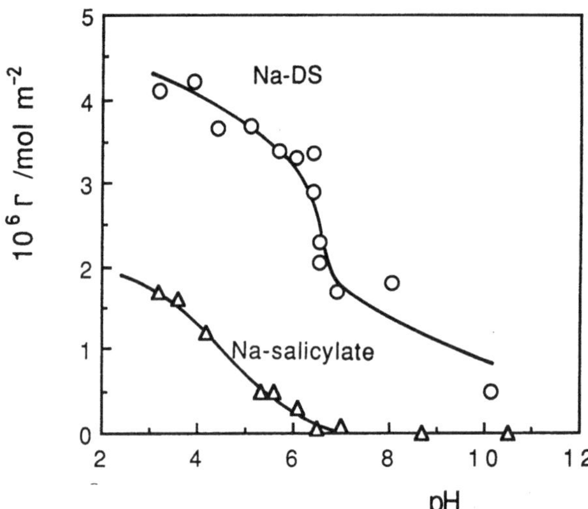


**Fig. 4** Surface concentration of dodecylsulfate (o) and salicylate ions ($\triangle$) as a function of pH at total concentrations of 0.01 mol dm$^{-3}$

($\Gamma_{max}$ is the maximum surface concentration, $a$ is area occupied by one molecule, and $K$ is the adsorption equilibrium constant as obtained on the basis of the Langmuir isotherm.)

The experiments were performed in the acidic region; pH = 4.6 for Ds$^-$ adsorption, and at pH = 4.2 for adsorption of salicylate ions. The interpretation of these data by the Langmuir isotherm is an approximation which should be further examined. At first, one may notice that the shape of the adsorption isotherm is of the Langmuir type. Secondly, the obtained values for occupied areas are reasonable. They correspond to the "flat" orientation of salicylate and to the "upright" orientation of dodecylsulfate adsorbed species. The more detailed analysis should take into account the effect of pH on the surface charge. At low pH, below the isoelectric point, the hematite surface is originally positive and it attracts negative organic ions promoting their adsorption. Therefore, at low pH and at high concentration of an organic substance, one may expect the surface to be completely covered so that maximum surface concentration

corresponds to the complete surface coverage and to the surface occupied by an adsorbed molecule. The effect of pH on adsorption equilibria of both anions is presented in Fig. 4. Below pH = 6 the surface concentration of Ds$^-$ is high, approaching the constant value; it agrees with the value of $\Gamma_{max}$ obtained from interpretation based on the Langmuir isotherm. Thus, it corresponds to the complete surface coverage, which is understandable because the initial NaDs concentration was sufficiently high. By increasing pH, the original hematite surface becomes negative, it repels negative Ds$^-$ ions and the adsorbed amount is markedly reduced. The same pattern of adsorption isotherm was obtained with salycilate ions. This finding agrees well with other observations on the adsorption equilibrium of organic ions on metal oxides [3, 4]. The effect of pH in these systems is twofold; at low pH the oxides are more positive which promotes adsorption of negative organic ions, but at the same time the weak acid is less dissociated so that the concentration of adsorbable anions is lowered which reduces the extent of adsorption. The latter isnot valued for relatively strong dodecylsulfonic acid.

# References

1. Dzombak DA, Morel FMM (1990) Surface Complexation Modeling, A Wiley-Interscience Publication, John Wiley and Sons, New York
2. Blesa MA, Kallay N (1988) Adv Colloid Interface Sci 28:111
3. Zhang Y, Kallay N, Matijević E (1985) Langmuir 1:201
4. Torres R, Kallay N, Matijević E (1988) Langmuir 4:706
5. Kallay N, Žalac S, Štefanić G (1993) Langmuir 9:3457
6. Definition of pH Scales, Standard Reference Values, Measurement of pH and Related Terminology (1985) Pure Appl Chem 57:531
7. Žalac S, Kallay N (1992) J Colloid Interface Sci 149:233
8. Simeon V, Ivičić N, Tkalčec M (1972) Z Phys Chem 78:1
9. Wagman DD, Evans WH, Parker VB, Halow I, Bailey SM, Schumm RH (1968) Selected Values of Chemical Thermodynamic Properties, NBS Technical Note, Institute for Basic Standards, National Bureau of Standards, Washington
10. De Keizer A, Fokkink LGJ, Lyklema J (1990) Colloid Surf 49:149
11. Fokkink LGJ, de Keizer A, Lyklema J (1989) J Colloid Interface Sci 127:116

Progr Colloid & Polym Sci (1994) 95:113—118
© Steinkopff Verlag 1994

A. Pohlmeier

# The kinetic spectrum method applied to the ion exchange of Cd(II) at Mg(II)-montmorillonite

A. Pohlmeier
Institute of Applied Physical Chemistry
(IPC),
Research Center Jülich (KFA),
52425 Jülich, FRG

**Abstract** In this work the ion exchange kinetics of $Cd^{2+}$ at $Mg^{2+}$-montmorillonite have been investigated. These kinetics are fast, i.e., the exchange in diluted suspension is complete after 30 s, therefore, the stopped flow method is applied. The data were analyzed by the kinetic spectrum method, because the reaction cannot be described by classical kinetic models. Two processes were observed: i) A fast one with a mean pseudo-first-order rate constant of 20 $s^{-1}$. This process is assigned to the exchange at easily accessible sites at the outer surface. ii) A group of slow processes with a distribution of pseudo first order rate constants between 0.1 $s^{-1}$ and 3 $s^{-1}$. These processes are assigned to the exchange in the interlayer space and have a mean energy and entropy of activation of 20 kJ $mol^{-1}$ and $-180$ J $mol^{-1}$ $K^{-1}$, respectively. The rate controlling step is the hindered diffusion at the internal surfaces of the montmorillonite.

**Key words** kinetic spectrum — cadmium(II) — montmorillonite — kinetics — stopped flow

## Introduction

Ion exchange reactions at mineral electrolyte interfaces belong to the most important processes that control the distribution of toxic metal ions between liquid and solid phases in the environment. While the equilibrium properties (adsorption isotherms, adsorption edges, equilibrium constants, etc.) have been widely investigated, less information is available on the kinetics of such processes. Most work in the latter field has been focused on the kinetics in the slow time domain (minutes to days). In general, the processes observed here are controlled by diffusion in the macropores of highly aggregated or compacted colloid particles or in concentrated suspensions [1—4]. The reactions observed are slow because of long diffusion paths. However, many authors report on fast reactions, whose equilibria are established instantaneous-ly compared to their accessible time domain. These fast reactions at the solid — electrolyte interface with half lives of less than 1 min have hardly been examined. By use of the pressure jump [5] and the stopped flow techniques [6] it was possible to investigate fast ion exchange processes at clay minerals. However, only few investigations have been performed on the fast exchange kinetics of heavy metals with layer-silicates in suspension.

Compared to reactions in solution, the interpretation of kinetic data at solid-electrolyte interfaces is often difficult because of the heterogenity of suspended particles. In general, the kinetic data recorded (concentration as a function of time) are interpreted by use of a special model like first- and second order kinetics, Elovich's equation or a $\sqrt{t}$ law [1]. These models imply some special experimental constraints, e.g., a limited concentration and time range. In Elovich's equation an exponentially de-

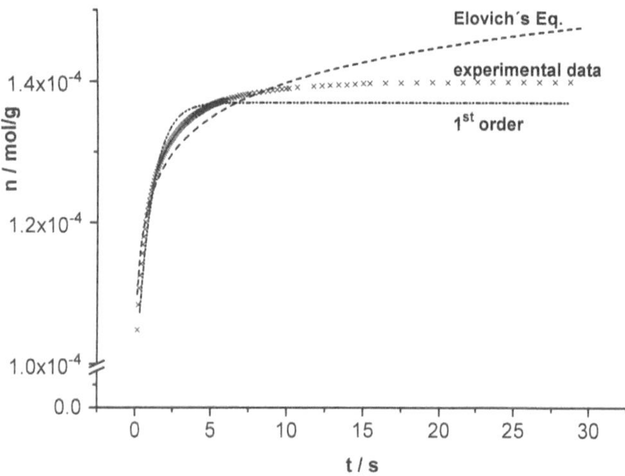

Fig. 1 Amount exchanged of $Cd^{2+}$, $n$, ($[Cd^{2+}]_{total}$: $1.5 \times 10^{-5}$ mol $dm^{-3}$) at $Mg^{2+}$-montmorillonite (0.05 g $dm^{-3}$). Crosses are experimental data, the lines are least-square-fits with Elovich's equation and the pseudo-first-order rate law, respectively

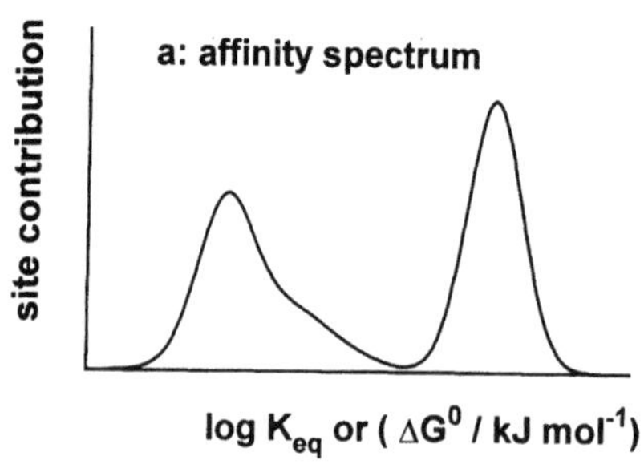


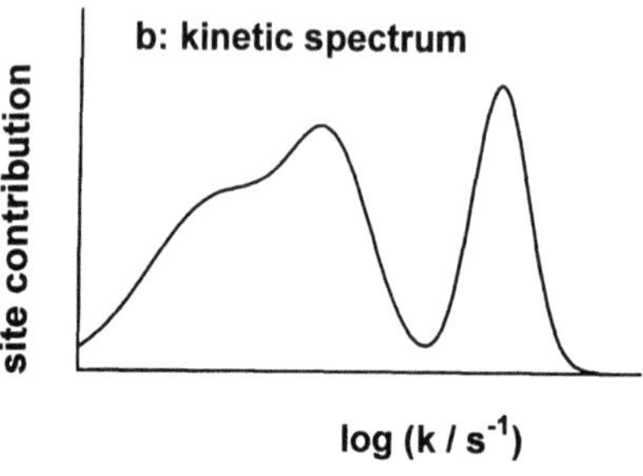


Fig. 2 Schematic drawing of the affinity spectrum (a) and the corresponding kinetic spectrum (b) for an arbitrary system

creasing affinity is assumed that depends on the degree of occupancy of surface binding sites, and the $\sqrt{t}$ law requires a uniform space geometry, i.e., pores of equal shape and size. In reality, those conditions are rarely met. As an example, Fig. 1 shows the reaction of $1.5 \times 10^{-5}$ mol $dm^{-3}$ $Cd^{2+}$ with 50 mg $dm^{-3}$ $Mg^{2+}$-montmorillonite ($2.4 \times 10^{-5}$ mol exchange sites per $dm^3$) at pH = 6.7 and $T = 25\,°C$. Crosses are experimental data (see below) and the lines show the best fits of Elovich's equation and a pseudo-first-order reaction, respectively. It can be clearly seen that in both cases the experimental data cannot be described correctly. The interpretation with Elovich's equation shows greater deviations at longer times, whereas the first-order law shows systematic deviations over the whole time range. Other fitting attempts like a second-order reaction and the $\sqrt{t}$-law also fail, but are not shown here.

In this work a method of data evaluation is applied that does not require an *a priori* choice of a definite model. In 1983, Shuman and Olsson proposed to apply the so-called kinetic spectrum method on the reactions of metals with dissolved humic material [7]. The basic idea is that binding sites on a homologous complexant [8] (like humic acid or suspended colloid particles) are heterogeneous. The system should not be described with a definite, discrete affinity (expressed by an equilibrium constant), but rather with a spectrum of affinities. That means at the surface of the homologous complexant there exists as very large number of exchange sites and each of them has its own affinity to a certain metal ion. The contribution of each site to the total occupancy can be plotted versus its affinity, expressed as the logarithm of its specific equilibrium constant or Gibb's function. This plot is call-

ed the affinity spectrum. As an example, Fig. 2a shows such an affinity spectrum for an arbitrary system[8].

For the kinetics in such a system this means each site should be described by its own rate constant, $k_i$. This rate constant implies, for example, its special energy of activation or its accessibility in porous media. Figure 2b shows the kinetic spectrum of an arbitrary system. The changes of this spectrum can be investigated as a function of temperature, pH or concentrations to give further information and knowledge about a certain system. As an example, in this work the ion exchange reaction of $Cd^{2+}$ with $Mg^{2+}$-montmorillonite is investigated and interpreted for the first time by the kinetic spectrum method.

## Theory

The surface of the montmorillonite particles is thought to consist of many different ion exchange sites. At each site "$L_i$" a bimolecular exchange reaction takes place:

$$Cd^{2+} + Mg^{2+}L_i \leftrightarrows Cd^{2+}L_i + Mg^{2+} . \qquad (1)$$

Under pseudo-first-order conditions the kinetics at each site are described by:

$$n_i(t) = n_i^0 (1 - \exp(-k_i t)) , \qquad (2)$$

where $n_i(t)$ is the concentration of site $L_i$ occupied with $Cd^{2+}$ at time $t$, $n_i^0$ is the total concentration of site $L_i$ —

$Cd^{2+}$ at equilibrium, and $k_i$ is the apparent rate constant at this site. It depends on concentrations, temperature, etc. For reasons of simplicity, Eq. (2)

$$\Delta n_i(t) = n_i^0 \exp(-k_i t) \tag{3}$$

is rearranged to Eq. (3) with $\Delta n_i(t) = n_i^0 - n_i(t)$. Of course, the exchange reaction $Cd^{2+}$-$Mg^{2+}$ at a single site $L_i$ is not observable, but rather only the total change at all existing sites.

Therefore, Eq. (3) must be converted to Eq. (4), which describes the total time-dependent change of concentration of $Cd^{2+}$ bound to the montmorillonite. The sum runs over all

$$\Delta n(t) = \sum_i^n \Delta n_i(t) = \sum_i^n n_i^0 \exp(-k_i t) \tag{4}$$

accesible sites $L_i$. If this number is high enough, the sum may be replaced by an integral, Eq. (5). Here, $\Delta n(t)$ is the experimentally observable quantity, i.e., the

$$\Delta n(t) = F \int_0^\infty A(k) \exp(-k_i t)\, dk \tag{5}$$

concentration of $Cd^{2+}$ bound to the montmorillonite. $F$ is a proportionality factor and $A(k)$ represents the distribution of sites, occupied by $Cd^{2+}$, as a function of the rate constant $k$, specific for each single site. If $A(k)$ is plotted versus $k$ or $\log k$, this plot is termed the kinetic spectrum of the observed reaction. In view of mathematics, $\Delta n(t)$ is the Laplace transform of the distribution function $A(k)$, which is unknown up to now. $A(k)$ can be principally obtained by performing an inverse Laplace transform on $n(t)$. Unfortunately, this is not trivial. Because each set of experimental data contains unknown noise components ($\varphi$), Eq. (5) has to be written in terms of:

$$\Delta n(t) = F \int_0^\infty A(k) \exp(-k_i t)\, dk + \varphi. \tag{6}$$

This problem is ill posed in contrast to Eq. (5): an infinite number of solutions exists, all satisfying Eq. (6). This number has to be reduced by introducing certain logical and statistical constraints. The CONTIN software [9—11] implies such conditions like non-negative values for $A(k)$ and uses the regularization method to find the simplest solution for Eq. (6). It requires a very accurate setof $\Delta n(t)$-data with a minimum of noise. The stopped flow method used in this study (see below) easily allows to accumulate many single runs and thereby improves the signal-to-noise ratio. Thus, it is possible to employ the CONTIN software to obtain the kinetic spectrum.

## Materials

$Mg^{2+}$-montmorillonite was prepared by fractionation (sedimentation), oxidation of organic material, ion ex-

change and washing from Ca-bentonite (Südchemie AG, Germany). Its $Mg^{2+}$-content is 99%, the BET-surface is 70 $m^2\,g^{-1}$ and the cation exchange capacity is $9 \times 10^{-4}$ $mol\,g^{-1}$. A stock suspension of 2 $g\,dm^{-3}$ was prepared. $Mg(NO_3)_2$, $Cd(NO_3)_2$ and the transition metal indicator pyridyl-2-azo-resorcinol (PAR) was purchased from Fluka Chemie AG, Switzerland. All solutions and suspensions were slightly buffered with $5 \times 10^{-4}$ mol $dm^{-3}$ of the non-complexing buffer 4-(2-hydroxyethyl)-piperazine-1-ethane-sulfonate (HEPES) at pH = 6.7. The stopped flow apparatus is a conventional one with a minimum mixing time of 10 ms. the photometric detection system consists of a VIS-lamp, a monochromator, photodiodes, a differential amplifier and a personal computer. Suspensions of 0.05 $g\,dm^{-3}$ montmorillonite were mixed with solutions containing $Cd(NO_3)_2$ and the indicator PAR, and the optical density at a certain wavelength is recorded in the flow-through mixing cell and a reference cell containing the equilibrated solution.

## Method

In the experiments presented here, the stopped flow method is used for measuring the kinetics of ion exchange of $Cd^{2+}$ at $Mg^{2+}$-montmorillonite. As $Cd^{2+}$ and $Mg^{2+}$ ions have nearly the same mobility in aqueous solution, the reaction cannot be detected by electrical conductivity. Therefore, photometric detection was chosen for this reaction. For this purpose a $Cd^{2+}$-sensitive indicator (PAR) reaction is coupled to the exchange reaction observed (Eq. (1)). The indicator itself does not react with $Mg^{2+}$ or the montmorillonite, as tested in separate experiments. Equations (1) and (7) describe the reaction observed, when a solution containing $Cd^{2+}$ plus indicator is mixed with a suspension of $Mg^{2+}$-montmorillonite.

$$Cd^{2+} + Mg^{2+}L_i \leftrightharpoons Cd^{2+}L_i + Mg^{2+} \tag{1}$$

$$Cd^{2+} + 2\,PARH^- \leftrightharpoons Cd\text{-}PAR + PARH^- + H^+$$

$$\leftrightharpoons Cd(PAR)_2^{2-} + 2\,H^+ \tag{7}$$

The absorption maximum of the Cd-PAR complexes is at 496 nm, that of the free indicator $PARH^-$ is at 412 nm. Under the conditions employed here, the equilibria of the reactions in Eq. (7) lie to about 90% on the left side and the rate is fast with a pseudo-first-order rate constant of about 35 $s^{-1}$, i.e., all reactions in the time range greater than 0.03 s can be monitored photometrically at 496 or 412 nm. As $Cd^{2+}$ vanishes from the electrolyte phase due to the reaction in Eq. (1), the equilibrium of the reactions in Eq. (6) is shifted to the left side, and the optical density at 496 nm decreases.

As an example, Fig. 3 shows the UV/VIS spectra of a fresh solution ($1.5 \times 10^{-5}$ mol $dm^{-3}$ $Cd^{2+}$ plus

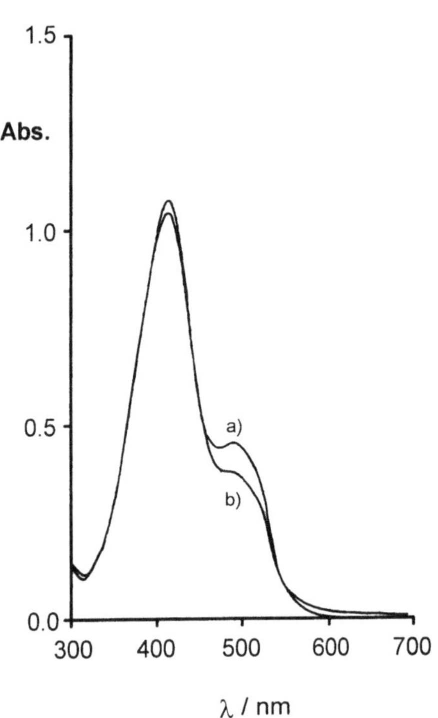

Fig. 3 UV/VIS spectra of $1.5 \times 10^{-5}$ mol dm$^{-3}$ Cd$^{2+}$ with 0.05 g dm$^{-3}$ Mg$^{2+}$ montmorillonite and a indicator (PAR) concentration of $2.5 \times 10^{-5}$ mol dm$^{-3}$. a) at the start of the ion exchange reaction (recorded without montmorillonite); b) at equilibrium

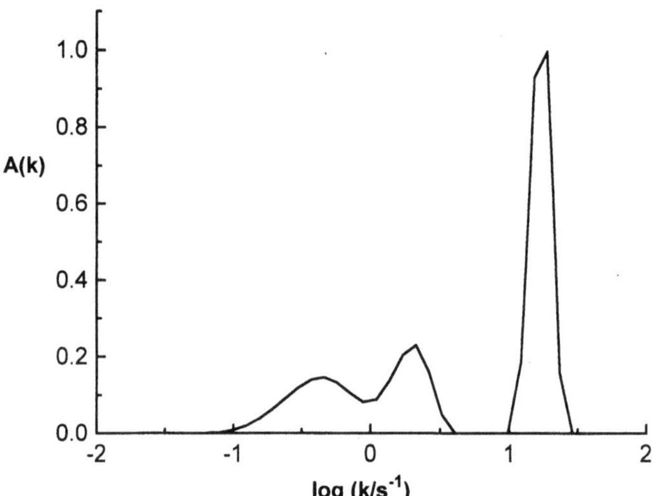

Fig. 4 Kinetic spectrum, obtained by inverse Laplace transform of the experimental data in Fig. 1, by use of the CONTIN software. $T = 25\,°C$

$2.5 \times 10^{-5}$ mol dm$^{-3}$ PAR at pH $= 6.7$ without montmorillonite) and the equilibrated mixture ($1.5 \times 10^{-5}$ mol dm$^{-3}$ Cd$^{2+}$ plus $2.5 \times 10^{-5}$ mol dm$^{-3}$ PAR at pH $= 6.7$ with 50 mg dm$^{-3}$ montmorillonite), recorded with a montmorillonite suspension of 50 mg dm$^{-3}$ as reference. These spectra correspond to the start and to the end of a stopped flow experiment. So, if the absorbance at 496 nm is recorded in the experiment, the amount Cd$^{2+}$ exchanged at the montmorillonite "$\Delta n$" can be calculated as a function of time. (See the example in Fig. 1). The obtained functions are now transformed according to Eq. (6) by the CONTIN software. It can be proven that the time functions have to be faded out nearly completely, otherwise no correct transformation could be performed.

## Results and discussion

Figure 4 shows the kinetic spectrum obtained for the reaction in Fig. 1. The distribution function $A(k)$ is plotted versus the logarithm of the pseudo-first-order reaction rate constant ($\log k$) that is characteristic for the exchange kinetics at each site. According to Sparks [1], exchange reactions at suspended particles with an inner surface or a porous structure could be controlled by three types of processes: firstly, the diffusion through the solution to the outer surface; secondly, the diffusion through the pores or along the inner surface, and thirdly, the exchange reaction itself. The last process is too fast to be rate determining, because ligand exchange reactions at the Cd$^{2+}$-ion are controlled by the water exchange in the first hydration shell, a process in the microsecond range. In our case the Mg$^{2+}$-montmorillonite is not completely desaggregated even in dilute suspension, i.e., no single platelets exist, but rather packages of several single sheets; they have an internal surface and a very narrow interlayer space. Many exchange sites exist in this interlayer space that are not easily accessible, so the second process could be rate determining, too, if the exchange reaction itself is too fast.

In Fig. 4 two groups of processes can be distinguished clearly: a fast one with a mean rate constant of 20 s$^{-1}$ and a more complex group with rate constants between 0.1 s$^{-1}$ and 3 s$^{-1}$. The fast process is nearly in the same order of magnitude as the decomplexation reaction of the Cd$^{2+}$ indicator complex, Eq. (6). It is attributed to the exchange of Cd$^{2+}$ at easily accessible outer surface binding sites of the suspended particles. This process is controlled by diffusion through the bulk solution surrounding the suspended particles, and it is therefore very fast. It could not be resolved further due to the lower time limit caused by the Cd$^{2+}$-indicator reaction. The slower processes are assigned to the binding at sites in the interlayer space, i.e., the second type of process as quoted above. Exchange reactions at these sites are much slower than at sites at the outer surface, because diffusion paths are long and the mobility in the interlayer space is hindered. The last argument implies an activated diffusion.

Figure 5 shows kinetic spectra of only the slower processes at different temperatures; the peaks of the fast process do not change with temperature and are clipped for

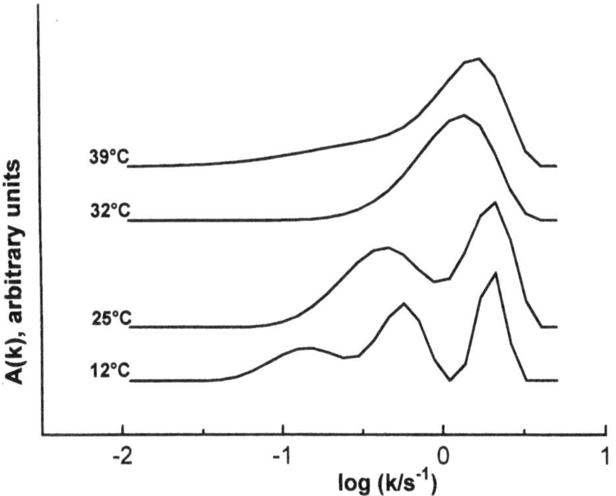

**Fig. 5** Kinetic spectra for the slow processes at the reaction of 1.5 × 10⁻⁵ mol dm⁻³ $Cd^{2+}$ with 0.05 g dm⁻³ $Mg^{2+}$ montmorillonite at different temperatures

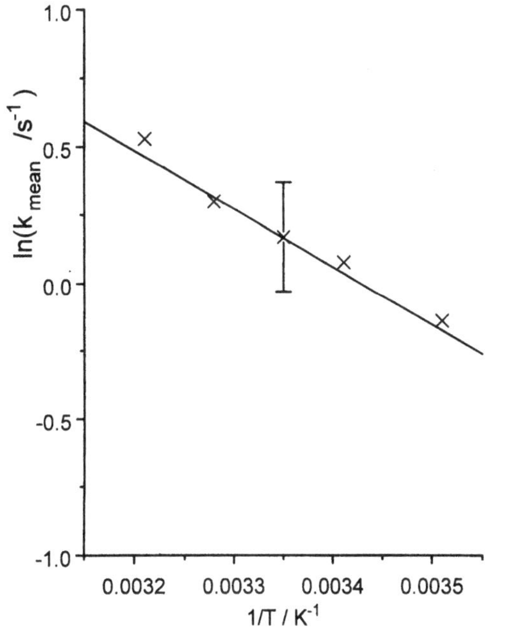

**Fig. 6** Arrhenius plot of the weighted mean rate constants of the slow processes. For $T = 25\,°C$ the error bar indicates the confidence range of the mean rate constant

reasons of lucidity. A significant temperature dependence is recognizeable, though the different processes could not be separated clearly. Therefore, a weighted mean rate constant $k_{mean}$ was calculated that describes all slow processes. In Fig. 6, ln $k_{mean}$ is plotted versus $T^{-1}$, yielding a mean energy of activation of 20 kJ mol⁻¹ and mean entropy of activation of −180 J K⁻¹ mol⁻¹. The first value is greater than expected for free diffusion which is about 10 kJ mole⁻¹, but much smaller than for ligand exchange reactions at transition metal ions in aqueous solution (70 kJ mol⁻¹). This proves that the slow processes really describe the exchange at the sites in the interior of the montmorillonite particles. The rate-controlling step is a hindered diffusion through the interlayer space. The large negative value of the activation entropy supports this view of a hindered diffusion through the interlayer space.

Our results are consistent with those of J. Crooks et al. [6], who found for the exchange of alkali ions at montmorillonite similar values for the activation parameters. But these authors analyzed the kinetics by fitting a single exponential function to the data obtained in a limited time range, and therefore they could not differentiate between various processes. Higher relaxation rates in the millisecond range, obtained by the pressure-jump technique, were also reported by Tang and Sparks [5] for the exchange of alkali and earth alkali ions at montmorillonite at higher concentrations of solid and cations. They evaluated the kinetic data with a single exponential function, too, and did not investigate the temperature dependence.

Summarizing our results, it can be said that:

i) the exchange reaction of $Cd^{2+}$ at montmorillonite is fast (complete in less than 30 s) and can be investigated in dilute suspensions by the stopped flow method with photometric detection.
ii) the kinetic runs are complex, i.e., they cannot be described with classical kinetic models like a $\sqrt{t}$ law, Elovich's equation or pseudo-first order kinetics.
iii) the kinetics can be evaluated by the kinetic spectrum method using the inverse Laplace transform of the time dependent functions.
iv) the exchange reaction of $Cd^{2+}$ at $Mg^{2+}$-montmorillonite consists of two processes: the fast exchange at outer surface sites is controlled by free diffusion in the bulk phase, and the slow exchange at sites in the interlayer space which is controlled by hindered diffusion at the internal surfaces. The mean energy and entropy of activation have been determined for this slow process as 20 kJ mol⁻¹ and −180 J mol⁻¹ K⁻¹, respectively, thus supporting the view of hindered diffusion.

## References

1. Sparks DL, Suarez DL (eds) Rates of Soil Chemical Processes, Soil Science Society of America Special Publication Nr 27, Madison
2. Gutzman DW, Langford CH (1993) Environm Sci Technol 27:1388—1393
3. Bruemmer GW, Gerth J, Tiller KG (1988) J Soil Sci 39:37—52
4. Barrow NJ, Gerth J, Bruemmer GW (1989) J Soil Sci 40:437—450
5. Tang L, Sparks DL (1993) Soil Sci Soc Am J 57:42—46
6. Crooks JE, El-Daly H, El-Sheikh MY, Habib AFM, Zaki AB (1993) Int J Chem Kinet 25:161—168
7. Olson DL, Shuman MS (1983) Anal Chem 55:1103—1107
8. Buffle J (1988) Complexation Reactions in Aquatic Systems: An Analytical Approach, Ellis Horwood Ltd, Chichester
9. Provencher S (1982) Comp Phys Comm 27:213—227
10. Provencher S (1982) Comp Phys Comm 27:229—242
11. Stanley BJ, Bialkowski SE, Marshall DB (1993) Anal Chem 65:259—267

Progr Colloid & Polym Sci (1994) 95:119—124
© Steinkopff Verlag 1994

B. D. Struck
I. Sistemich
R. Pelzer

# An approximative method for the transformation of kinetic sorption data from batch-experiments to transport processes in model soil columns

Dr. B. D. Struck (✉) · I. Sistemich ·
R. Pelzer
Institut für Angewandte Physikalische
Chemie (IPC),
Forschungszentrum Jülich GmbH,
52425 Jülich, FRG

**Abstract** In connection with the mathematical modeling of the transport of pollutants in soil the selection of data from literature plays an important role. In the case of sorption of pollutants mainly partition coefficients and rate constants from batch-experiments are available. For the transformation of these sorption parameters to the transport of pollutants in soil, however, the dependence of these parameters from the concentration of the adsorbent in the batch-experiment and in soil must be considered.

In the present paper, for the relatively frequent case of a pollutant sorbing only partially reversibly, the transformation of these parameters is described by a model regarding the numerical effect of this transformation on the transport of pollutants in precharged and non-precharged soil. Model assumptions were made in order to extend the equilibrium case to the non-equilibrium case. As an example, sorption of 3-chlorophenol in a model soil column from Na bentonite was chosen. Results show that a considerable influence of adsorbent concentration on the proportion of reversibly and irreversibly adsorbed species and, therefore, also via this effect on the transport velocity in the model soil column is to be expected.

**Key words** Batch — transport — sorption — 3-chlorophenol — bentonite — model

## Introduction

Batch-experiments are well suited for the measurement of sorption kinetics on defined soil components such as oxides or clay minerals. For the transformation of kinetic data from these batch-experiments to transport processes in model soil columns attention has to be paid to the dependence of the kinetic sorption parameters on the concentration of the suspended particles.

In this context, Di Toro et al. [1] made batch-experiments on the equilibrium of sorption of hexachloro-biphenyl on montmorillonite and constructed a sorption model. They additionally considered the partial reversibility of this sorption process. A. M. Rodrigo and P. C. Chan [2] found partial reversibility also for sorption of 3-chlorophenol on Na bentonite.

In the studies presented the model developed for sorption equilibrium [1] was extended also to sorption kinetics. Furthermore, an approximative procedure was outlined for the transformation of equilibrium and kinetic sorption data from batch-experiments to transport processes in model soil columns.

## Nomenclature

$a$: suffix "adsorption"
$A$: constant
$b$: suffix "batch"

$B$:  constant
$c$:  adsorbate concentration
d:  suffix "desorption"
$D$:  dispersion coefficient
irr:  suffix "irreversible"
$Jv$:  percolation rate of soil water (by Darcy)
$k$:  rate constant
$K$:  partition coefficient
$m$:  adsorbent mass
max: suffix "maximum value"
$n$:  specific pollutant amount adsorbed
rev:  suffix "reversible"
s:  suffix "model soil column"
$t$:  time
$V$:  volume of the aqueous suspension
$z$:  axial coordinate of the model soil column
$\theta$:  porosity
0:  suffix "initial value"

## Model assumptions for sorption kinetics

In order to transform physicochemical sorption parameters from batch-experiments to transport processes in model soil columns the following model assumptions were made:

i) Sorption is only partially revesible;
ii) The adsorption rate of the adsorbate is reciprocally proportional to adsorbent concentration;
iii) The desorption rate of the adsorbed species is independent of adsorbent concentration;
iv) The partition coefficient of the irreversibly adsorbed species is independent of adsorbent concentration;
v) The ratio of the irreversibly adsorbed species amount and the total adsorbed species amount is constant at constant adsorbent concentration.

## Determination of sorption parameters by batch-experiments

Batch-experiments for sorption equilibrium

*General Considerations* For batch-experiments it is preassumed that the reversible sorption is in equilibrium and that irreversible adsorption corresponds to these equilibrium conditions. Therefore, the corresponding partition coefficients for total, reversible and irreversible sorption are

$$K = n/c \quad K_{\mathrm{rev}} = n_{\mathrm{rev}}/c \quad K_{\mathrm{irr}} = n_{\mathrm{irr}}/c , \tag{1}$$

with

$$n = n_{\mathrm{rev}} + n_{\mathrm{irr}} \tag{2}$$

$$K = K_{\mathrm{rev}} + K_{\mathrm{irr}} . \tag{3}$$

Due to the model assumptions ii) and iii) (viz. also Eqs. (12—15)) the partition coefficient $K_{\mathrm{rev}}$ for reversible sorption is given by

$$K_{\mathrm{rev}} = A/(m/V) . \tag{4}$$

$A$ is a constant independent of adsorbent concentration.

From Eqs. (3) and (4) results in a linear equation for the dependence of the partition coefficient on the concentration $m/V$ of the adsorbent, viz. Fig. 1 in logarithmic scale.

$$K = A/(m/V) + K_{\mathrm{irr}} . \tag{5}$$

$K_{\mathrm{irr}}$ is independent of adsorbent concentration regarding model assumption iv). Equations (4) and (5) were published in [1].

**Fig. 1** The partition coefficients of several chloroorganic compounds decreases with increasing adsorbent concentration. Temperature: 20—25 °C; refs. [1—5]

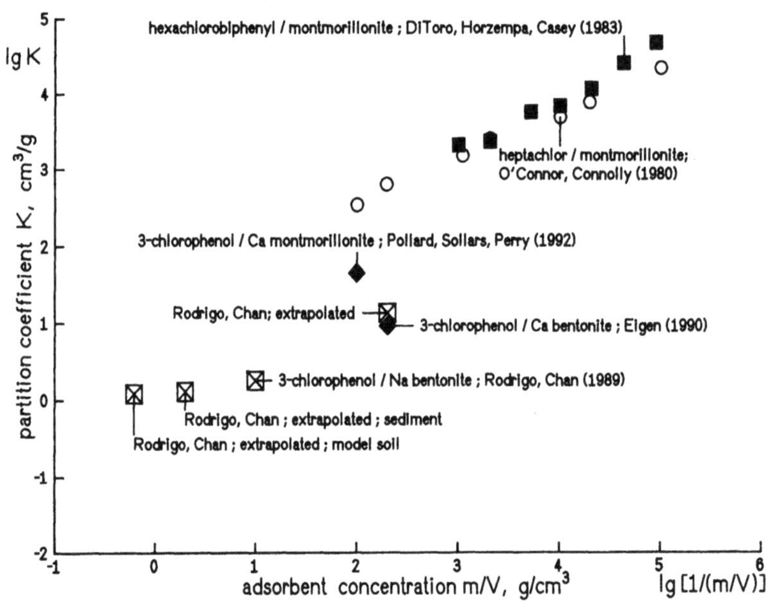

Furthermore, model assumption v), also considering Eq. (1), provides in the case of constant adsorbent concentration

$$n_{irr}/n = K_{irr}/K = B \; . \tag{6}$$

$B$ is a constant independent of adsorbate equilibrium concentration, but it depends on the concentration of the adsorbent. Equation (6) was verified for the sorption equilibrium of 3-chlorophenol and other substances on Na bentonite by [2]. In the following, Eq. (6) is also applied to non-equilibrium conditions (assuming $B$ as independent of adsorbate concentration) as an approximation.

*Experimental determinations* Via batch-adsorption-experiments the respective partition coefficient $K_b$ can be calculated from the linear Henry adsorption isotherm. The irreversibly adsorbed specific amount $n_{irr,b}$, related to the equilibrium concentration $c_b$ in the adsorption experiment, results from subsequent desorption experiments. According to Eqs. (1), (3) and (4), it is

$$K_{irr,b} = n_{irr,b}/c_b \tag{1}$$

$$K_{rev,b} = K_b - K_{irr,b} \tag{7}$$

$$A = K_{rev,b} \cdot m_b/V_b \; . \tag{8}$$

Consideration of the model assumption iv) yields

$$K_{irr} = K_{irr,b} \; . \tag{9}$$

Batch-Experiments for sorption kinetics

*General considerations* The reversible sorption is assumed to be in non-equilibrium; the irreversible adsorption corresponds to this condition.

The equation for the related sorption kinetics is derived as follows. Derivating Eq. (2) yields

$$\partial n/\partial t = \partial n_{rev}/\partial t + \partial n_{irr}/\partial t \; . \tag{10}$$

For environmentally low adsorbate concentrations first order kinetics according to Eq. (11) can be suggested.

$$\partial n_{rev}/\partial t = k_{a,rev} \cdot c - k_{d,rev} \cdot n_{rev} \; . \tag{11}$$

Model assumption ii) states that the adsorption rate is reciprocally proportional to adsorbent concentration. Therefore,

$$k_{a,rev} = A_a(m/V) \; , \tag{12}$$

where $A_a$ is a constant, independent of adsorbent concentration.

The desorption rate $k_{d,rev}$ is independent of adsorbent concentration according to model assumption iii).

In the following, all transformations of sorption parameters will be described in terms of partition coefficient and of desorption rate for reasons of comparison between the equilibrium and the non-equilibrium case.

For the partition coefficient $K_{rev}$ of reversible sorption, Eq. (13) is valid in general:

$$K_{rev} = k_{a,rev}/k_{d,rev} \; , \tag{13}$$

and with Eq. (12) results in

$$K_{rev} = (A_a/k_{d,rev})/(m/V) \; . \tag{14}$$

Abbreviating

$$A = A_a/k_{d,rev} \tag{15}$$

provides Eq. (4) for the partition coefficient $K_{rev}$ of reversible sorption as a function of adsorbent concentration.

Continuing the derivation of the differential equation for sorption kinetics under model conditions, the combination of Eqs. (11), (13), and (2) yields

$$\partial n_{rev}/\partial t = k_{d,rev} \cdot K_{rev} \cdot c - k_{d,rev} \cdot n + k_{d,rev} \cdot n_{irr}, \tag{16}$$

and by insertion of Eq. (16) into Eq. (10),

$$\partial n/\partial t = k_{d,rev} \cdot K_{rev} \cdot c - k_{d,rev} \cdot n$$
$$+ k_{d,rev} \cdot n_{irr} + \partial n_{irr}/\partial t \; . \tag{17}$$

From the kinetic adsorption curve for 3-chlorophenol on Na bentonite (Fig. 2), it is realized that there is neither an instant adsorption step for the irreversibly adsorbing adsorbate indicated by a steep initial rise of the kinetic curve, nor a two-step curve indicating faster irreversible adsorption than reversible adsorption. Therefore, the irreversibly adsorbed amount can be regarded as steadily proportional to the total amount adsorbed all along the kinetic curve. This fact suggests, as an approximation, the application of Eq. (6) also in this kinetic theory.

From Eq. (6), it follows that

$$\partial n_{irr}/\partial t = (K_{irr}/K) \cdot \partial n/\partial t \; . \tag{18}$$

Inserting Eqs. (16) and (18) into Eq. (10), also including Eq. (6), results in

$$\partial n/\partial t = k_{d,rev} \cdot K_{rev} \cdot c - k_{d,rev} \cdot n$$
$$+ k_{d,rev} \cdot (K_{irr}/K) \cdot n + (K_{irr}/K) \cdot \partial n/\partial t. \tag{19}$$

Summing up terms with $\partial n/\partial t$ and $n$ in Eq. (19) provides

$$\partial n/\partial t = k_{d,rev} \cdot [K_{rev}/(1 - (K_{irr}/K))] \cdot c$$
$$- k_{d,rev} \cdot n \; . \tag{20}$$

Combination of Eqs. (20) and (3) leads to

$$\partial n/\partial t = k_{d,rev} \cdot K \cdot c - k_{d,rev} \cdot n \; , \tag{21}$$

for the total sorption in the kinetic adsorption experiment.

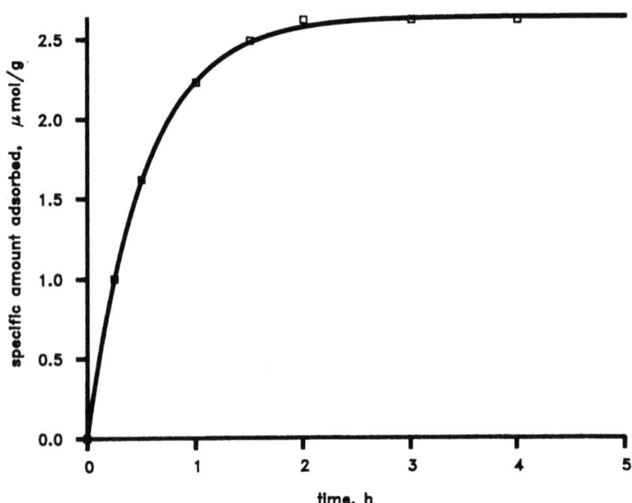

**Fig. 2** Adsorption of 3-chlorophenol on Na bentonite, 25 °C [2]

*Experimental determinations* If the rate constants $k_{a,b}$ and $k_{d,b}$ corresponding to

$$\partial n/\partial t = k_{a,b} \cdot c - k_{d,b} \cdot n \tag{22}$$

are measured in this experiment, then Eqs. (23) and (24) for the calculation of model parameters are obtained by comparison of Eqs. (21) and (22):

$$k_{d,rev} = k_{d,b} \tag{23}$$

$$K_b = k_{a,b}/k_{d,b} \; ; \tag{24}$$

$K_{irr,b}$ is determined by desorption experiments and will be inserted into Eqs. (7—9).

## Model assumptions for the soil column

i) Within the surveyed interval of environmentally relevant low percolation rates of soil water the model soil column is treated with respect to sorption as a suspension with an adsorbent concentration equal to soil density.

ii) The amount of irreversibly adsorbed pollutant is always proportional to the maximum total amount adsorbed at this location.

iii) The constants of the convection-dispersion-equation correspond to those of a natural soil.

## Model equations for pollutant transport in the soil column

Equilibrium conditions for sorption

For the transport in the model soil column it is assumed that reversible sorption is in the equilibrium state and that irreversible adsorption corresponds to this condition.

The differential equation (30) or (31) for sorption at variable equilibrium concentration of the adsorbate in pore water is derived employing the model assumption ii) for the transport in the soil column formulated analogous with Eq. (6), thus,

$$n_{irr} = (K_{irr}/K_s) \cdot n_{max} \; , \tag{25}$$

where $n_{max}$ is maximum specific total amount of pollutant adsorbed at the respective location $z$.

Moreover, for the model soil column, we can write, in accordance with Eq. (1),

$$K_s = n_{max}/c_{max} \; , \tag{26}$$

where $c_{max}$ is maximum adsorbate concentration in pore water at the respective location $z$:

$$n_{irr} = K_{irr} \cdot c_{max} \tag{27}$$

$$n_{rev} = K_{rev,s} \cdot c \; , \tag{28}$$

and in accordance with Eqs. (4) and (8):

$$K_{rev,s} = A/(m_s/V_s) = K_{rev,b} \cdot (m_b/V_b)/(m_s/V_s) \; . \tag{29}$$

Equation (10) and (27—29) yield

$$\partial n/\partial t = K_{rev,b} \cdot (m_b/V_b)/(m_s/V_s) \cdot \partial c/\partial t + K_{irr} \cdot \partial c_{max}/\partial t \; , \tag{30}$$

or by insertion of Eq. (26) into Eq. (30),

$$\partial n/\partial t = K_{rev,b} \cdot (m_b/V_b)/(m_s/V_s) \cdot \partial c/\partial t + (K_{irr}/K_s) \cdot \partial n_{max}/\partial t \; . \tag{31}$$

Non-equilibrium conditions for sorption

Reversible sorption is assumed to be in non-equilibrium state, irreversible adsorption corresponds to this condition. In this case sorption is described by Eq. (32), which is derived by insertion of Eq. (25) into Eq. (17):

$$\partial n/\partial t = k_{d,rev} \cdot K_{rev,s} \cdot c - k_{d,rev} \cdot n$$
$$+ k_{d,rev} \cdot (K_{irr}/K_s) \cdot n_{max}$$
$$+ (K_{irr}/K_s) \cdot \partial n_{max}/\partial t \; . \tag{32}$$

Inserting

$$K_s = K_{rev,s} + K_{irr} \; , \tag{33}$$

and inserting Eq. (29) into Eq. (32) provides

$$\partial n/\partial t = k_{d,rev} \cdot [K_{rev,b} \cdot (m_b/V_b)/(m_s/V_s) \cdot c - n$$
$$+ (K_{irr}/(K_{rev,b} \cdot (m_b/V_b)/(m_s/V_s) + K_{irr})) \cdot n_{max}]$$
$$+ (K_{irr}/(K_{rev,b} \cdot (m_b/V_b)/(m_s/V_s) + K_{irr})) \cdot \partial n_{max}/\partial t \tag{34}$$

for sorption in non-equilibrium state.

## Convection-dispersion-equation

Equation (35) shows the applied type of convection-dispersion-equation about the transport of a pollutant in the model soil column. The term on the lefthand side describes the change of adsorbate concentration with time at a certain location $z$. The first term on the righthand side calculates the dispersion, the second term calculates the convection, and the third term the sorption of the transported pollutant at the respective location.

$$\partial c/\partial t = D \cdot \partial^2 c/\partial z^2 - (Jv/\theta) \cdot \partial c/\partial z$$

$$- [(m_s/V_s)/\theta] \cdot \partial n/\partial t . \tag{35}$$

## Relevant conditions for the model calculations

The kinetic parameters measured by batch-experiments [2] for the sorption of 3-chlorophenol on Na bentonite at 25°C (viz. the subsequent list of data and Fig. 2) are transformed to the transport in a model soil column. The model soil column is assumed to consist of Na bentonite particles, to have the porosity of natural soil, and to be saturated with water. There is a 1-day injection of aqueous 3-chlorophenol into the column. Model calculations are made for a non-precharged column and a column precharged by irreversibly adsorbed 3-chlorophenol. Observation time is 1.5 days. The experimental conditions for the calculations have model character with respect to the transfer of data and do not consider possible experimental difficulties in the soil column, e.g., by swelling of Na bentonite.

Model calculations comprised the solution of the system of the differential Eqs. (31) and (35) for the case of sorption equilibrium as well as (34) and (35) for the kinetic case (non-equilibrium). Calculations were performed by computer programs in FORTRAN 77 with the aid of the subroutine D03PGF from the NAG-library on the IBM-computer system ES/9600-620 of the research center. The subroutine D03PGF employs Gear's method for the numerical solution of differential equations.

The differential quotient $\partial n_{max}/\partial t$ in the Eqs. (31) and (34) was considered approximately by resetting its value after each completed step of the numerical iteration procedure. It was set $\partial n_{max}/\partial t = \partial n/\partial t$ if $n_{max}$ was increased by the foregoing step and

$$\partial n_{max}/\partial t = 0$$

if $n_{max}$ was not increased by the foregoing step.

Calculations were based on the following data.

$K_{rev,b}$: .63 cm$^3$/g [2]

$K_{irr}$: 1.18 cm$^3$/g [2]

$k_{d,rev}$: 1.79 1/h (calculated by data of [2])

$m_b/V_b$: 0.1 g/cm$^3$

$m_s/V_s$: 1.62 g/cm$^3$

$Jv$: 3 cm/d

$D$: 0.5 cm$^2$/d

$\theta$: 0.4 cm$^3$/cm$^3$

$T$: 25°C

$c_0$: 2.2 $\cdot$ 10$^{-4}$ g/cm$^3$

## Results

Results describe the application of the model assumptions on the transport process. Figures 3—6 show the numerical differences of pollutant distribution in the column which arise when the dependence of the partition coefficient on the concentration of the adsorbent (Fig. 1) is considered or not. Results comprise the equilibrium and kinetic case. Application of the mentioned correction for adsorbent concentration increases the amount of irreversibly adsorbed pollutant in the non-precharged column in comparison with neglection of this correction (Figs. 3, 4). The concentration of the adsorbate in pore water becomes correspondingly diminished and the percolated soil depth lengthened (Fig. 5). In the precharged column the percolated soil depth is clearly increased in the case of correction.

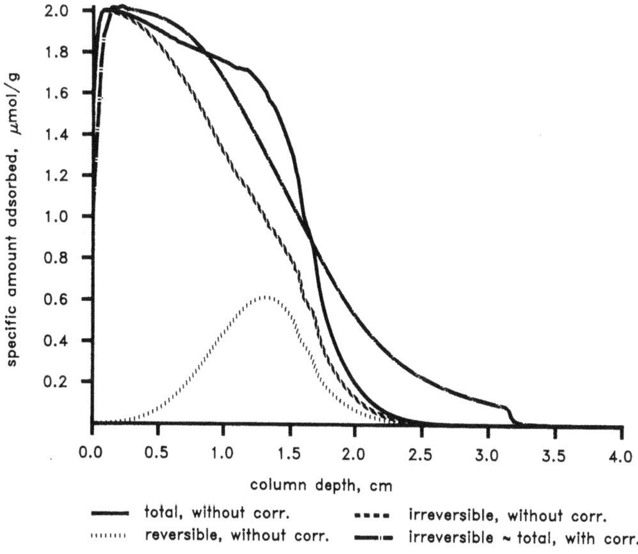


**Fig. 3** The distribution of adsorbed 3-chlorophenol in a model soil column from Na bentonite not precharged before adsorbate insertion (1 day) for the case of sorption equilibrium after 1.5 days. "Without correction" and "with correction" means without and with regard of differing adsorbent concentrations, respectively

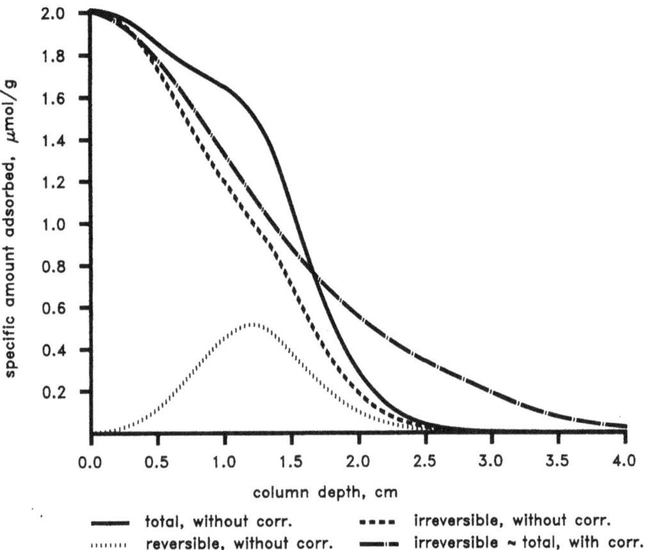



**Fig. 4** The distribution of adsorbed 3-chlorophenol in a model soil column from Na bentonite not precharged before adsorbate insertion for the case of sorption non-equilibrium after 1.5 days

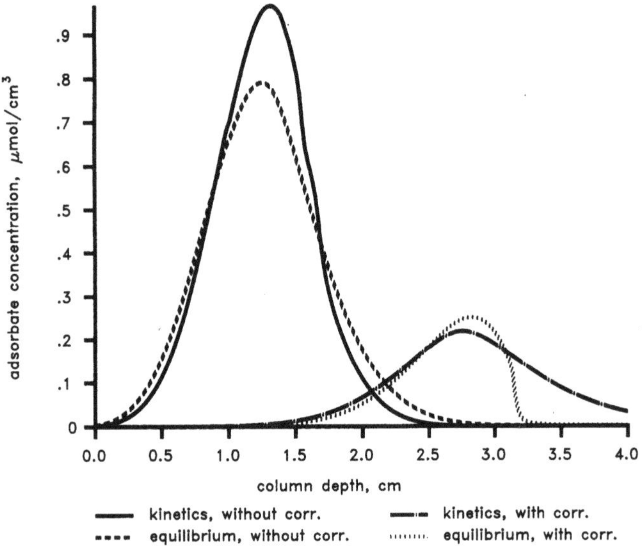



**Fig. 5** The distribution of 3-chlorophenol in the pore water of a model soil column from Na bentonite not precharged before adsorbate insertion for the case of sorption equilibrium and non-equilibrium ("kinetics") after 1.5 days

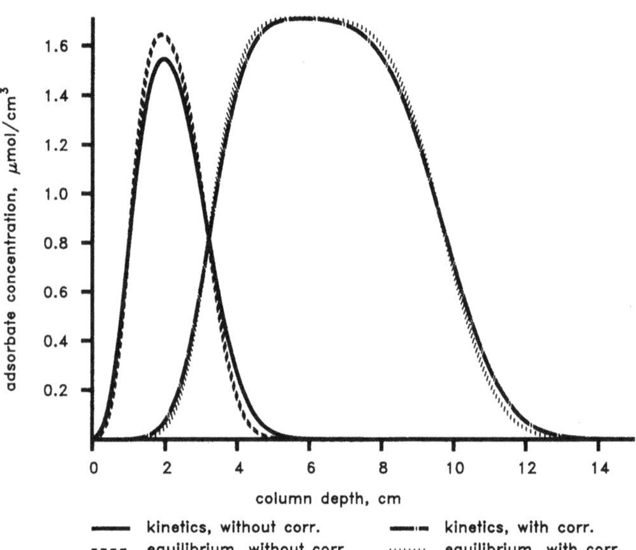



**Fig. 6** The distribution of 3-chlorophenol in the pore water of a model soil column from Na bentonite irreversibly precharged by the pollutant for the case of sorption equilibrium and non-equilibrium ("kinetics") after 1.5 days

Comparison of pollutant distribution in the non-precharged column in the case of sorption equilibrium and non-equilibrium provides only little differences without the correction (Figs. 3—5). Application of the correction leads in the kinetic case to an acceleration of adsorbate transport in pore water in comparison with the equilibrium case (Fig. 5). Furthermore, in the kinetic case a more extended distribution of the irreversibly adsorbed pollutant in the column is consequently observed (Figs. 3, 4).

## Conclusions

The dependence of the kinetic sorption parameters on the concentration of the adsorbent in the suspension of the batch-experiment is not negligible. Neglecting this concentration dependence for reversible sorption leads to wrong estimations of the transport velocity of the pollutant in sediment or soil. Therefore, environmentally relevant batch-experiments have to be made at adsorbent concentrations true to nature. If this is not possible, corrective calculations under the model conditions described could be performed as an approximation.

## References

1. Di Toro DM, Horzempa LM, Casey MC (1983) Adsorption and Desorption of Hexachlorbiphenyl, U.S. Department of Commerce, National Technical Information Service, PB83-261677
2. Rodrigo AM, Chan PC (1989) Sorption Processes of Bentonite with Liquid Organics, Proceedings 43rd Purdue Industrial Waste Conference, Lewis Publishers Inc., Chelsea, Michigan 48118, USA, pp 115—121
3. Eigen H (1990) Zur Wechselwirkung aromatischer Kohlenwasserstoffderivate mit Schichtsilikaten, Diplomarbeit, Forschungszentrum Jülich, Institut für Angewandte Physikalische Chemie
4. O'Connor DJ, Connolly JP (1980) Water Research 14:1517—1523
5. Pollard SJT, Sollars CJ, Perry R (1992) Carbon 30:639—645

Progr Colloid & Polym Sci (1994) 95:125—129
© Steinkopff Verlag 1994

T. Sobisch
L. Kühnemund
H. Hübner
G. Reinisch
J. Krägel

# Physicochemical aspects of in situ surfactant washing of oil contaminated soils

T. Sobisch (✉) · L. Kühnemund ·
H. Hübner · G. Reinisch
Adlershofer Umweltschutztechnik- und
Forschungsgesellschaft,
Rudower Chaussee 5,
12484 Berlin

J. Krägel
Universität Potsdam
Institut für Festkörperphysik

**Abstract** The paper reports on lab-scale column studies for in situ surfactant washing of oil-contaminated soils. The displacment efficiency of nonionic surfactants and their blends was tested on small columns. The kinetics of oil displacement and surfactant sorption-desorption were studied. Further, the viscosity behavior of emulsions of varying compositions was investigated. No single optimum HLB value was found for the displacement of a selected oil. It depends on the surfactant pair chosen. Optimum surfactant composition is distinctly different for weathered and unaltered diesel oil. The application of single nonionic surfactants leads to a substantial reduction of soil permeability or even to clogging. This is due to formation of viscous emulsions. This can be avoided by using surfactant blends. This way, the complex requirements for in-situ application, i.e., maintenance of permeability, low surfactant losses, and low residual levels of oil can be achieved.

**Key words** in situ soil washing — nonionic surfactant mixtures — mineral oil emulsions — viscosity — sand

## Introduction

One of the main restricting factors in the remediation of soils is the low aqueous solubility of several organic compounds. It is connected with a resistance to mobilization by conventional pump-and-treat measures. Therefore, the residual concentration of mineral oils in sandy soils is in the range of 3—20 g/kg soil [1]. To further reduce this concentration by conventional extraction-injection techniques large volumes of water have to be treated to remove small quantities of contaminants over a period of years [2].

In situ application of surfactants to solve these problems has been under discussion in recent years. Despite encouraging lab-scale results [3—11] a number of field tests was unsuccessful [12—16]. A key problem of surfactant enhanced soil cleanup is the formation of viscous emulsions inhibiting flow through contaminated zones. To overcome these viscosity problems combined polymer surfactant systems were adopted [17, 18] to increase the viscosity of the displacing fluid. Our approach was to avoid the formation of viscous emulsions by using surfactant blends [19—20]. This paper summarizes lab-scale displacement studies and investigations on the viscosity behavior of mineral oil emulsions for selected surfactant blends.

## Experimental

### Materials

Surfactants used were commercial polyoxyethylene alkylether (A1, A2, A3, A4 — HLB 10.5, 11.6, 12.5, 13.2, B1, B2 — HLB 16.2, 17.9; C — HLB 9.7) and their blends. Diesel oil was supplied by a petrol station. The sand with a density of 2.62 g/cm³ and a main particle fraction ranging from 0.1 to 1 mm was collected from an uncontaminated site.

### Small scale column tests

Small columns with an inner diameter of 17 mm were packed with dry sand resulting in sand columns with an approximate volume of 10 cm³ and an average porosity of 24%. Oil was added onto the top of these columns and flushed after 15 min discontinuously with portions of 1) 1 w% surfactant solution and 2) 100 cm³ water. Surfactant solutions and water moved through the columns simply by gravitation. The oil/surfactant ratio was 1. Flow rate was monitored and the residual oil concentration in soil was determined.

### Kinetics

60 g of dry sand was packed into two 15 cm ID glass columns resulting in an average porosity of 35%. Flow was controlled by gravity.

The sand-packed columns were doped with 1.8 g oil. After 1 week the columns were eluted with 1) 3 dm³ water, 2) 180 cm³ 1 w % surfactant solution (first column A1 alone, second column A1/B1 with HLB 12.2) and 3) 1 dm³ water. Surfactant and oil concentration in the eluat fractions and flow rate was monitored.

### Analytical procedures

The content of mineral oils in soils and eluat fractions were determined according to the standard German method (Deutsche Normen: DIN 38 409, Teil 18).

The concentration of the individual surfactants in the aqueous samples from the columns was measured by using isocratic HPLC equipped with a refractive index detector (30°C), and a RP-8 column (5 μm, 250 × 4 mm ID). Aqueous samples (20 μl) were injected either directly or after a preconcentration step on SPE octadecyl (C18) disposable columns and then eluted with acetonitrile/water (80/20).

### Viscosity measurements

Emulsions were prepared from diesel oil and water at low shear rates by adding water or 1 w% aqueous surfactant solutions to surfactant/oil mixtures. The viscosity was measured varying surfactant/oil and surfactant/surfactant ratios as well as water content. For emulsions of higher viscosity the flow behavior and the viscosity function was determined by the rotational rheometer Rheotest 2 (MLW Medingen) equipped with concentrical cylinder geometry. In the range of low viscosity the apparent viscosity was measured using an Ubbelohde capillary viscosimeter (K: 0.0999 and 0.005). To monitor the dependence of viscosity on water content the values of apparent viscosity at a shear rate of 300 cm⁻¹ were chosen from the measurements with the rotational rheometer and compared with the values obtained by capillary viscosimetry.

All measurements and the preparation of emulsions were carried out at room temperature (23 ± 1°).

## Results and discussion

Figure 1 shows the results of the small scale column tests. No value was obtained for pure nonionic C because formation of a viscous emulsion caused clogging. For surfactants A1—A4 as well as for the three different surfactant pairs an optimum HLB value was found for effective oil displacement. However, it is interesting to note that the optimum values differ significantly from each other. The comparison with previous results [19, 20] shows that for the same pair of surfactants the course of residual oil saturation with composition is very different for weathered diesel oil, i.e., the blend C/B1 with HLB 11.7 displaced only 49.5% of the oil in this investigation, whereas it produced the best result for weathered oil (2% residual oil). As can be seen from Figs. 1 and 2, efficient oil displacement as well as acceptable flow rates during surfactant flushing and water rinsing could only be achieved using selected surfactant combinations. This is the case using blends A1/B1 (HLB 12.2) and A1/B2 (HLB 11.2) with residual oil values and flow rates during the surfactant washing and the water rinsing step of 9.3%, 100 ml/h, 150 ml/h and 5.2%, 100 ml/h, 100 ml/h, respectively.

Figure 3 comprises the kinetics of oil displacement (Fig. 3a), surfactant sorption/desorption (Fig. 3b) and the ratio of surfactant to oil on column (Fig. 3c). A relatively small volume of surfactant solution displaces a great amount of oil. With surfactant A1 alone the residual oil

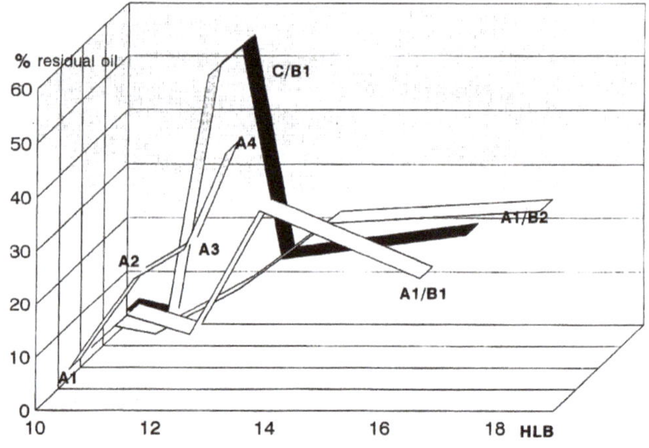


**Fig. 1** Residual oil saturation as function of blend composition

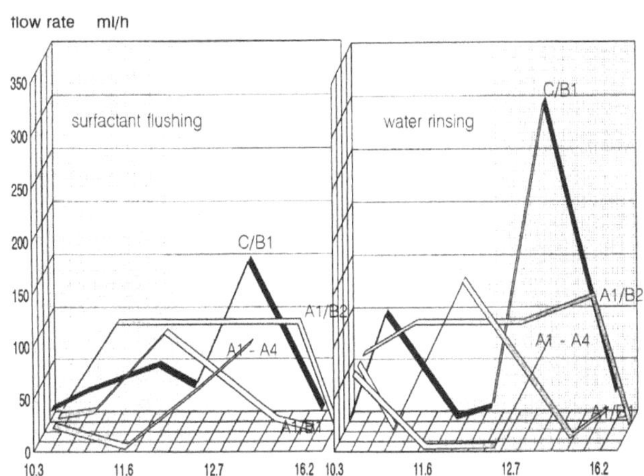


**Fig. 2**  Flow rate as function of blend composition

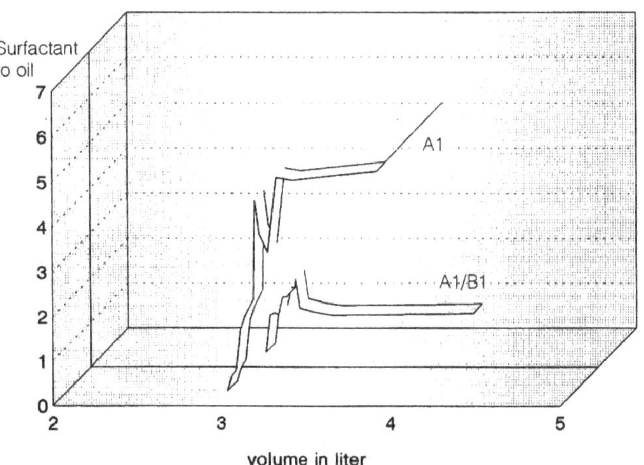


**Fig. 3c**  Variation of surfactant/oil ratio during displacement

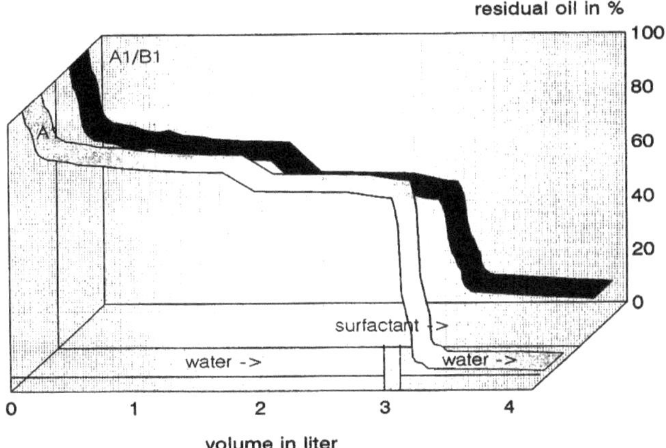


**Fig. 3a**  Kinetics of oil displacement

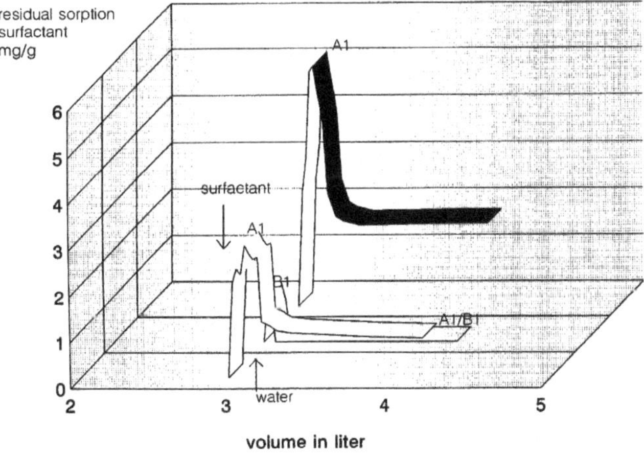


**Fig. 3b**  Kinetics of surfactant sorption

concentration (230 mg/kg) was lower than in the case of the blend A1/B1 with HLB 12.2 (990 mg/kg). On the other hand, the residual sorption in case of the first column (1700 mg/kg) was substantially higher than in the case of the surfactant blend (850 mg/kg). The ratio of surfactant-to-oil on the first column rose well beyond a value of 1, whereas on column two the value approached 1. Further, it is important to note that during surfactant flushing the permeability of the first column reduced by 75%, whereas for the second column permeability reduced only by 25%. Thus, despite different experimental procedures the same permeability behavior was observed as in the case of small scale column tests (Fig. 2).

Figure 4 shows the change of the apparent viscosity with water content for different surfactant/oil ratios in the case of the polyoxyethylene alkylether with the HLB 10.5. At high surfactant/oil ratio (1:2) gel formation occurs at low water content resulting in high viscosity values. Upon dilution the gel structure breaks down and the emulsion formed exhibits a second viscosity maximum at higher water content.

No gel formation was observed for the surfactant/oil ratio 1:4. Viscosity goes through a maximum between 25 and 30% water content. Between 15 and 25% water content phase separation occurred immediately after stopping agitation. Thus, viscosity could not be measured in this region.

For surfactant/oil ratios < 1:4 the maximum viscosity is shifted towards lower water content by lowering the surfactant/oil ratio. For these low surfactant/oil ratios total lack of stability was found between 0 and 25% water content.

The effect of blending on viscosity of emulsions containing 1 w% surfactant is shown in Fig. 5. Emulsions formed are unstable below 35% water for HLB 11—16 and below 20% for HLB < 11. For the blend (HLB 10.7) with a very low content of the more hydrophilic surfactant the viscosity maximum is shifted towards lower

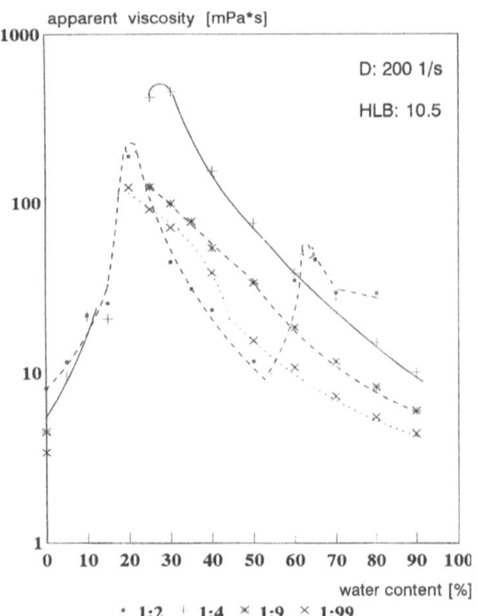


**Fig. 4** Viscosity as function of surfactant/oil ratio

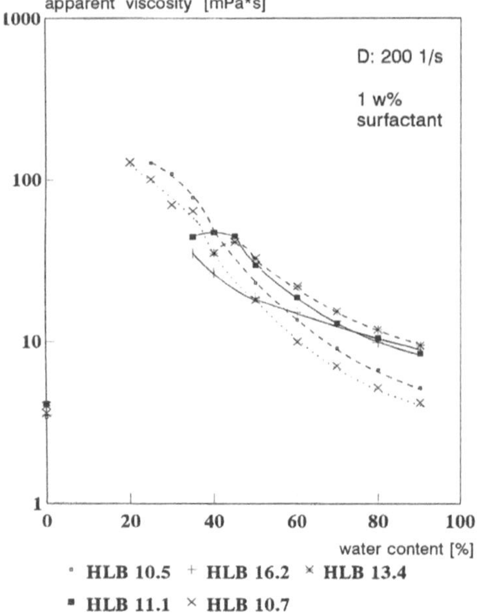


**Fig. 5** Viscosity as function of HLB

water content, whereas for blends with higher HLB it is shifted to the opposite direction. This behavior could be correlated with the changes in the aggregation of the oligoether chains at the micelle/water interface upon blending [21].

Comparing the results of displacement and viscosity investigations no simple correlation could be found. This is not surprising if one keeps in mind the complexity and dynamic nature of the displacement process. Viscosity plays a key role, however, concentration of water, surfactant and oil as well as droplet size distribution in emulsions changes continuously throughout displacement with time as well from place to place. Earlier investigations [20] already showed that the apparent viscosity is very sensitive to changes of the surfactant/oil ratio. As can be seen from Fig. 4 high viscosities are observed at high surfactant/oil ratios ($\geqslant 1:4$), especially in the oil-rich region. It is supposed that the flow behavior of the oil or emulsion film on the sand surface is influenced mainly by the viscosity behavior observed in the range of lower water contents. Somewhat unexpectedly, it was observed that the surfactant/oil ratio indeed approached a value of 1 for blend A1/B1 and a value even higher when A1 was used alone during displacement (Fig. 3c). In the case of surfactant A1 the high ratio of surfactant to oil on the column observed during the kinetic investigations should hinder further oil displacement. Actually, a substantial decrease in permeability was observed which corresponds to the formation of a viscous gel.

To discuss the results of oil displacement studies obtained with different surfactant blends on the basis of the viscosity behavior additional viscosity measurements have to be carried out at higher surfactant/oil ratios than already obtained (Fig. 5). This work is in progress [21].

## Conclusions

No single optimum HLB value was found for the displacement of a selected oil. It depends on the surfactant pair chosen.

Optimum surfactant composition is distinctly different for weathered and unaltered diesel oil. The application of single nonionic surfactants leads to a substantial reduction of soil permeability or even to clogging. This is due to formation of viscous emulsions. This can be avoided by using surfactant blends. This way, the complex requirements for in-situ application, i.e., maintenance of permeability, low surfactant losses and low residual levels of oil can be achieved.

## References

1. Lund CN (1991) Veröffentlichungen des Instituts für Bodenmechanik und Felsmechanik der Universität Fridericiana in Karlsruhe, Heft 119: Beitrag zur biologischen in situ-Reinigung kohlenwasserstoffbelasteter körniger Böden, Dissertation

2. Mackay DM, Cherry JA (1989) Environ Sci Technol 23:630—636

3. Fountain JC, Klimek A, Beikirch MG,

Middleton TM (1991) J Hazard Mat 28:295—311

4. Ellis WD, Payne JR, Mc Nabb GD (1985) EPA/600/2-85/129, U. S. Environmental Protection Agency 85 pp

5. Ellis WD, Payne JR, Tafuri AN, Freestone FJ (1984) In: Proc of Hazardous Material Spills Conference, Nashville, pp 116-124

6. Ang CC, Abdul AS (1991) Ground Water Monit Rev 11:121—127

7. Gannon OK (1988) PhD: Environmental reclamation through use of colloidal foam flotation, in-situ soil aeration and in-situ surfactant flushing, UMI Disseration Services Order Number 8910849

8. Abdul AS, Gibson TL (1991) Environ Sci Technol 25:665—671

9. Rickabaugh J, Clement S, Lewis RF (1986) Proc 41st Purdue Industrial Waste Conference 41:377-382

10. Gannon OK, Bibring P, Raney K, Ward JA, Wilson DJ (1989) Sep Sci Technol 24:1073-1094

11. Texas Research Institute (1979) API Publication 4317

12. Chawla RC, Diallo MS, Cannon JN, Johnson JH, Porzucek C (1990) 1989 International Symposium on Solid/Liquid Separation: Waste Management and Productivity Enhancement, pp 355—367

13. Texas Research Institute (1985) API Publication No 4390

14. Nash JH (1987) EPA/600/2-87/110, PB-146808

15. Nash JH, Traver RP (1989) In: Kostecki PT, Calabrese EJ (eds) Petroleum contaminated soils, Vol. 1. Lewis Publishers, p 157

16. Downey DC (1990) Military Engineer 82:(n 53) 22—24

17. Sale T, Piontek K, Pitts M (1989) In: The proceedings of the petroleum hydrocarbons and organic chemicals in ground water: Prevention, detection and restoration, NWWA/API conference 1989, pp 487—503

18. Pouska GA, Trost PB, Day M (1989) In: Proc 6th National RCRA, Superfund Conference and Exhibition, New Orleans, pp 423—430

19. Sobisch T, Reinisch G, Hübner H, Karg A, Niebelschütz H (1993) In: Arendt F, Annokee GJ, Bosman R, van den Brink WJ (eds) Contaminated Soil '93, Proceedings of the Fourth International KfK/TNO Conference on Contaminated Soil, 3—7 May 1993, Berlin, Germany, Volume II. Kluwer Academic Publishers, Dordrecht/Boston/London, pp 1453—1454

20. Sobisch T, Krägel J, Reinisch G (1993) Proceedings of the First World Congress on Emulsions — Paris 19—22 oct. 1993, 3—30 307

21. Sobisch T, Krägel J, in preparation

Progr Colloid & Polym Sci (1994) 95:130—134
© Steinkopff Verlag 1994

G. Hollederer
W. Calmano

# Changes of the density of charge on mineral soil components by adsorption of some metabolites of hydrocarbons

G. Hollederer (✉) · W. Calmano
Technische Universität Hamburg-Harburg
Arbeitsbereich Umweltschutztechnik
Eissendorfer Straße 40
21073 Hamburg

**Abstract** The adsorption on clay minerals and sesquioxides of some polar degradation products of naphthalene and alkylated benzenes was investigated by $^{14}$C-tracer experiments. Surface charge density of the solids was measured by titration with sodium polyethene sulfonate and polydiallyl-dimethyl-ammonium chloride at pH-range 4—7. Adsorption of organic anions reduced the positive charge on oxidic surfaces and increased the density of negative charge on clay minerals, respectively. The increase of the density of charge on clay minerals indicated an effective bonding mechanism. Changes of the density of charge were due to the adsorption of anions. The method is suitable for screening the interaction between polar metabolites of hydrocarbons and mineral soil components.

**Key words** Adsorption — clay minerals — hydrocarbons — particle charge — sesquioxides

## Introduction

In recent years a large number of hydrocarbon-contaminated sites has been discovered. As a consequence, microbial clean-up technologies have been developed. It has been shown, however, that the restoration of sites contaminated with lubricating oils or other complex mixtures is very complicated [1]. The biodegradation of the substances is often limited to secondary formation of alcohols and acids. There is little information about the interaction of polar substances with soils that have a low carbon content, although the importance of mineral soil components as adsorbents of organic chemicals is now generally accepted [2, 3]. The aim of the paper is to describe changes of surface characteristics of mineral adsorbents especially of the density of charge by adsorption of polar aromatic degradation products.

## Methods and materials

Model components of soils

The major constituents of soils are sesquioxides and clay minerals. Iron oxide (goethite) was synthesized by a method described by Gerth [4] and manganese oxide (manganite) was prepared by a method described by McKenzie [5]. The clay mineral montmorillonite was obtained from Wyoming, USA, while kaolinite was obtained from Ward Corp., USA. Organic substances were destroyed by a pretreatment of the clay minerals with $H_2O_2$. The surfaces were loaded with calcium chloride and adjusted to pH 7.

All minerals were characterized by x-ray-diffractometry. The components were stored as a suspension in the dark at 5 °C. Table 1 shows some important characteristics of the components. The measurement of the specific surface area was done by $N_2$-

**Table 1** Characteristics of the soil components

| | Particle size (μm) | Spec. surface area (m$^2$/g) |
|---|---|---|
| Manganite | 0.10 × 10 | 13.9 |
| Goethite | 0.03 × 0.80 | 62.7 |
| Montmorillonite | 0.20— 2.00 | 67.2 |
| Kaolinite | 1.00— 10.0 | 7.9 |

adsorption and calculation of the BET-Isotherm. The particle sizes were measured by scanning and transmission electron microscopy (SEM, TEM).

## Metabolites of hydrocarbons

2-hydroxybenzoic acid (salicylic acid) and 1,2-dihydroxybenzene (catechol) were used as catabolite products of naphthalene, benzoic and 4-methylbenzoic acid were used as degradation products of alkylated aromatic substances, which were important constituents of lubricating oils. The biodegradation of naphthalene by bacteria has been described [6], and aromatic acids have been found in groundwater that has passed a site contaminated by alkylated benzenes [7]. The substances were obtained from Sigma Chemie, FRG.

## Experiments with $^{14}$C-tracers

Experiments with radioactive chemicals were performed to receive ensured information about the distribution of the metabolites in the suspension. The experiments were carried out at different pH values. In the same way the concentration of calcium chloride was varied between 1.3 and 3.2 μmole/l. The concentration of the metabolites was 9 μmole/l and that of the solids was 300 mg/l. The vials which were used during this assay had a volume of 6 ml and were kept under constant shaking for 4 days at 21°C. After this time, the solid phase was separated by centrifugation. Measurement of radioactivity was done by β-scintillation counting. The comparison of the radioactivity in the aqueous phase with a solid free variant was used to describe the association of organic compounds with the mineral soil components. All experiments were done in triplicate.

## Measurement of the charge density

The density of charge of the soil components was measured by a "particle charge detector" constructed by Mütek, FRG. The suspension was filled into a cylindrical testing apparatus and a large fraction of the solids was adsorbed on its wall. As a function of pH and characteristics of the soil constituents these solids are charged except at the point of zero charge. Charge equalization is achieved by the sorption of counterions. During the measurement this equilibrium is disturbed by a laminar flow generated by a piston stroke. The spatial shift of the counterions generates an electrical potential, which can be recorded and counted by detectors at the ends of the testing arrangement. The suspension was titrated with a polyelectrolyte and in this way the charge of the solids was compensated. When the counterionic space charge cloud disappears and the value of the electrical potential drops to zero, the density of charge $\sigma$ can be calculated as follows:

$$\sigma = \frac{V \cdot c(\text{PE}) \cdot F}{m \cdot A/m} , \tag{1}$$

where $V$ is the volume of the titrating solution, $c(\text{PE})$ is the concentration of the polyelectrolyte, $F$ is the Faraday constant (9,64845 · 10$^4$ C/mole), $m$ is the mass of the particles, and $A/m$ is the specific surface area of the solids.

Preliminary tests have shown a decrease of titration speed at the end of the measurement, and much time is needed to obtain a result. Sporadically, the reaction between the polyelectrolyte and the surfaces was not finished, and a second titration step was necessary. Because of these disadvantages the samples were not titrated in this way. Instead, the solids were treated with an excess of polyelectrolyte. The mass of the non reacting polyelectrolyte was determined by back-titration with the opposite kind of polyelectrolyte. In Eq. (1) the term $V \cdot c(\text{PE})$ has to be substituted as follows:

$$V \cdot c(\text{PE}) = V_0 \cdot c(\text{PE}_0) - V_T \cdot c(PE_T) , \tag{2}$$

where $V_0$ is the volume of the polyelectrolyte, which was used in excess, and $c(\text{PE}_0)$ is its concentration. $V_T$ is the volume of the polyelectrolyte solution which was used for back-titration, and $c(\text{PE}_T)$ is its concentration. In relation to the sign of the charge of the soil compounds counterionic and isoionic polyelectrolytes were used. The sign of $V \cdot c(\text{PE})$ is identical with the sign of the charge at the surface if $\text{PE}_0$ is the anionic and $\text{PE}_T$ the cationic polyelectrolyte. Sodium polyethane sulfonate was used as anionic polyelectrolyte and polydiallyl-dimethyl-ammonium chloride was used as a cationic one.

The experiments were carried out as follows: 3 mg of a mineral soil component were transferred with a pipette as a suspension into 20-ml glass vials. 5 μmole of polyelectrolyte, 0.3 μmole of calcium chloride and different amounts of hydrochloric acid (HCl) and sodium hydroxide (NaOH), respectively, were pipetted into the vials, which were filled up to 10 ml with demineralized water. The vials were shaken for 1 week in the dark at room temperature. All experiments were done in triplicate.

## pH and metabolites

An interaction between the metabolites and the polyelectrolytes was investigated by titrating solid free controls at pH 5 and 8, respectively. Results that differ significantly from the values obtained from back-titrating the counterionic polyelectrolyte could be observed in the alkaline controls. Therefore, the following experiments were carried out at lower pH values. Some data were obtained from alkaline controls that did not contain any metabolites.

The concentration of hydronium ions was adjusted to pH values between 3 and 7. The concentration of the metabolites was 500 μmole/l. The concentration of 4-methylbenzoic acid was 250 μmole because of the limited solubility. The polar organic chemicals existed as anions with the exception of 1,2-dihydroxybenzene.

## Calcium chloride assay

In the presence of high concentrations of calcium chloride the spatial extent of the electric field around the soil colloids was reduced significantly. The recording and counting of the electrical potential was difficult and was partly not possible. Beyond this, the experimental results dependend on the concentration of calcium chloride. The reason for this might be an interaction between calcium chloride and the polyelectrolytes or a modification of the solid surfaces.

These preliminary studies have shown that the concentration of calcium chloride should not significantly exceed 30 μmole/l, if a solid concentration of 300 mg/l is chosen. The tests were carried out at this concentration and were compared with experiments without calcium chloride to investigate the effects of this electrolyte.

## Results

### Tracer assay

Figure 1 shows the adsorbed mass ($m_{ad}$) related to the total mass ($m_t$) of the metabolites in the tests. The results were derived from the comparison of the radioactivity of the supernatant and the radioactivity of the solid free control. Sorption of less than 3% of the total mass was not significant.

As a unitary result, it was found that the sorbed mass increased with decreasing pH values. Protonation of the surfaces of the solids with an increased affinity to anions might be the reason for this, rather than a hydrophobic effect. The pH-values were some orders above the $pk_s$-values of the acids. There was no interaction between any soil component and the substance with the highest octanol-water-partition coefficient $P_{ow}$, which is 2.27 for 4-methylbenzoic acid. This result emphasizes the polar properties of the mineral soil particles in the suspension.

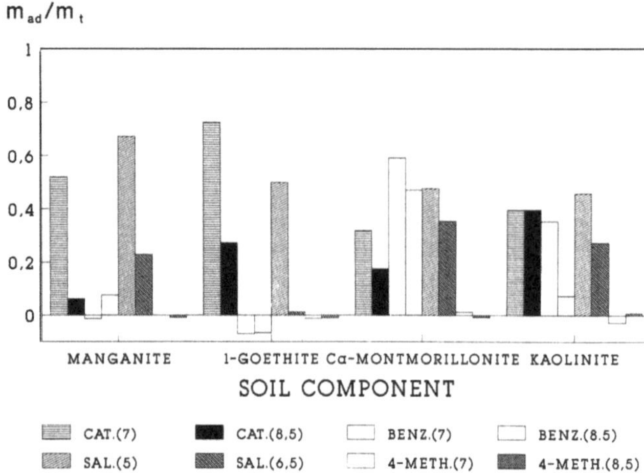



**Fig. 1** Sorption of the metabolites on the solids at different pH values, [14]C-tracer assay

There was no difference in partitioning of the metabolites at different concentrations of calcium chloride (1.3 and 3.2 μmole/l, respectively) that lies above the range of variations of these measurements.

The interactions between goethite and benzoic acid were unexpected and a plausible interpretation cannot be given here.

## Density of charge

### pH value

The results of the measurements carried out with an excess of cationic electrolyte and with calcium chloride are shown in Figs. 2—3. An increase of the hydronium ions concentration in the aqueous phase is coupled with an increase of the positive charge of the particles. This can be derived from the slope of the regression line. This result was expected because of the interaction of the hydronium ions with the surfaces of the soil components. The value of the slope is higher for sesquioxides in comparison with clay minerals. The surface of montmorillonite is negatively charged because of the permanent charge.

### Interactions with the metabolites

As demonstrated in Fig. 2, organic substances interact with the positively charged surface of manganite. The positive charges were reduced in the order 4-methylbenzoic acid, 2-hydroxybenzoic acid, 1,2-dihydroxybenzene. If goethite is used as an adsorbing substance, only the reduction of the positive charge by sorption of 1,2-dihydroxibenzene is significant. Beyond that, both oxides react with the phenolic compound 1,2-dihydroxibenzene, forming green-colored complexes.

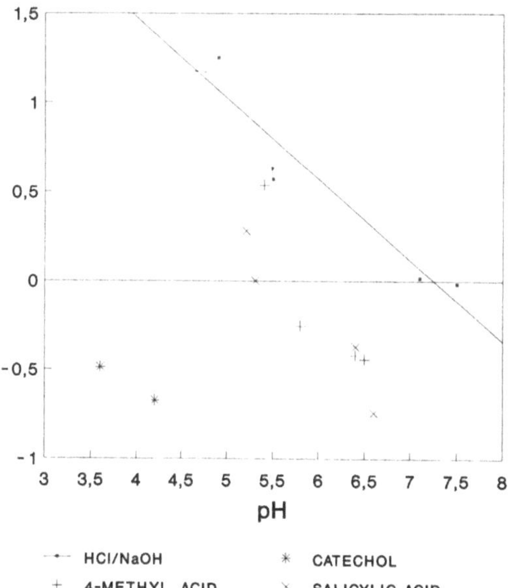



**Fig. 2** Changes of the charge density $\sigma$ on manganite at different pH values in the presence of metabolites

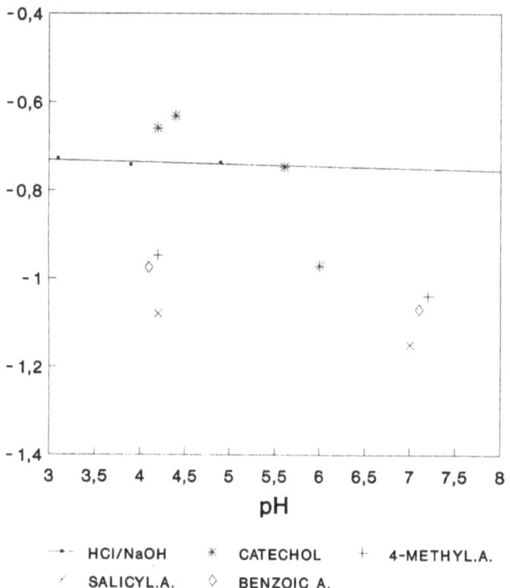



**Fig. 3** Changes of the charge density $\sigma$ of on montmorillonite at different pH values in the presence of metabolites

The density of charge is nearly constant over the whole range of pH if clay minerals are tested (Fig. 3). The negative charges increase in the order 1,2-dihydroxybenzene, 4-methylbenzoic acid, benzoic acid, 2-hydroxybenzoic acid. The change of the charge density strongly depends on the pH if 1,2-dihydroxibenzene is the metabolite.

No change of the density of charge was seen in the experiments carried out with the organic anions and kaolinite as adsorbent. All experiments performed at an excess of anionic polyelectrolyte show high values of the mean error, and therefore changes in the density of charge were not significant.

## Calcium chloride assay

The presence of calcium chloride at the concentration of 30 μmole/l reduced the order of magnitude of the mean error of the measuring points around the regression line of the hydronium ions adsorption by HCl and NaOH, respectively. At least the variance of the values was the result of the instability of the system during pH measurement and when no additional electrolyte was present. As a rule, more than 10 minutes were required to get values that could be regarded as reliable.

## Discussion

Changes of the density of charge observed during the adsorption of some metabolites of hydrocarbons on mineral soil components were compared with the results derived from partitioning experiments with radiochemicals. Some hints at the interaction between 1,2-dihydroxybenzene and related compounds, respectively, with iron ions [8] or Al-OH groups [9] were found in the literature. Sigg and Stumm [10] emphasized the reactivity between metal hydroxides and those chemicals forming ring-shaped surface complexes like 2-hydroxybenzoic acid.

The adsorption of organic anions reduces the positive charge of the oxidic surfaces of manganite and goethite and increases the density of negative charge of montmorillonite, respectively. While the neutralization of positive charges on protonated surfaces by adsorbing anions seems to be plausible because of electrostatic reasons, changes in the density of charge indicate an effective bonding mechanism if — as in the case of montmorillonite — Coulomb forces were surmounted.

Changes of the density of charge were due to the adsorption of anions. The measurements of the density of charge were semiquantitative, because the interaction between metabolite and polyelectrolyte has to be taken into account. The developed method could be useful to obtain some quick information about the interaction between organic substances and soil components without using radiotracer tests, which are more sophisticated when contaminated sites have to be investigated to develop clean up strategies.

**Acknowelegement** This study was supported by the Deutsche Forschungsgemeinschaft.

## References

1. Hollederer G, Hofmann R, Filip Z (1993) In WaBoLu-Heft 1/92, Berlin, pp 1—99
2. Kango RA, Quinn JG (1989) Chemosphere 19:1269—1276
3. Bouchard DC, Enfield CG, Piwoni MD (1989) Soil Sci Soc Am and Am Soc of Agronomy, SP22:350—356
4. Gerth J (1985) Ph D Thesis, Kiel
5. McKenzie RM (1971) Mineral Mag 38:493—502
6. Cerniglia CE (1984) In: Atlas RM (ed) Petroleum Microbiol. Macmillan Publ Comp, New York, pp 99—128
7. Cozzarelli IM, Eganhouse RP, Baedecker M (1990) J Environ Geol Water Sci 16:135—141
8. Patrovsky V (1970) Collect Czechoslovak Chem Commun 35:1599—1604
9. McBride MB, Wesselink LG (1988) Environ Sci Technol 22:703—708
10. Sigg L, Stumm W (1991) In: Sigg L, Stumm W (eds) Aquatische Chemie. Teubner, Stuttgart, pp 331—371

Progr Colloid & Polym Sci (1994) 95:135—138
© Steinkopff Verlag 1994

J. Thieme
J. Niemeyer
P. Guttmann
T. Wilhein
D. Rudolph
G. Schmahl

# X-ray microscopy studies of aqueous colloid systems

Dr. J. Thieme (✉) · P. Guttmann ·
T. Wilhein · D. Rudolph · G. Schmahl
Universität Göttingen
Forschungseinrichtung Röntgenphysik
Geiststraße 11
370373 Göttingen

J. Niemeyer
Universität Göttingen
Institut für Bodenwissenschaften
von-Siebold-Straße 4
37075 Göttingen, FRG

**Abstract** X-ray microscopy is capable of imaging objects with a higher resolution than light microscopy. The reason is the shorter wavelength of x-rays compared to visible light. Up to now, the smallest structures that can be visualized are about 30 nm in size. It is possible to image objects directly in their aqueous environment without preparational procedures like drying, staining, etc. In the wavelength range of 2.4 nm $\leqslant \lambda \leqslant$ 4.5 nm water absorbs x-radiation weakly compared to the absorption of other materials like alumino silicates, iron oxides or organic materials. This yields a natural contrast mechanism for the imaging of structures in their aqueous environment. Studies of aqueous colloidal systems demonstrate the abilities of x-ray microscopy. The aggregation of hematite particles under the influence of divalent cations ($Ca^{2+}$) is investigated. The dependence of the size and the shape of the aggregates on the amount of added $Ca^{2+}$ is directly visualized.

**Key words** X-ray microscopy — aqueous colloidal systems — hematite

## Introduction

X-rays within the wavelength range of 0.3 nm—5 nm are best suited for x-ray microscopy studies of wet colloical systems. The wavelength range between the K absorption edges of oxygen at $\lambda = 2.34$ nm and carbon at $\lambda = 4.38$ nm is especially interesting because in this wavelength range the radiation is weakly absorbed by water but strongly absorbed by iron oxides, silicates, organic matter, etc., resulting in a good amplitude contrast of wet specimens [1]. Due to the much shorter wavelength, x-ray microscopy provides higher resolution than optical microscopy. Most importantly, x-ray microscopy has the potential for imaging hydrated specimens with high resolution [2, 3]. Photoelectric absorption and phase shift are the two diminating processes for the contrast. In the mentioned wavelength range the absorption cross-sections for x-rays are about one order of magnitude smaller than the cross-sections for electrons in the energy range in which electron microscopy is performed. These absorption cross-sections for soft X-rays are appropriate for high-resolution investigation of specimens in a water layer of up to 10 µm thickness.

## The x-ray microscope at the electron storage ring BESSY

The Forschungseinrichtung Röntgenphysik has developed two x-ray microscopes with zone plate optics, an x-ray microscope and a scanning x-ray microscope. The two microscopes are installed at the electron storage ring BESSY in Berlin utilizing synchrotron radiation. Figure 1 shows the scheme of the x-ray microscope at BESSY.

For x-ray microscopy two types of zone plate optics are needed. The object has to be illuminated by a condenser zone plate, which has to collect as much

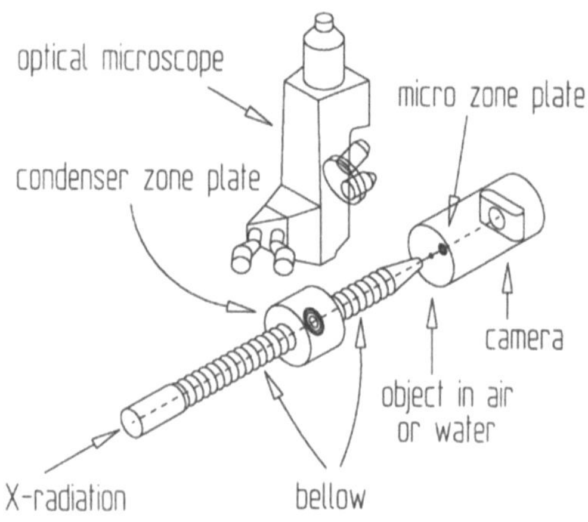


**Fig. 1** Scheme of the x-ray microscope at BESSY

radiation as possible from the x-ray source. The object is imaged by a high-resolution zone plate. As the size of object structures that can be imaged is nearly equal to the width of outermost zone of this zone plate, this zone width $dr_n$ has to be as small as possible. Zone widths in the range of $dr_n = 20$—$50$ nm are achieved [4]. The images of the hematite particles shown below are made with a zone plate with $dr_n = 30$ nm.

The condenser zone plate is mounted in a bellow which ends in an $Al_2O_3$-foil transparent to x-radiation. This foil separates the vacuum of the bellow from air. The object field is limited by a pinhole with a diameter of about 20 µm. The object is mounted in an environmental chamber — not shown in Fig. 1 — where it can be prepared in its natural state. The next vacuum compartment contains the micro zone plate and the camera. The micro zone plate as the high resolution x-ray objective

generates a magnified image of the object in the image field. The enlarged x-ray image is recorded by a CCD-camera. An optical microscope on a swivel-mount is installed to adjust and to prefocus the object. Therefore, the bellow containing the condenser zone plate can be bent downwards [5]. Figure 2 shows a photograph of the x-ray microscope at BESSY.

## X-ray microscopy studies of aqueous colloidal systems

The aggregation of hematite particles with a diameter of nominal 80 nm by $CaCl_2$ was investigated. 150 mg freeze dried hematite was suspended in 10 ml bidistilled water by ultrasonic treatment. To 1 ml aliquots different amounts of a 1% $CaCl_2 \cdot 2\,H_2O$ solution were added. In Fig. 3 primary hematite particles suspended by ultrasonic treatment in bidistilled water are shown. As can be seen, the primary particles are completely dispersed. The addition of $CaCl_2$ leads to an aggregation of the particles. The size and the shape of the aggregates depend on the amount of added $Ca^{2+}$. In Figs. 4, 5, and 6 the influence on the aggregation is shown after the addition of 20 µl, 80 µl, and 200 µl of the $CaCl_2$ solution, respectively. As can be expected, the size increases with the $Ca^{2+}$-content of the hematite suspension. The internal structure of the aggregates can be visualized by x-ray microscopy without any preparational procedures. This

**Fig. 2** The x-ray microscope at BESSY

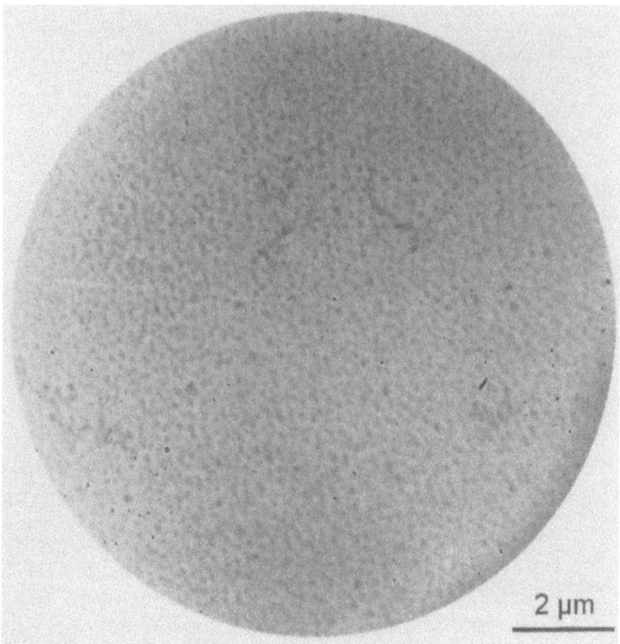


**Fig. 3** Completely dispersed primary hematite particles. 150 mg freeze-dried hematite was suspended in 10 ml bidistilled water by ultrasonic treatment

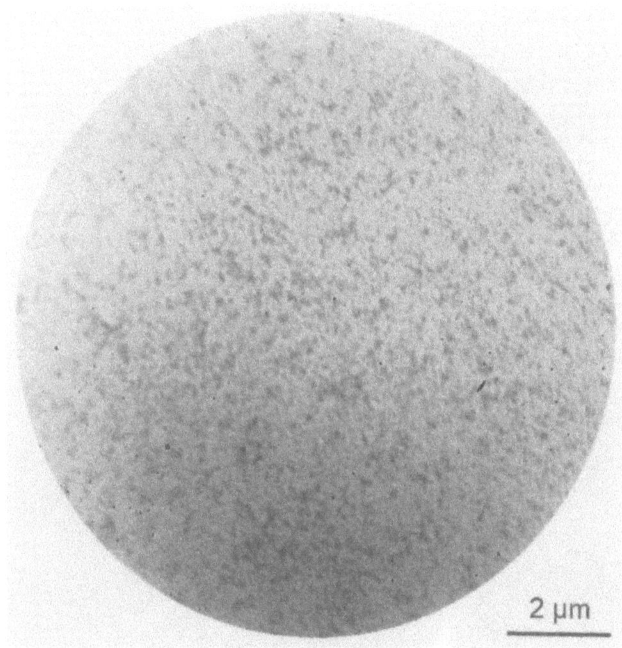

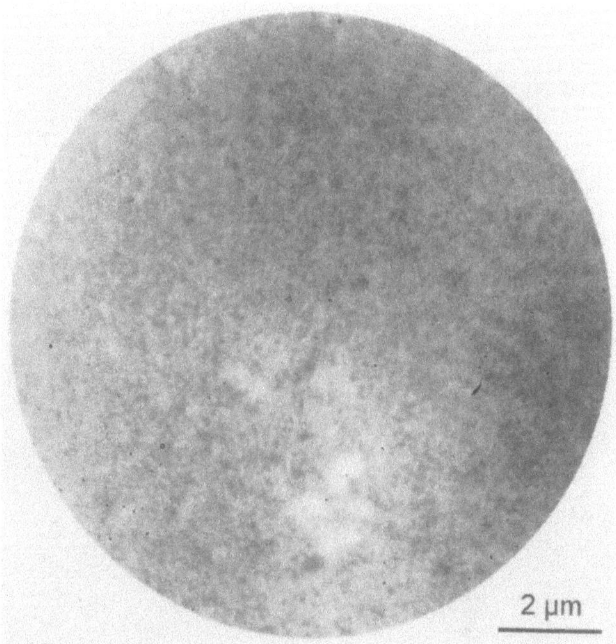

**Fig. 4** Primary hematite particles (see Fig. 3) after the addition of 20 µl of a 1% $CaCl_2 \cdot 2\ H_2O$ solution to a 1 ml aliquot

**Fig. 6** Primary hematite particles (see Fig. 3) after the addition of 200 µl of a 1% $CaCl_2 \cdot 2\ H_2O$ solution to a 1 ml aliquot

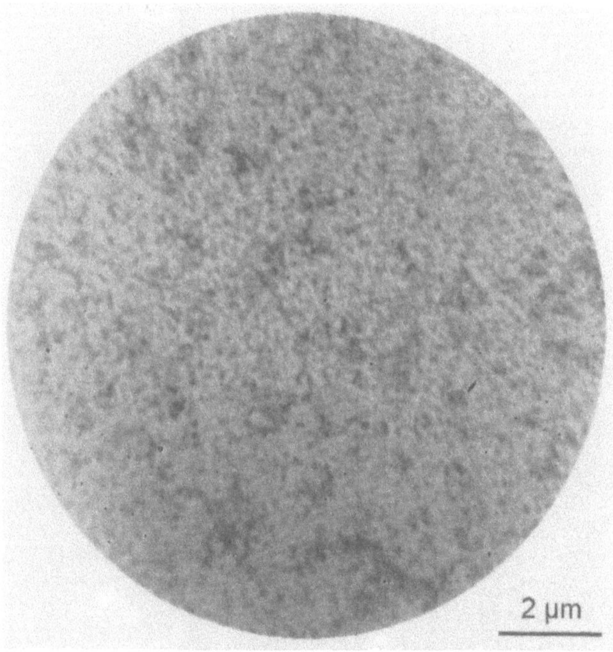

will lead to a better understanding of the underlying aggregation mechanisms. In addition, the analysis of these images can lead to a confirmation of the proposed mechanisms like cluster-cluster-aggregation CCA and diffusion-limited-aggregation DLA [6, 7].

**Acknowledgements** Special thanks to Prof. Dr. H. Sticher and Dr. M. Borkovec, Institute of Terrestrial Ecology, ETH Zürich, Switzerland, who provided the hematite. The current research is being funded by the German Federal Minister of Research and Technology (BMFT) under contract number 05 5MGDXB6.

**Fig. 5** Primary hematite particles (see Fig. 3) after the addition of 80 µl of a 1% $CaCl_2 \cdot 2\ H_2O$ solution to a 1 ml aliquot

# References

1. Wolter H (1952) Ann Phys 10(6): 94—114
2. Thieme J, Guttmann P, Niemeyer J, Schneider G, David C, Niemann B, Rudolph D, Schmahl G (1992) Nachr Chem Techn Lab 40(5):562—563
3. Rudolph D, Schneider G, Guttmann P, Schmahl G, Niemann B, Thieme J (1992) In: Michette AG, Morrison GR, Buckley CJ (eds) X-Ray Microscopy III. Springer, Berling, pp 392—396
4. Thieme J, David C, Fay N, Kaulich B, Medenwaldt R, Hettwer M, Guttmann G, Kögler U, Schneider G, Rudolph D, Schmahl G (1993) In: X-Ray Microscopy IV, to be published
5. Schmahl G, Rudolph D, Niemann B, Guttmann P, Thieme J, Schneider G, David C, Diehl M, Wilhein T (1993) Optik 93(3):95—102
6. Viscek T (1989) Fractal Growth Phenomena. World Scientific, Singapore
7. Meakin P (1988) Ann Rev Phys Chem 39:237—267

Progr Colloid & Polym Sci (1994) 95:139—142
© Steinkopff Verlag 1994

J. Niemeyer
J. Thieme
P. Guttmann
T. Wilhein
D. Rudolph
G. Schmahl

# Direct imaging of aggregates in aqueous clay-suspensions by x-ray microscopy

J. Niemeyer (✉)
Universität Göttingen
Institut für Bodenwissenschaften
von-Siebold-Straße 4
37075 Göttingen

J. Thieme · P. Guttmann · T. Wilhein ·
D. Rudolph · G. Schmahl
Universität Göttingen
Forschungseinrichtung Röntgenphysik
Geiststraße 11
37073 Göttingen

**Abstract** The applicability of x-ray microscopy for the direct investigation of clay aggregates in water is demonstrated. As coagulating agents $Ca^{2+}$, an anionic and a cationic detergent were used. The internal structures of the aggregates formed by the addition of the surfactant could be clearly visualized.

**Key words** X-ray-microscopy — clay, aggregates — detergents — suspensions

## Introduction

Clay minerals influence a great variety of chemical, biochemical, and microbiological processes in soils and sediments [1]. These colloids bind cations, act as carriers of radionuclides, and influence the growth and metabolic activity of soil microorganisms, to give but a few examples [2]. In the soil, the clay minerals are associated with a great variety of other materials, like different iron oxides, aluminium aquoxides, and humic substances [3]. This association leads to the formation of larger aggregates. Due to the reduced accessibility of the active parts inside the aggregates, the turnover rates of chemicals are reduced [4].

To obtain a measure for the influence of the geometry on the accessibility of the centers, a direct visualization of these aggregates and their internal structure is highly desirable. To avoid samples under investigation being damaged by preparation procedures, these aggregates should be investigated in their natural aqueous environment. Due to the size of the colloidal particles, a direct visualization with the light microscope is not possible. All electron microscopic techniques require that the water be removed from the sample [5]. Since these processes can influence the object, a number of different methods has been developed to reduce this risk. On the other hand, x-ray microscopy permits the direct investigation of clay particles in the natural aqueous environment.

## Methods

The principles of x-ray microscopy are presented in detail by Schmahl et al. [6] and in the accompanying article in this volume [7].

*Preparation of the monocationic clay sample:*

Na-montmorrilonite was prepared by washing the clay (after removing particles larger than 2 μm), three times with 1 M solution of sodiumchloride. Excess salt was removed by washing the clay with distilled water until the supernatant was free of chloride. The salt free clay was freeze dried and resuspended in distilled water. The coagulating agents were added according to their desired concentration.

## Results

In Figs. 1A—C, the influence of increasing amounts of $Na^+$ on the delamination of Wyoming montmorillonite suspensions in bidestilled water is shown. In Fig. 1A, structures within a 0.1% suspension (weight by weight) can be seen. Besides some larger particles, many platelets of 70—100 nm thickness are observable. As can be expected, the treatment with $Na^+$ leads to a decay of the larger particles, while the platelets are still present with the various forms of contacts (Fig. 1B). The addition of

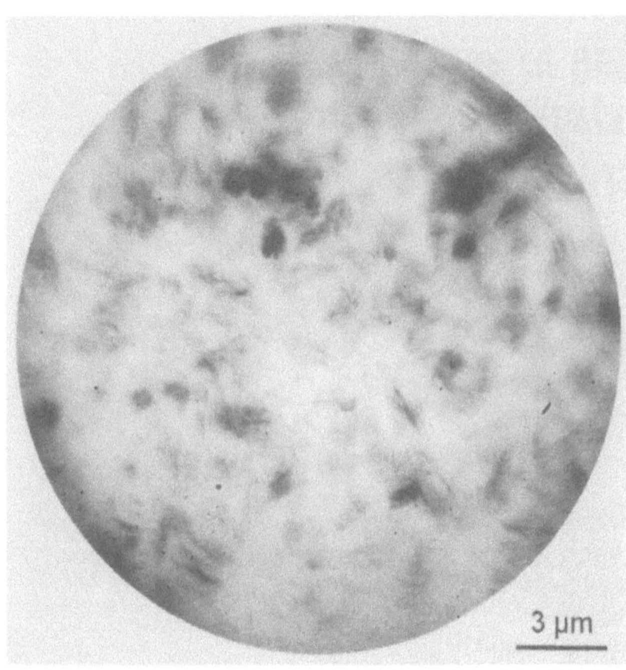

**Fig. 1A** Structures within 0.1% suspension of Na$^+$-montmorillonite (Upton, Wyoming)

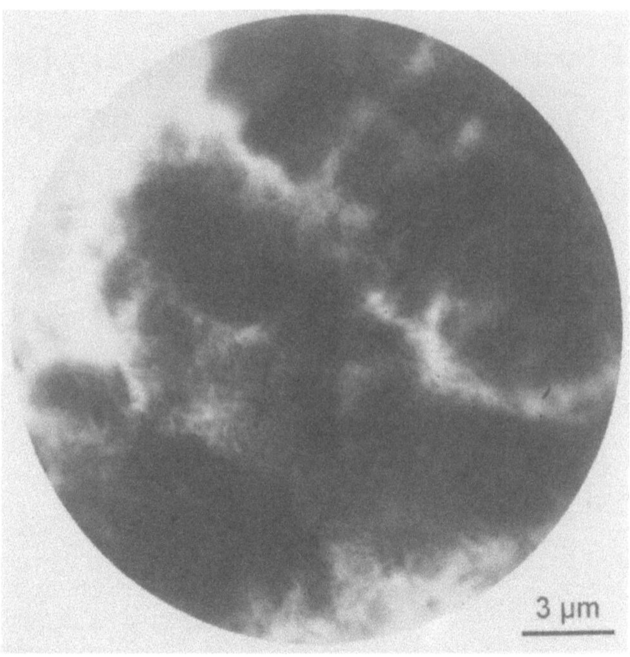

**Fig. 1C** Structures within 0.1% suspension of Na$^+$-montmorillonite (Upton, Wyoming) after the addition of Na$^+$, exceeding the critical coagulation concentration

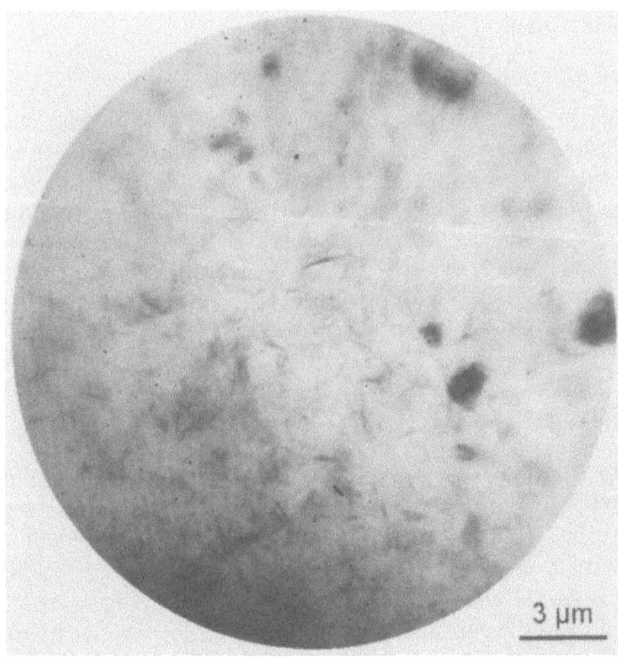

**Fig. 1B** Structures within 0.1% suspension of Na$^+$-montmorillonite (Upton, Wyoming) after the addition of Na$^+$

further amounts of Na$^+$, of more than the critical coagulation concentration, leads to a complete decay of the originally observed platelets into smaller particles. These small particles (Fig. 1C), consisting of a few elementary sheets, are the building units of flocs, which are observable with the naked eye.

It is known that the addition of cationic detergents to clay suspensions leads to a coagulation of dispersed particles. In Figs. 2A—C, the influence of a cationic detergent, hexadecyltrimethylammonium bromide (CTB), on a 0.1% clay suspension is demonstrated. This time, a Na-montmorillonite from Moosburg, Bavaria was chosen. In the original suspension, the clay particles form loose aggregates, as can be seen in the center of Fig. 2A. The addition of a small amount of the detergent (5% weight by weight), Fig. 2B, results in a compaction of the particles. Further addition of the detergent causes the formation of very dense structures, Fig. 2C. Here, the amount of added detergent is doubled, so that the compaction effect is clearly visible.

Detergents are used as soil conditioners. To investigate their influence on the inner structure of the aggregates formed from a naturally Ca$^{2+}$-coated montmorillonite, the effects of CTB or an anionic detergent sodium dodecylbenzene-sulfonate (LAS) were investigated. The compaction of the large aggregates obtained is demonstrated in Figs. 3A—C. The aggregates

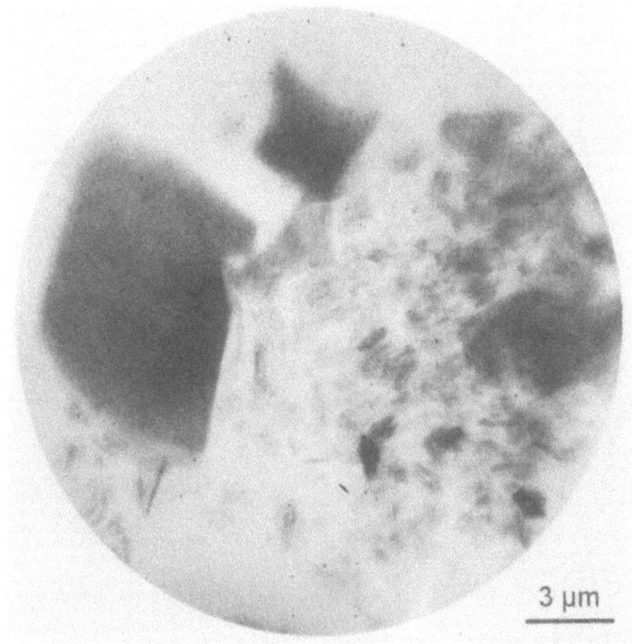

Fig. 2A Structures within 0.1% suspension of Na$^+$-montmorillonite (Moosburg, Bavaria)

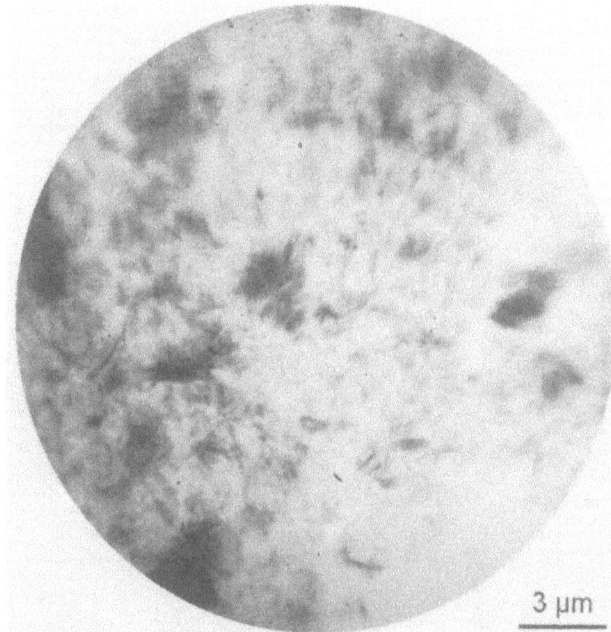

Fig. 2C Structures within 0.1% suspension of Na$^+$-montmorillonite (Moosburg, Bavaria) after the addition of 10% (weight by weight) CTB

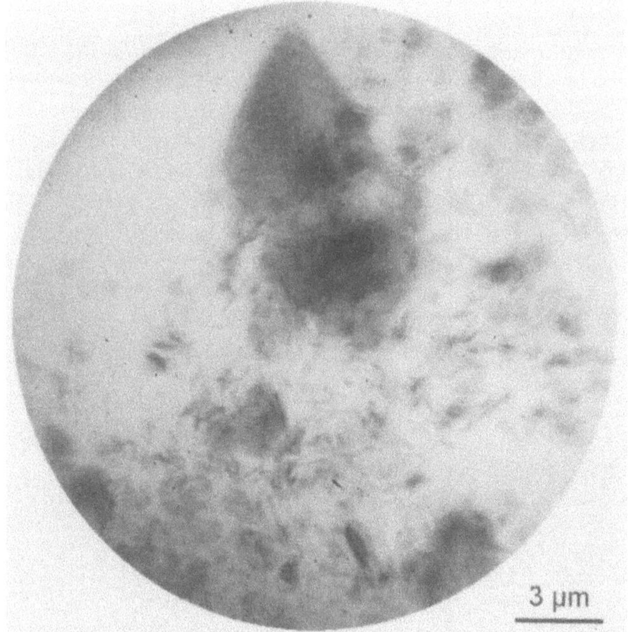
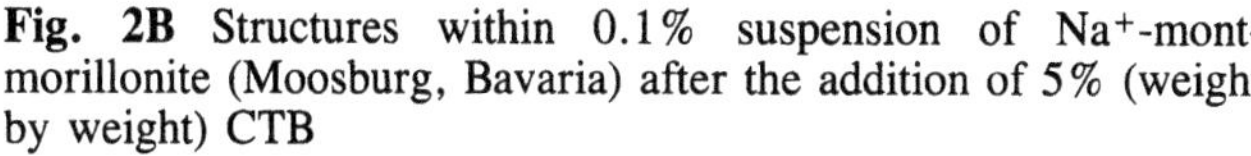

Fig. 2B Structures within 0.1% suspension of Na$^+$-montmorillonite (Moosburg, Bavaria) after the addition of 5% (weight by weight) CTB

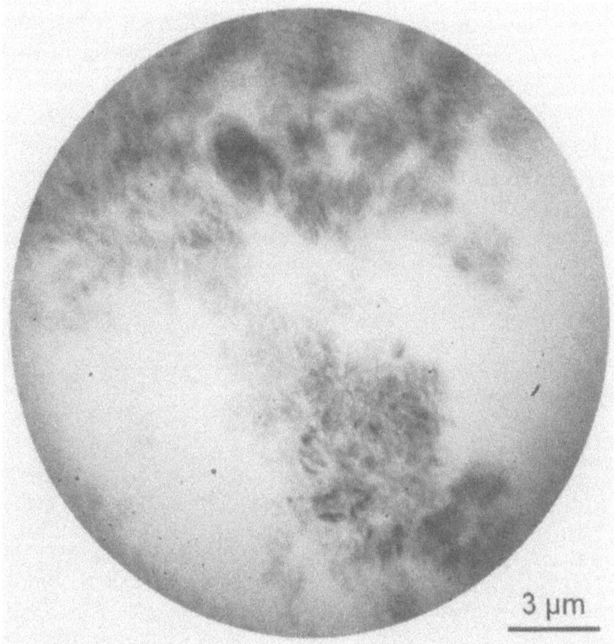

Fig. 3A Structures within 0.1% suspension of a naturally Ca$^{2+}$-coated montmorillonite (Moosburg, Bavaria)

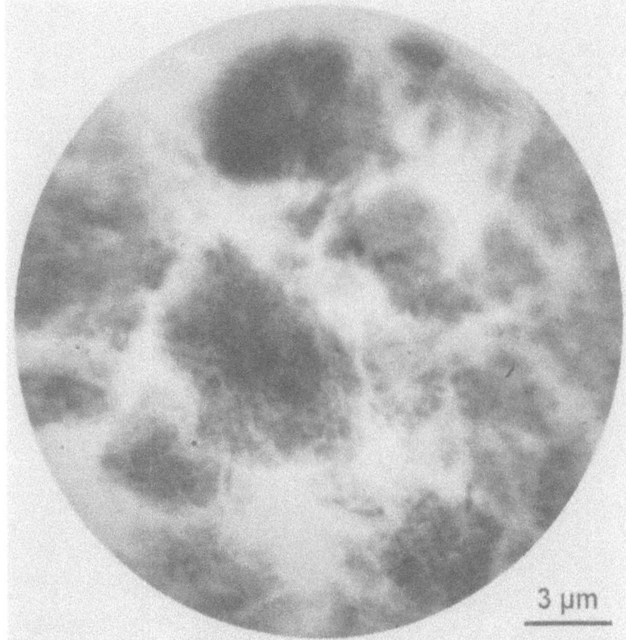



**Fig. 3B** Aggregates formed after the addition of CTB to 0.1% suspension of a naturally Ca$^{2+}$-coated montmorillonite (Moosburg, Bavaria)

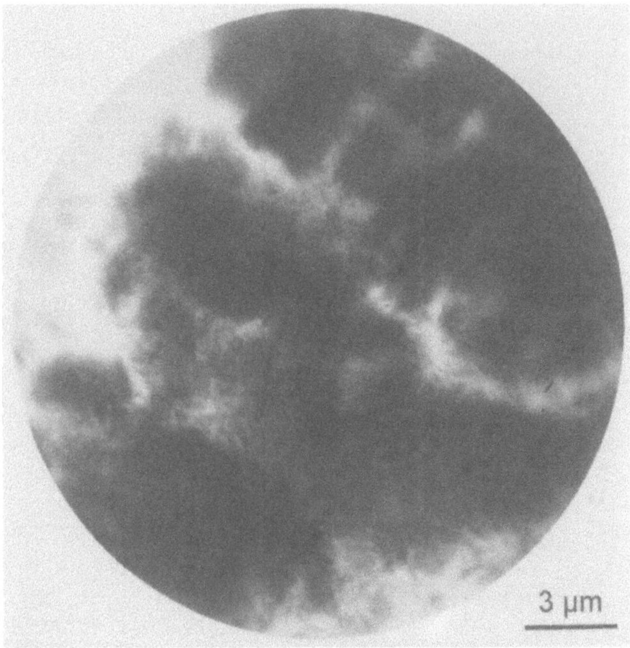

**Fig. 3C** Aggregates formed after the addition of LAS to 0.1% suspension of a naturally Ca$^{2+}$-coated montmorillonite (Moosburg, Bavaria)

observed in 1% clay suspension, Fig. 3A, have a very loose and open structure. This could result in an erosion due to wind or running water. The influence of the addition of CTB is clearly visible in Fig. 3B. The aggregates formed have a very compact and dense internal structure. In contrast to this, the addition of the anionic detergent, which still causes a coagulation of the suspension, also results in very large aggregates, but now with a more open structure, Fig. 3C. In soils, this type of structure is desirable, because it allows a good exchange of substances with the surrounding soil water.

## Summary

The behavior of montmorillonite suspension on the addition of sodium, on a cationic, and on an anionic detergent could be directly observed with x-ray microscopy. The internal structures of the aggregates formed by the addition of the surfactants could be clearly visualized. Therefore, it can be expected that x-ray microscopy will become a very useful tool in the investigation of coagulation phenomena.

## References

1. Sposito G (1989) The chemistry of soils. Oxford University Press
2. Mills WB, Liu S, Fong FK (1991) Ground Water 29:199—208; Michel R, Chenu C (1992) In: Stotzky G, Bollag J-M (eds) Soil biochemistry. Marcel Dekker, New York
3. Oades JM (1990) In: De Boodt MF, Hayes MHB, Herbillon A (eds) Soil colloids and their association in aggregates. Plenum Press, New York, pp 463—484
4. Burns RG (1990) loc cit pp 337—361
5. Vali H, Bachmann L (1988) J Colloid Interface Sci 126:278—291
6. Schmahl G et al. (1993) Optik 90:95—102
7. Thieme J et al., this volume

Progr Colloid & Polym Sci (1994) 95:143—152
© Steinkopff Verlag 1994

E. Koglin
B. Laumen
E. Borgarello
G. Borgarello

# Competitive and displacement adsorption of cetylpyridinium chloride and p-nitrophenol on charged surfaces: A surface-enhanced Raman microprobe Scattering (SERS) study

E. Koglin (✉) · B. Laumen
Institute of Applied Physical Chemistry
(IPC)
Research Center Jülich (KFA)
52425 Jülich, FRG

E. Borgarello · G. Borgarello
Colloid Science Group
Eniricerche S.p.A.
20097 Milano, Italy

**Abstract** Micro-surface-enhanced Raman scattering (micro-SERS) has been used to study, in situ, the competitive and displacement adsorption of cetylpyridinium chloride (CPC) and p-nitrophenol (PNP) on charged silver micro particle surfaces. The investigations have shown that this surface spectroscopy is an extremely powerful technique for monitoring the interfacial behavior of CPC and PNP, for assessing adsorbate orientation, displacement kinetics, competitive adsorption and for probing the effect of various environmental factors on the adsorbate/substrate interaction. Considerations of surface selection rules suggest that the CPC surfactant molecule is adsorbed with the ionic head group towards the charged surface and the long hydrophobic tail is directed away from the surface. Coadsorbed halide ions are very important for the binding in the first monolayer. The positively charged nitrogen atom of the pyridinium group is surrounded by chloride ions and this pyridinium/chloride surface complex is strongly adsorbed even on the positively charged surface. The p-nitrophenol is transformed to p-nitrophenolate when adsorbed on the charged surface. Therefore, the p-nitrophenol is bound to the surface through the ionized Ph-O$^-$ group by releasing its proton and is oriented perpendicularly to the surface with the $NO_2$ group extending into solution. The competitive coadsorption measurements have shown that the adsorption process depends greatly on the surface charge, the relative concentration of the CPC/PNP mixture and the counterion concentration. The kinetic displacement studies have demonstrated that the cationic surfactant is able to remove pre-adsorbed p-nitrophenol in a second-time scale.

**Key words** Micro-SERS — solid/liquid interface — cationic surfactants — organic pollutants — in situ determination — first monolayer adsorption

## Introduction

The synergistic coadsorption of surface-active substances and organic pollutants with soils or other natural adsorbents in aqueous environment is a mechanism of fundamental importance in determining the fate and distribution of these substances in the environment. In addition, even small amounts of surfactants produce major effects with respect to the transport and deposition of pollutants in soils and sediments. Another major environemental problem and technological challenge is the removal of hazardous chemicals from contaminated soil. The in situ investigation of the adsorption kinetic of organic pollutants, the interaction with surfactants, and the replacement of contaminants by means of surfactants is an open field, still largely unexplored, mainly because of a lack of an appropriate experimental technique for determining the adsorption process in the first monolayer. Adsorption of single surfactants is already a complex process involving several physico-chemical factors. The phenomena become more complicated when different types of molecules compete for adsorption at a solid/liquid interface.

A new spectroscopic method that appears to be promising for answering some basic questions regarding surfactant adsorption, coadsorption, and displacement reactions on charged model surfaces is Surface-Enhanced Raman Scattering (SERS) spectroscopy [1—6]. This enables researchers to characterize in situ the chemical identity, structure, orientation, and chemical reaction of species adsorbed at surfaces.

The purpose of this contribution is to utilize the SERS microprobe analysis for obtaining fundamental information on the adsorption mechanism of cationic surfactants, p-nitrophenol, coadsorption, and the removal of this priority pollutant in the first adsorption layer.

The CPC and PNP molecules were chosen because of the following reason: i) Cationic modified clay adsorbents (organo-clays) are eminently suitable for the removal of organic contaminants from ground water and industrial waste water. Therefore, to determinte the optimal use of organo-clays for environmental application the fundamental understanding of adsorption processes of cationic surfactants on charged surfaces should be first addressed. ii) Nitroaromatic compounds are widely used as herbicides and insecticides, as intermediates in the synthesis of pesticides and dye molecules, as solvents, and as explosives. They have been found to be ubiquitous environmental pollutants, particularly in subsurface environments. iii) The cationic surfactant and the nitrophenol display an opposite polarity in the adsorbed state so that the competition between the two molecules will be important to study and to explain.

The model substrate for the adsorption studies consisted of a two-dimensional colloid-like layer of randomly distributed submicroscopic metallic bumps (nanostructured surfaces) on a flat conductive silver surface.

## Experimental

Cetylpyridinium chloride $(C_{16}H_{33}H^+C_5H_5Cl^-)$ and 4-nitro-phenol were delivered by FLUKA-Chemie AG in the purest form. All other chemical reagents were of analytical quality from E. Merck (Darmstadt, FRG).

A SERS apparatus consists basically of a laser excitation source, the potential-controlled electrochemical cell, the optics for collecting the surface scattering, the computercontrolled spectrometer (double or triple monochromator), the photon-counting electronic or multichannel detection system and the display unit.

Raman microprobe spectroscopy has become an established technique in the past decase for particulate and microcontaminant analysis, as well as for molecular identification of the constituents of spatially heterogeneous natural and synthetic materials. To our knowledge, there are few reports in the literature of surface Raman microprobe spectra which invoke electrode/SERS spectroscopy.

The use of a microscope as an integral part of the SERS optical system provides precise control over focusing the laser light. For objectives with high angular aperture, the intensity of the incident laser radiation decrease dramatically outside the focal-point region. Using a laser Raman microscope to obtain vibrational information about electrode processes offers unique capabilities in terms of precise placement of the laser beam and in terms of rejection of significant Raman scattering outside the region of the laser focal point (distinction between the adsorbed state and solution molecules).

Our spectroelectrochemical cell consisted of a quartz glass cylinder with a diameter ca. 1 cm and a capacity of ca. 1 cm$^3$ of solution. The working electrode was a polycrystalline Ag metal disk, ca. 2 mm in diameter, enclosed in a Teflon holder. This metal electrode was prepared before each experiment by mechanical polishing, ultrasonic treatment and electrochemical cleaning by $H_2$ evolution.

The electrochemical equipment consisted of a potentiostat (PAR, model 173) using a three-electrode system and a function generator (PAR, model 175) as programmer for the ORC. Furthermore, a digital Coulomb meter was connected to measure the charge transfer during the ORC, which indicates the degree of metal dissolution and recrystallization of the electrode.

Surface-enhanced Raman signals from the electrode surface can be observed using different pretreatment procedures: a) in situ roughening. The surface is pre-treated by running the ORC in the electrolyte solution containing the molecule studied. b) ex situ roughening. The electrode roughening procedure is carried out in the electrolyte solution in the absence of the surfactant/nitrophenol molecule and the solution is then added to the optoelectrochemical cell. After both roughening procedures the silver surface is composed of randomly distributed nodular deposits in a nm-scale. To avoid conformational changes of large molecules during the electrochemical cycle procedure b) should be preferred. Potential measurements were made with respect to the saturated calomel reference electrode (SCE).

Micro-electrode/SERS spectra were obtained with an Instrument S. A. MOLE-S-3000 spectrometer. The MOLE-S-3000 is a new, fully computerized, triple spectrometer system with multichannel (E-IRY 1024) acquisition of data. The excitation sources were the lines of an argon-ion laser or an He-Ne laser (Spectra physics, model 2020-03 and 127). The actual laser focus used in obtaining the reported spectra was ca. 1 μm. The more modern and ideal instrumentation for electrode-chemical micro-SERS investigations is the multichannel detection system. In the study of the potential dependence of SERS during an ORC (Oxidation Reduction Cycle) at 5 mV s$^{-1}$, for example, the multichannel system allows rapid acquisition of several Raman peaks every 5—10 s. Furthermore, since the multichannel system covers a spectral range of ca. 1600 cm$^{-1}$ with a 600 grooves mm$^{-1}$ holographic grating, variations in the relativge intensities of micro-SERS peaks, shifts in the Raman frequencies, and changes in the line shape of SERS bands can all be observed at the same time the surface potential is linearly ramped.

## Results and discussion

### Raman and SERS Studies of CPC and PNP

#### Raman microprobe analysis of crystallized CPC

In order to utilize micro-SERS spectra to study surfactant/organic pollutant adsorption processes on charged surfaces, it is necessary to have reasonable good assignments of the Raman vibrational modes. For comparison purposes the microprobe spectrum of a microparticle of CPC is shown in Figs. 1, 2. The assignments are based on extensive comparisons with vibrational spectra of long chain molecules (polymers) and surfactants with different head groups (7—9).

Prominated head group ($-N^+C_5H_5$ pyridinium ring)-related bands in the crystalline CPC molecule are the $\nu_{6b}(A_1)$ in plane ring deformation vibration at

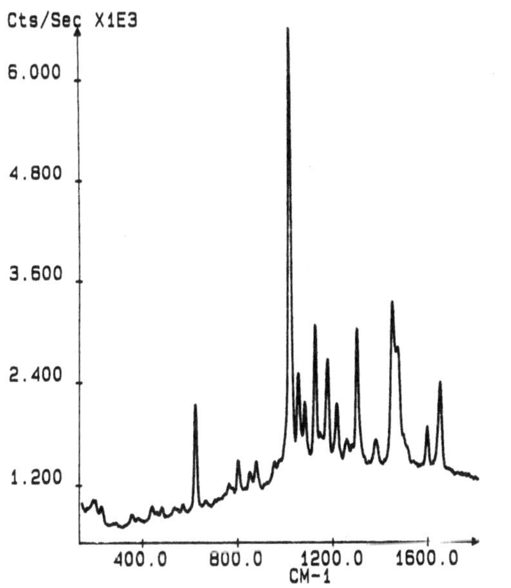

**Fig. 1** Raman microprobe analysis of cetylpyridinium chloride (CPC): Raman spectrum of a microparticle, particle size about 20 μm · 30 μm, $\lambda_0 = 514.5$ nm, 6 mW (at sample), beam spot 1 μm diameter, integration time 2 s, number of readings 30

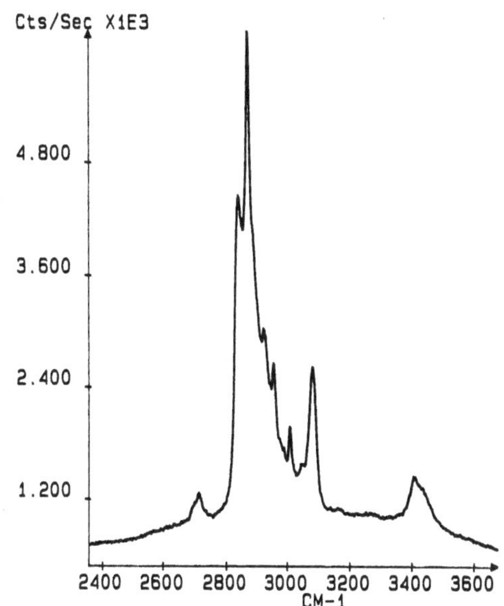

**Fig. 2** Raman microprobe analysis of cetylpyridinium chloride (CPC) in the C—H frequency range. Experimental conditions as in Fig. 1

651 cm$^{-1}$; the intense $\nu_1(A_1)$ symmetrical and trigonal ring breathing mode at 1028 cm$^{-1}$; the $\nu_{8a}(A_1)$ ring stretching vibration at 1638 cm$^{-1}$; and the CH ring stretching mode at 3084 cm$^{-1}$. The characteristic tail group vibrations are located around 1062, 1132, 1301, 1458, 2852 and 2881 cm$^{-1}$. The bands at 1062 and 1132 cm$^{-1}$ have been assigned to C—C symmetric and anti symmetric vibrations, respectively and represent the trans (T) conformation of the $C_{16}$ alkyl chain bound to the pyridinium head group. The only band of strong intensity in the mid-frequency range which can be unequivocally assigned to the CH$_2$ group is the twisting motion at 1301 cm$^{-1}$. All the other bands for CH$_2$ vibrations are in ranges that overlap those of the hydrophobic CH$_3$ terminal vibrations. The strong CH$_2$ scissoring motion at 1458 cm$^{-1}$ is also in an area where CH$_3$ bending and CH$_2$ bending vibrations both contribute. But for the CPC long-chain surfactant molecule this dominant band would be automatically assigned to CH$_2$ scissors. The bands at 2852 and 2881 cm$^{-1}$ have been assigned to CH$_2$ symmetric and anti symmetric modes respectively for different long-chain molecules with a hydrophobic (CH$_2$)$_n$-tail.

As a result, head group "markers" are at 1028, 1638 and 3084 cm$^{-1}$ and the "fingerprints" for the hydrocarbon chain are the bands at 1301, 1458, 2852 and 2881 cm$^{-1}$.

#### Micro-SERS spectroscopy of CPC

The potential dependence, the influence of the counterion concentration, and the substance concentration of the micro-SERS spectra will be discussed in sequence.

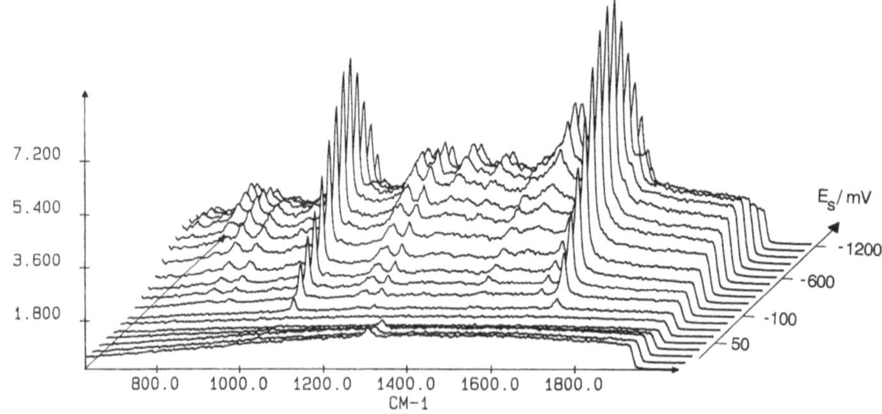

**Fig. 3** Micro-SERS spectra recorded as the Ag electrode surface is cycled through an ORC at 5 m V s$^{-1}$ from —0.6 V to +0.2 V and back to —1.2 V in 0.1 M KCl plus 2 · 10$^{-5}$ M cetylpyridinium chloride. Laser excitation line 514.5 nm, laser power 10 mW, laser spot focus 1 μm

*Potential dependence:* The use of an optical multichannel analyzer for recording micro-SERS spectra not only has the advantage of fast recording of a single spectrum, but also allows successive spectra to record as a function of electrode potential for three-dimensional recordings of SERS intensity-wavenumber-potential plot. Figure 3 shows the micro-SERS spectra in the wavenumber range between 400 cm$^{-1}$ and 2000 cm$^{-1}$ at different potentials. There are two intensity maxima of the SERS vibrational modes in the potential range from 0.0 V to —1.0 V. The potential of 0.0 V corresponds to a very positively charged surface and the potential at —0.6 V is the neutral surface (point of zero charge —p.z.c.—). The first maximun is around 1030 cm$^{-1}$ and this SERS band can be attributed to the enhanced $v_1(A_1)$ ring breathing mode of the pyridinium head group. The second SERS maximum is at 1626 cm$^{-1}$ and is assigned to the $v_{8a}(A_1)$ ring stretching vibration.

Comparing the relative intensities of the micro-SERS spectra and the microprobe normal Raman spectrum, we can clearly see that the enhancement of the pyridinium ring vibrations and the hydrocarbons modes are quite different. The characteristic tail vibration at 1301 cm$^{-1}$ (chain $CH_2$ twisting motion) is completely absent in the adsorbed state. The $CH_2$-scissoring vibration at 1452 cm$^{-1}$ shows a very low intensity. For a better understanding of these intensity changes in the ORC-SERS spectra, the spectrum at —0.6 V is shown as an example in Fig. 4. On the basis of the short-range sensitivity of the SERS enhancement (10), we can conclude that the pyridinium head group is attached to the surface, leaving the hydrocarbon chain directed away from the surface. Perhaps the most striking evidence for this molecular orientation with the pyridinium ring adsorption at the neutral surface is the SERS behavior in the $v$(C—H) region, shown in Fig. 5. The pyridinium CH-ring stretching mode at 3080 cm$^{-1}$ is more enhanced relative to other SERS bands in this spectral range. The strong intensity of this CH-vibration, the enhanced ring breathing mode, and the intense ring stretching band at 1626 cm$^{-1}$

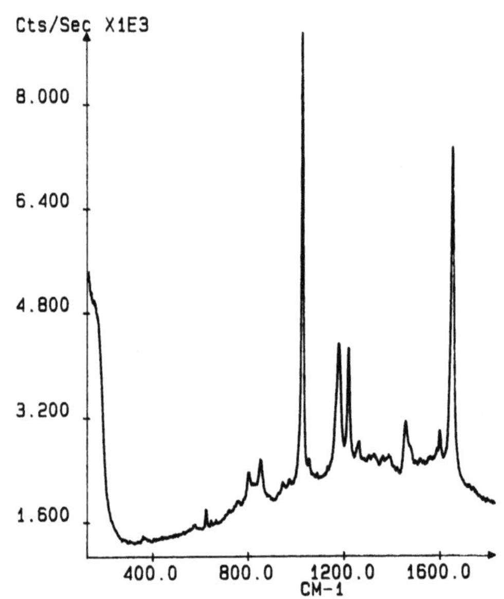

**Fig. 4** Micro-electrode SERS spectrum of cetylpyridinium chloride. Ag electrode, ex situ roughening, $E_s$ of —0.6 V vs. SCE, concentration 5 · 10$^{-5}$ M, $\lambda_0 = 514.5$ nm, 10 mW, integration time 1.5 s, number of readings 60

suggest that the pyridinium head group is in a standing configuration (edgewise orientation) [3, 1]. These results are in agreement with the investigations of Sun et al. [8].

By changing the potential to more negative values, e.g., —1.0 V the most intensive band in the SERS spectrum is now the mode at the frequency of 1518 cm$^{-1}$ (cf. Fig. 6). This band can also be assigned to a totally symmetric stretching vibration $v_{19a}(A_1)$. The strong enhancement of this mode at the negatively charged surface can be explained by the surface resonance Raman selection rules (electronic SERS contribution). At this negative potential the Fermi level of Ag metal comes in the region where the laser incident energy ($\lambda_{ex} = 514.5$ nm) can give this energy to an electron and a charge transfer from the metal to the affinity level of the adsorbed CPC

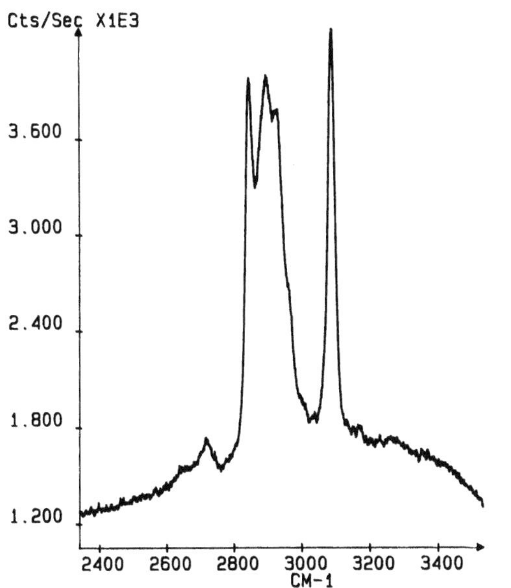

**Fig. 5** Micro-SERS spectrum of cetylpyridinium chloride in the C—H frequency range. Experimental conditions as in Fig. 4

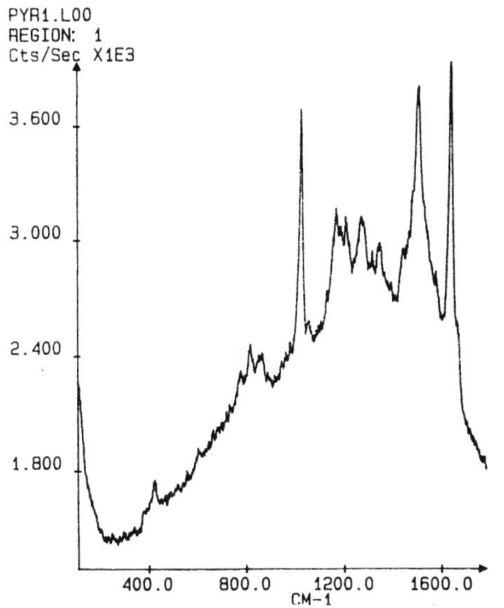

**Fig. 6** Micro-SERS spectrum of cetylpyridinium chloride. Surface potential —1.0 V vs. SCE. Other experimental conditions as in Fig. 4

molecule is possible. In this case, we have a strong enhanced intensity about the A-term resonance mechanism [11], especially for this $v_{19a}(A_1)$ charge transfer band (CT band). This electronic resonance is a short-range phenomenon which is characteristic of the adsorbed CPC surface complex stabilized by coadsorbed $Cl^-$ ions.

### The effect of counter ions:

In order to further elucidate the nature of the CPC/halide/complex, the effect of chloride concentration on

the adsorption process was studied. The cetylpyridinium concentration was kept constant at $10^{-3}$ M and the chloride concentration was varied from 0.01 M KCl to 1 M KCl aqueous solutions. The ex situ pretreatment was carried out in a very low electrolyte concentration and this pretreatment solution was replaced by purified water. Increasing the specifically adsorbing $Cl^-$ concentration from 0.1 M to 1 M affects the adsorption of the active surface complex and markedly changes the SERS spectra in the whole potential range from —0.1 to —1.2 V. Considering the CPC SERS spectrum in the 1 M KCl solution, the CT band at 1518 $cm^{-1}$ is significantly enhanced at the positive, neutral and negative charged surface. Rojhantalab and Richmond (12) observed in SHG (optical second harmonic generation) experiments that by increasing the concentration of $Cl^-$, the surface charge density increase while the maximum in the surface excess charge versus potential curve shifts to negative potentials. These SHG results are in agreement with our SERS investigations: at high concentrations of the halide supporting electrolyte ions the SER-scattering of the characteristic CT band is strongly increased. Therefore, it is possible that the chloride acts as a bridge for electron transfer between the metal surface and the CPC molecule in the charge transfer type enhancement mechanism [13].

Besides this charge-transfer effect, the orientational change of the adsorbed molecules might be correlated with the position of the energy difference between the ground state of the CT-complex and the charge-transfer band [11]. This orientational change in the adsorbed state at high $Cl^-$ concentration and at a positive surface charge results in intense SERS bands of the two characteristic $(CH_2)_n$-tail vibration: the $CH_2$ twisting mode and the $CH_2$ scissoring vibration. Under these conditions the CPC headgroup is in close proximity to the surface and C—C bonds of the long chain stretch tortuously on the surface with some parts of the chain close to the surface.

### The effect of CPC concentration:

The SERS spectra of CPC molecules from an aqueous solution containing 0.1 M KCl have been obtained over the concentration range of $1 \cdot 10^{-3}$ to $1 \cdot 10^{-5}$ M. This concentration range encompassed the CPC's critical micelle concentration (CMC) which is known to be $2.4 \cdot 10^{-4}$ M. No strong structural adsorption changes appear to take place with increasing the surfactant concentration from the monomer to micelle conformation. That means that the CPC's packing/orientation in the first monolayer does not change with increasing surface coverage. The onset of micellization does not appear to cause any discontinuity in the SERS spectra in the potential range from —0.1 to —1.2 V. These results are in agreement with the SERS investigations on cetyl trimethylammonium bromide (CTAB) [14].

## Raman microprobe analysis of crystallized PNP

The microprobe Raman spectrum of p-nitrophenol will be discussed first, to provide assignments for the surface species in the first monolayer. Figure 7 shows the Raman spectrum of a microparticle of PNP. Traditional notation is used to describe the vibrations of a substituted phenole molecule. Stretching vibrations are denoted by $v$, and in-plane bending or deformation by $\beta$. Symmetric deformation modes involving the $-NO_2$ substituent are denoted by $\delta$. A brief discussion of several characteristic bands is given below. The most intense feature in the spectrum appears at 1337 cm$^{-1}$ and is assigned as the $v_s(NO_2)$ stretch of the nitro group on the benzene ring which is strongly electron withdrawing and rotates freely about the Ph$-NO_2$ bond [15, 16]. The asymmetric $v_{as}(NO_2)$ stretching vibration of the nitro group occurs at 1504 cm$^{-1}$. The strong band at 872 cm$^{-1}$ is due to the deformation of the $NO_2$ group, and its shoulder at 809 cm$^{-1}$ (not shown in Fig. 7) is attributed to a ring-breathing mode. The $v(C-C)$ ring stretching mode is located at 1593 cm$^{-1}$. Another characteristic C$-$C stretching vibration with a very low intensity is located at 1458 cm$^{-1}$. The most important vibration of the Ph$-OH$ moiety is the $A_1$ symmetric stretching mode at 1286 cm$^{-1}$. This band shifts up to about 20 cm$^{-1}$ in an alkaline medium, indicating the phenoxide ion vibration $v(Ph-O^-)$.

## Micro-SERS spectroscopy of PNP

There are many changes in the PNP micro-SERS spectra as the surface is changed between positive, neutral, and

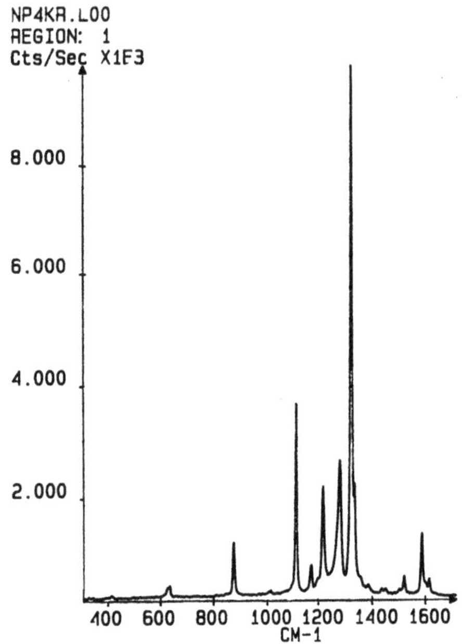



**Fig. 7** Raman microprobe analysis of p-nitrophenol: Raman spectrum of a microparticle, particle size about 10 μm · 20 μm, $\lambda_0$ = 514.5 nm, 1 mW (at sample), beam spot 1 μm diameter, integration time 1 s, number of readings 10

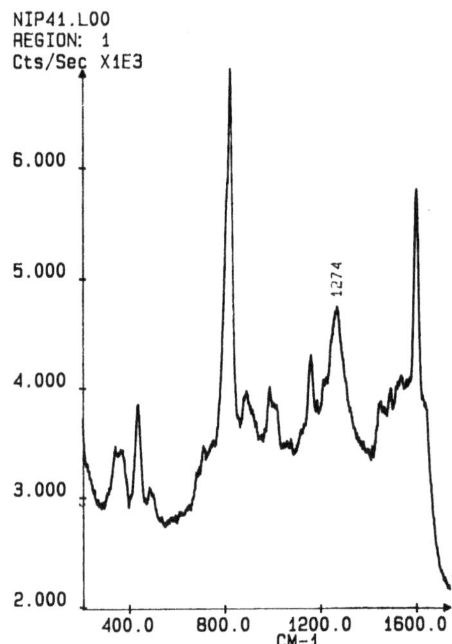



**Fig. 8** Micro-SERS spectrum of p-nitrophenol. Ag electrode, ex situ roughening, $E_s$ of $-0.6$ V vs. SCE, concentration 2 · $10^{-3}$ M, $\lambda_0$ = 514.5 nm, laser power 4 mW, laser spot focus 1 μm, integration time 1.5 s, number of readings 60

negative charges. For example, as the potential is changed from $-1.3$ V to $-0.7$ V, the ring breath mode at 848 cm$^{-1}$ and the ring stretching vibration at 1582 cm$^{-1}$ decrease in intensity. At $-0.3$ V the band at 1465 cm$^{-1}$ (very weak $v(C-C)$ stretching mode at 1458 cm$^{-1}$ in the microprobe Raman spectrum) begins to dominate the SERS spectrum, particularly at $-0.01$ V. As an example for a characteristic SERS spectrum of PNP the spectrum at $-0.6$ V (potential of zero charge-pzc-) is shown in Fig. 8. Competitive investigations of protonated and deprotonated nitrophenol micro-SERS measurements have shown that the PNP molecule is transformed to p-nitrophenolate when adsorbed from the neutral 0.1 M KCl aqueous solution. The SERS spectrum obtained from the solution that was adjusted to pH 12 with KOH shows the same vibration bands as in neutral solution. In Table 1 the surface Raman bands are compared with the Raman microprobe vibrations and the solution Raman characteristics at pH 5 and pH 12.

The preferential binding of the Ph$-O^-$ group to the surface over the $NO_2$ group is also demonstrated by the complete absence of the $v(NO_2)$ vibrational band (cf. Fig. 8). Therefore, under these conditions the nitrophenolate ring would be oriented perpendicularly to the surface with the $-NO_2$ group extending into solution. The strong increase of the ring breathing mode at 848 cm$^{-1}$ and the intense ring stretch SERS band at 1582 cm$^{-1}$ support this end-on orientation.

**Table 1** Vibrational data (wavenumber cm$^{-1}$) and corresponding assignments for p-nitrophenol (PNP)

| Micro Ra crystal | PNP in KCl 0.1 M, pH 5 | PNP in KCl 0.1 M, pH 12 | SERS Ag, —0.6 V | SERS Ag, —1.1 V | Vibrational assignment |
|---|---|---|---|---|---|
| — | — | 370 w | 373 w | 375 m | |
| 438 m | 457 w | — | 464 w | 467 m | ring vibration |
| 636 s | 640 w | 634 w | 636 m | 631 w | C—H o.-p. deformation |
| **809 w** | — | **822 sh** | **848 vs** | **847 vs** | **ring breathing mode** |
| 872 s | 868 s | 858 vs | — | — | $\delta(NO_2)$ |
| 968 w | 1009 m | 998 w | 1020 m | 1026 m | i.p. ring deformation |
| 1172 m | 1174 m | 1176 m | 1162 vw | 1164 m | $\beta(C—H)$ |
| **1286 s** | **1290 s** | **1294 vs** | **1278 s** | **1268 s** | $\nu(Ph—O^-)$ |
| 1337 vs | 1344 vs | 1338 m | 1346 vw | — | $\delta_s(NO_2)$ |
| 1504 m | 1508 m | 1529 m | — | — | $\nu_{as}(NO_2)$ |
| **1593 s** | **1598 s** | **1586 w** | **1582 s** | **1581 s** | **ring stretch** |

Intensities: m = medium, w = weak, s = strong, vw = very weak, vs = very strong, sh = shoulder

The coadsorbed K$^+$ cations are very important for the strong p-nitrophenolate binding at the negative charged surface. The negative charged Ph—O$^-$ group is surrounded by potassium ions and this phenolate/K$^+$ complex can strongly adsorb on the negatively charged surface.

On the basis of the above results, the micro-SERS "marker" is at 1465 cm$^{-1}$ for the positively charged surface and the "fingerprints" at the neutral and negative surface are the SERS bands at 848 cm$^{-1}$ and 1582 cm$^{-1}$. Together, these three bands provide an excellent indication of the structural rearrangement possibilities in the surfactant/nitrophenol monolayer system.

Simultaneous Adsorption: Competition between CPC and PNP

Adsorption of single surfactants or nitrophenol is already a complex process involving several physico-chemical factors. The phenomena become even more complicated when different types of molecules compete for adsorption at the interface. With respect to competitive adsorption, distinction has to be made between simultaneous adsorption and displacement reactions. SERS spectroscopy is probably one of the most sensitive and chemically specific methods for investigating the processes at the solid/liquid interface because the signals of the coadsorbed CPC and PNP molecules can be observed simultaneously in the first monolayer.

Synergetic coadsorption of cetylpyridinium chloride and p-nitrophenol have been studied as a function of CPC and PNP in aqueous KCl solutions, from 1000:1 to 1:1000, the surface charge of the substrate from positive to negative charged surfaces, and the presence of different concentrations of the Cl$^-$ counterions in the solution.

*Adsorption from CPC:PNP Mixtures*

Surface-enhanced Raman scattering is a surface phenomenon and only molecules close to the substrate surface show enhanced Raman spectra. In a mixture there will be competition for surface sites. If one species is more highly attracted to the surface than another the latter will not show a SERS spectrum. For this reason, we can use SERS spectroscopy for identifying and quantifying the individual adsorbed compound in the first monolayer of mixtures of CPC and PNP. The ex situ surface roughening procedure is carried out in a 0.1 M KCl electrolyte solution in the absence of other molecules and the binary mixture of CPC/PNP is then added to the optoelectrochemical cell. The detailed analysis of the system containing mixtures of CPC and PNP can be summarized as follow:

— For a mixture containing 10$^{-3}$ M each of CPC and PNP in a 0.1 M KCl solution it was found that there is no measurable amount of PNP at a negatively charged surface and that, instead, the CPC molecules are the dominant chemisorbed species.

— At a neutral or positive surface, the competitive adsorption of the two molecules becomes dependent on the coadsorbed Cl$^-$ ion concentration. With decreasing the KCl concentration to 0.01 M a decreasing in the intensity of CPC modes at 1030 cm$^{-1}$ and 1626 cm$^{-1}$ is observed. Simultaneously, new bands at 848 cm$^{-1}$ and 1582 cm$^{-1}$ develop, and this is ascribed to the corresponding vibrations of adsorbed nitrophenolate ions (cf. Fig. 9). Therefore, the coadsorbed halide ions are very important for the binding of cationic surfactants and organic pollutants in the first adsorption layer.

— Similar results are obtained if we study different relative ratios between the pyridinium ion and the p-nitrophenolate ions, from 1000:1 up to 1:1000 in

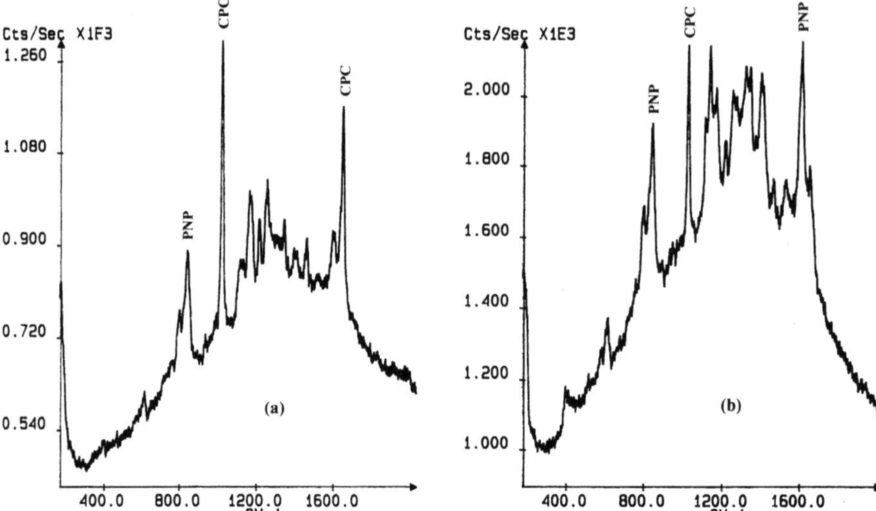



**Fig. 9** Micro-SERS spectra of CPC:PNP mixtures. Mixture containing $10^{-3}$ M each of CPC and PNP, ex situ surface roughening, $E_s$ of $-0.6$ V vs. SCE, laser excitation line 514.5 nm, laser power 8 mW, laser spot focus 1 μm.
a) SERS spectrum in a 0.1 M KCl supporting electrolyte solution
b) SERS spectrum in a 0.01 M KCl supporting electrolyte solution

moles per liter at different charged surfaces. As an example, in a mixture of $10^{-6}$ M of CPC and $10^{-3}$ M of PNP the pyridinium surfactant molecules are always very strongly adsorbed at high $Cl^-$ ion concentration in a potential range between $-1.2$ V and $-0.1$ V.

— If the KCl concentration is decreased to 0.01 M, also the adsorption of the CPC is reduced. At a positively charged surface the CPC signals have completely vanished and only the SERS bands of PNP appear. But, even at this very low CPC concentration of $10^{-6}$ M, in presence of 0.1 M of KCl the surfactant molecules are still adsorbed.

— All these experimental findings indicate the importance of $CPC/Cl^-/$surface complex formation in determining the role of stabilizing $Cl^-$ or other coadsorbed species on the positively charged metal surface.

Displacement reactions:
Competition between CPC and pre-adsorbed PNP

Adsorbed chemicals of environmental and biological interest can be desorbed by suitable displacers, either small molecules or surface active substances (surfactants).

To understand the details of such surface rearrangement following displacement reactions, it is first necessary to understand as fully as possible the kinetic behavior (on a second time scale) of molecules which are undergoing these surface desorption reactions in the first adsorption layer. Information from SERS experiments allows the temporal course of the displacement reaction and additional information about structures in the adsorbed state. Therefore, to examine the question of reorientation and desorption of PNP by CPC molecules, surface Raman spectra were collected in real time at different stages of adsorption.

The SERS activation of the Ag electrode surface was carried out ex situ, in pure supporting electrolyte (0.1 M

**Fig. 10** Micro-SERS spectra of displacement reactions of pre-adsorbed PNP by CPC.
a) Micro electrode SERS spectrum of adsorbed p-nitrophenol. Ag-electrode surface, ex situ roughening, $E_s = -0.6$ V vs. SCE, concentration $1.5 \cdot 10^{-3}$ M, $\lambda_0 = 514.5$ nm, laser power 5 mW, laser focus 1 μm, integration time 0.8 s, number of readings 10
b) Micro-SERS spectrum of the displacement reaction: 1 μl of $10^{-3}$ M CPC was added to the optoelectrochemical cell. The SERS spectrum was taken 15 s after injection of the displacing CPC molecules. Experimental conditions as in a)
c) Micro-SERS spectrum of the displacement process. The SERS spectrum was taken 90 s after injection of the CPC surfactant molecules. Experimental conditions as in a)

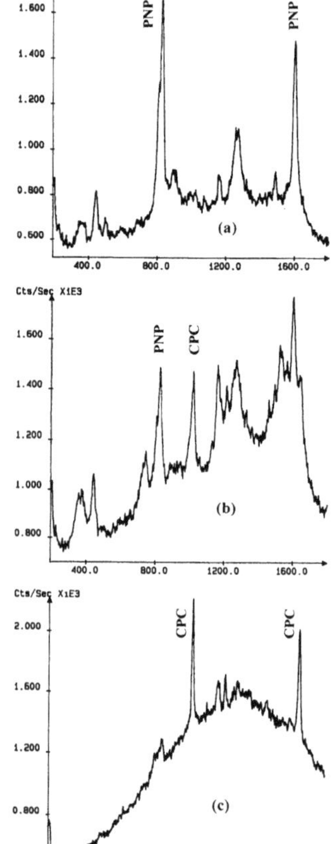



KCl). This electrolyte solution was removed and the same electrolyte with $10^{-3}$ M concentration of PNP was added. The SERS spectrum of PNP measured at $-0.6$ V is shown in Fig. 10a. 1 μl of a $10^{-3}$ M CPC solution was then added to the 1 ml optoelectrochemical cell volume. First excellent results can be obtained from SERS measurements even at early times. The first data point was taken 15 s after injection of the displacing CPC molecules (cf. Fig. 10b). The partial disappearance of the PNP bands at 848 $cm^{-1}$ and 1582 $cm^{-1}$ indicates a characteristic interaction between CPC and PNP on the surface. As well as the marked decrease in the PNP surface scattering intensity the coadsorption effect of CPC and PNP is clearly demonstrated by the strong scattering intensity of the CPC vibration modes at 1032 $cm^{-1}$ and 1624 $cm^{-1}$. CPC is able to displace a fraction of the pre-adsorbed PNP in the first 15 s after the injection. The pre-adsorbed PNP fraction decreases with increasing the rearrangement time, which suggests that CPC becomes more strongly attached at the surface. After 90 s a complete displacement of the pre-adsorbed PNP has already taken place (cf. Fig. 10c).

With the information obtained from this measurements of surface displacement kinetics, the question of which interaction dominates in the adlayer can be obtained. Physically, one can describe the desorption of the nitrophenolate ions through two main structural models. The first one in which the CPC molecule bonds with the pyridinium head group to the surface, but in which the hydrocarbon tails remain in the electrolyte solution, and in this state there is a competition with the PNP molecules. In the next step, the cationic surfactant molecules align and van der Waals interaction between adjacent tails becomes important until a complete CPC monolayer is formed: the pre-adsorbed nitrophenolate ions are completely removed from the first monolayer.

If the displacement time of PNP from the surface is also electrostatically determined, it would be expected that the PNP desorption kinetic can be strongly influenced by the electrode surface charge. It appears, therefore, that electrostatic considerations of the PNP adsorption on the charged metal surface play a central role in determining the desorption time of PNP by CPC. Electrochemical measurements show that ionic molecules like the nitrophenolate ions generally adsorb very weakly on an electrode surface of the p.z.c. These investigations are in agreement with the very fast desorption time of 90 s at this electrode potential in the SERS replacement experiments as shown in Fig. 10. After pretreating of the PNP adsorption at $-0.6$ V the surface potential was scanned in the negative direction from $-0.6$ V to $-1.2$ V. At this potential the negatively charged Ph—$O^-$ group of the PNP molecule is surrounded by potassium ions and this phenolate/$K^+$/Ag surface state is strongly adsorbed on the negatively charged surface. This strong PNP adsorption process results in a longer desorption time in the CPC displacement investigations: if the pre-adsorbed PNP molecules are adsorbed at the negative surface the complete replacement time takes 3 min.

The strong potential dependence of the desorption process can be observed when the potential is shifted to $-0.1$ V. At this highly positively charged substrate surface the adsorption of the nitrophenolate ions increase dramatically by the electrostatic attraction to the surface and we obtain a high population of nitrophenolate ions that are strongly attached to the surface. As mentioned before, the displacement of the pre-adsorbed PNP by the secondly added CPC is strongly influenced by the electrostatic interaction between the adsorbent surface and the respective surfactant. At positive potentials to the point of zero charge, cationic surfactants can desorb from the surface because of the electrostatic repulsion. Specifically, coadsorbed $Cl^-$ counterions near the positively charged surfactant head group are necessary to explain the strong binding of cationic surfactants on positively charged surfaces. Therefore, the adsorption, rearrangements, and desorption of PNP molecules from positively charged surfaces follow a more complicated displacement mechanism. The SERS desorption measurements at this surface charge are quantitatively similar to those at the neutral and negatively charged surfaces, but everything now occurs much more slowly: the CPC adsorbs at a much slower rate and the nitrophenolate ions leave the surface much more slowly. These slower desorption kinetics result in a complete replacement time of 9 min.

## Conclusions

In summary, the utility of micro-SERS spectroscopy for the evaluation of potential-dependent interfacial competititve and displacement reactions at charged surface has been demonstrated. The data obtained allow the determination of the chemical identity, structure, orientation, competitive and displacement adsorption of cationic surfactants and nitrophenol in the first adsorption layer. The examples of these measurements in the field of surfactants and organic pollutants reviewed in this article were selected to illustrate the sensitivity, molecular specificity of adsorption processes, accuracy, ease of substrate preparation, and manifold applications of Raman analysis. The spatial resolution of the laser microprobe, coupled with the $10^6$ enhancement of the Raman cross-section, means that picogram quantities of material localized to $\mu$m-sized surfaces areas can be detected and identified by SERS vibrational spectroscopy.

**Acknowledgment** The authors are indebted to M. J. Schwuger for his continuous encouragement and for valuable discussions. We also gratefully acknowledge S. Dohmen for her skillful assistance.

## References

1. Otto A (1983) In: Cardano M, Güntherodt G (eds) Light scattering in solids IV:89—411
2. Chang R, Laube B (1984) CRC Crit Rev Mater Sci 12:1—74
3. Moskovits M (1985) Rev Mod phys 57:83—804
4. Koglin E, Sequaris J-M (1986) In: Dewar M, Dunitz J, Hafner K, Heilronner E, Lehn J-M, Niedenzau H, Raymond K, Rees C, Vögtle F, Wittig G (eds) Topics in Current Chemistry 134:1—53
5. Garrell R (1989) Ana Chem 61:401—411
6. Otto A (1991) J Raman Spectrosc 22:743
7. Dendramis A, Schwinn L, Sperline E (1983) Surf Sci 134:675—688
8. Sun S, Birke R, Lombardi J (1990) J Phys Chem 94:2005—2010
9. Bunding K, Bell M, Durst R (1982) Chem Phys Let 89:54—58
10. Koglin E (1983) J de Physique, Colloque N° 10, C10:487—490
11. Creighton J (1988) In: Clark R, Hester R (eds) Spectroscopy of surfaces 2:37—64
12. Rojhantalab H, Richmond G (1989) J Phys Chem 93:3269—3275
13. Roy D, Furtak T (1984) J Chem Phys 81:4168—4175
14. Koglin E, Krug O (1992) Kiefer W, Cardona M, Schaack G, Schneider F, Schrötter H (eds) INCORS Würzburg '92, pp 650—651
15. Kumar K, Carey P (1975) J Chem Phys 63:3697—3707
16. Ni F, Thomas L, Cotton T (1989) Anal Chem 61:888—894

Progr Colloid & Polym Sci (1994) 95:153—160
© Steinkopff Verlag 1994

Hermann H. Hahn

# Chemical and physical aspects of coagulation

Hermann H. Hahn
Institut für Siedlungswasserwirtschaft
Universität Fridericiana
76128 Karlsruhe, FRG

**Abstract** It is the aim of this contribution to identify situations where predominantly physical variables affect or even control the coagulation process in a chemically controlled and optimized system. Observations in technical systems show that the mixing of coagulation chemicals can be slower or faster, depending upon the hydraulic characteristics of the mixing reactor. This leads to satisfactory destabilization in one case and unsatisfactory destabilization in another, in particular if metal ions or hydroxilized metal species are used. It has been shown that the energy dissipation in stirred real reactors leads to a nonhomogeneous distribution in terms of the absolute size of the (locally linear) velocity gradient, i.e., aggregation conditions. This will be superimposed by the characteristics of the coagulating chemicals, i.e., the ability to form stronger or less strong flocs. Data on the separation effectivity of geometrically different units for systematically varied boundary conditions of the coagulation process show that one type of floc is withheld more efficiently than another in the same separation unit and likewise that one kind of separation unit may be more efficient for a specific floc.

**Key words** Colloid destabilization with metal ions — colloid collision — shear flow — liquid-solid-separation — homogeneity and heterogeneity of coagulating systems

## Introduction

Coagulation, as described for instance by von Smoluchoski's concepts [1], is the result of a process-initiating group of more physicochemical reactions causing colloid destabilization and a second process step consisting of a group of predominantly physical transport phenomena. Investigations on the rate-limiting step have identified the transport reaction as controlling process in many instances. For other, very specific boundary conditions the destabilization reaction has been found to be the controlling one. Thus, in one instance coagulation has been described more in physical terms, and on other occasions it has been described as a predominantly chemically controlled process. This dichotomy of definitions and interpretations still exists today.

If coagulation is employed technically, for instance, for the improvement of the separation of solids (such as sedimentation, flotation or filtration) through aggregation, induced by technical means, then additional reaction steps and reaction parameters must be considered. These include the addition and mixing of the chemical reagent that must be dosed for the destabilization of the sol. Furthermore, desirable fluid velocity gradients must be created, mostly through controlled energy dissipation. And, finally, the relationship between the characteristics of the solids formed and the specifics of the large-scale separation reactor must be observed.

In the analysis of these reaction phenomena, one finds that there is a strong interaction of chemical and physical parameters, i.e., coagulation is then no longer a predominantly chemical process or is described mostly by physical parameters: coagulation in large-scale systems depends to a large extent on the simultaneous optimization of chemical and physical boundary conditions.

It is the aim of this discussion to identify situations where predominantly physical variables affect or even control the coagulation process. The fact that these phenomena have not been described extensively until now may be explained by the effect of optimizing or even maximizing chemical aspects: non-optimal physical boundary conditions for the reaction, resulting in a lower reaction rate, might be compensated by higher dosages of chemicals. Thus, the main objective of this contribution is the discussion of the "rate-controling" effects of physical parameters in a chemically controlled and optimized system.

## The interrelationship of chemical and physical phenomena in the effective mixing of coagulating reagents with the sol

Coagulation has been analyzed so far as the result of two major reaction steps, the destabilization of the colloids and the transport and adherence of the destabilized dispersed solids to form permanent aggregates. And this model suffices to explain all observations as long as coagulation is studied in a laboratory context, i.e., small-volume batch reactions.

As soon as coagulation is explored in a technical context, under conditions of continuous flow and in larger reactor vessels, then the addition and mixing of the destabilizing reagent must be considered an additional and possibly rate-controlling step.

This may be illustrated by the following example: If one wants to effectively coagulate a suspension under the conditions of a large (hydraulic) flux, one will have to be careful not to use too large a volume of liquid (i.e., water) for the dissolution and transport of the chemical[1]). To mix such disparate streams will require additional reaction time. It may also lead, under unfavorable reaction conditions, to situations where this step will be rate-limiting.

If addition and mixing of chemicals is thought of as an independent reaction step, then the following configurations of physical and chemical boundary conditions could be envisioned:

a) The coagulating chemical is formed in situ upon its entrance into the sol and its effectiveness depends upon the solution regime encountered, i.e., the homogeneous concentration of all solute constituents (e.g., hydrolyzing metal ions).
b) The destabilizing chemical is introduced already in its active form and the reaction is not dependent upon the mixing characteristics of the system in this instance (e.g., prepolymerized metal hydroxo-complexes).
c) The coagulant or flocculant is of such concentration and of such molecular size that its "interaction" with the colloid, for instance, its motion toward the colloid surface, is dependent upon turbulent transport as well as upon diffusion processes (high molecular weight organic polymers).
d) The interaction of the coagulating/flocculating chemical with the colloid is independent of turbulent transport phenomena and therefore cannot be improved by controlling the hydraulic regime (highly charged counterions).
e) The reaction of the destabilizing chemical with the colloidal surface is, from a point of view of available reaction time, irreversible (e.g., adsorption or organic polymers).
f) The surface-chemical reaction is to be considered reversible under conditions of practical application, i.e., non-homogeneous solution characteristics can be corrected or compensated later on (e.g., counterions).

For the most frequently encountered situations where hydrolyzed and hydroxo-complexing metal ions are used for the destabilization of aqueous sols, the situation, as it is to be envisioned, is shown in Fig. 1. The upper part describes schematically the theoretical situation: the first phase of the chemical mixing step leads to uniform hydroxilation of the metal ion. Subsequently, the metal-hydroxo-complex adsorbs - again uniformly — at the surface of the predominantly hydrophilic colloid (as encountered in aqueous systems).

The lower part of the figure describes the situation as it might be encountered under technical conditions where mixing and transport leads to non-homogeneous situations: First of all the hydroxo-complexation is non-uniform. This is "visualized" by showing mixed species of hydroxilated and hydrolyzed ions simultaneously. Second, the interaction of the colloid surface and the metal-hydroxo-complexes leads to insufficient destabilization in one case and re-stabilization in another, for reasons of less efficient transport. The overall effect is insufficient destabilization, even though the amount of chemicals added was sufficient on the basis of "theoretical" calculations. Such hpyotheses are confirmed by observations reported in the literature. Figure 2 illustrates with data by Masides et al. [2] the reversible or irreversible hydroxilation of aluminum as used in the coagulation of technical systems: depending upon the previous pH-regime, to

---

[1] In practical applications this can lead to situations where a suspension stream of 3000 l/h has to be mixed effectively with a chemicals stream of as little as 0.5 l/h, i.e. one thousands of the volume flow or less.

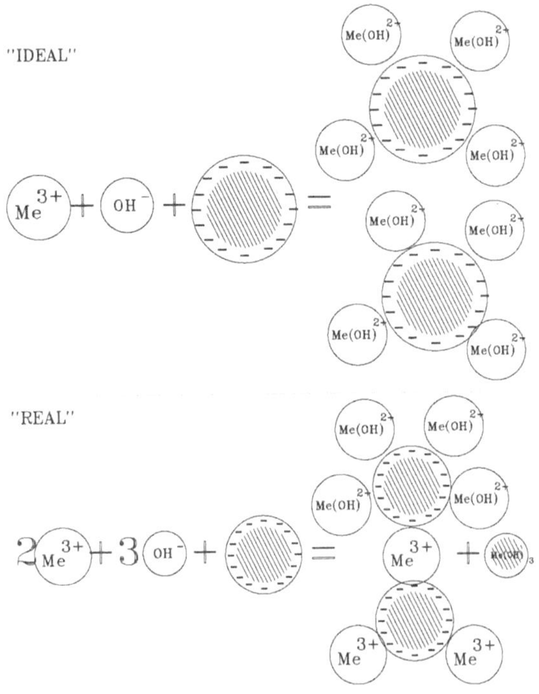



**Fig. 1** Schematic depiction of the consequences for colloid destabilization if non-homogeneous mixing of chemicals (in this instance, trivalent metal ions) occurs

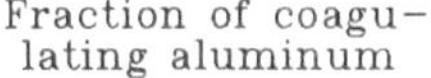

Fraction of coagu-
lating aluminum

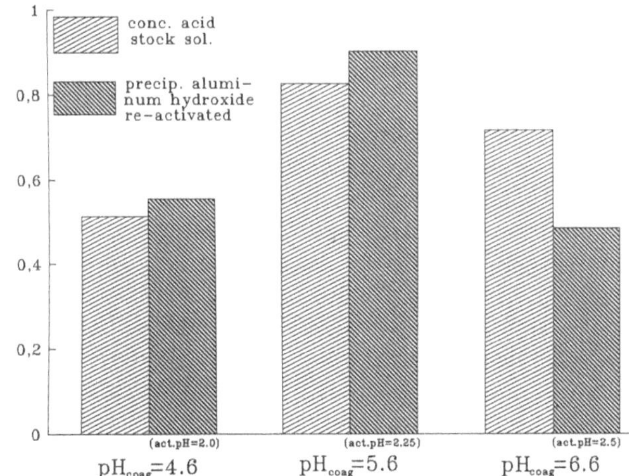



**Fig. 2** Coagulation efficiency, i.e., destabilizing efficiency of aluminum prepared under different pH regimes, illustrating the irreversible hydroxilation or hydroxide formation (after [2])

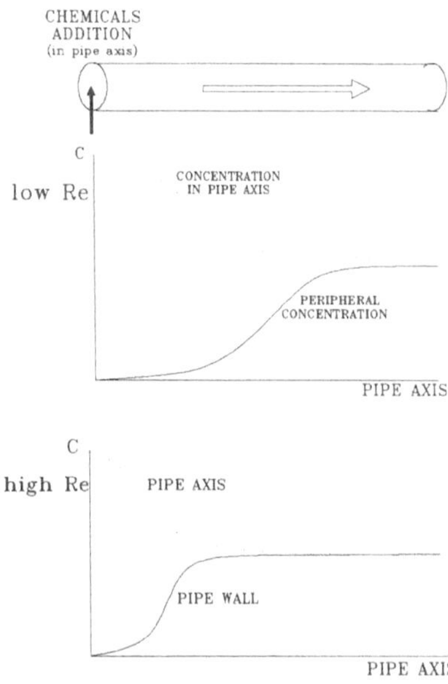



**Fig. 3** Mixing of chemicals in pipe flow, described by the change in concentration along the pipe axis and the pipe wall for two different hydraulic regimes

which the coagulant was exposed, the chemical is more or less effective in destabilizing water-borne colloids.

Observations on mixing in real systems, schematically described in Fig. 3 (after [3]) show that the mixing can be slower or faster, depending upon the hydraulic characteristics of the mixing reactor, denoted by lower or higher Reynolds' numbers. Here, the point of addition of chemicals has been designed such that the highly concentrated stock solution is added into the central flow trajectory of the pipe flow. The flow regime is laminar or turbulent, depending upon the hydraulic load of the system. Measurements have shown that for conditions of "normal" loading of the system, i.e., lower flow velocity and therefore lower Reynolds numbers, the time required for complete mixing is very large. The state of mixing can be described by the difference in concentration, for instance, between the peripheral and the central flow trajectory. If the so-called through-put, i.e., the flow per unit cross-sectional area, is increased, this leads to an increase in the flow velocity and an increase in the Reynolds number. Within limits there is an increase in the degree of turbulence within the system. The result is a shortening of the mixing time, the time to reach homogeneous concentration of for instance, metal ions or hydroxilyzed metal species in the system.

The actual effect of such different hydraulic boundary conditions in the stage of chemicals addition, i.e., more or less successful chemicals' mixing, is shown by the data described in Fig. 4 (after [4]). Here, the coagulation efficiency is described for two different chemicals dosing systems. The difference between the two systems is a solely geometrical one, as the schematic indicates.

In the instance of the dosing fixture "A", one needs an amount of some 40 mg/l prepolymerized metal-hydroxo-complexes coagulant to reach effective destabilization. This is recognized from measurements on electrophoretic mobility as well as from observations on the removal rate (measured as particle number, or turbidity

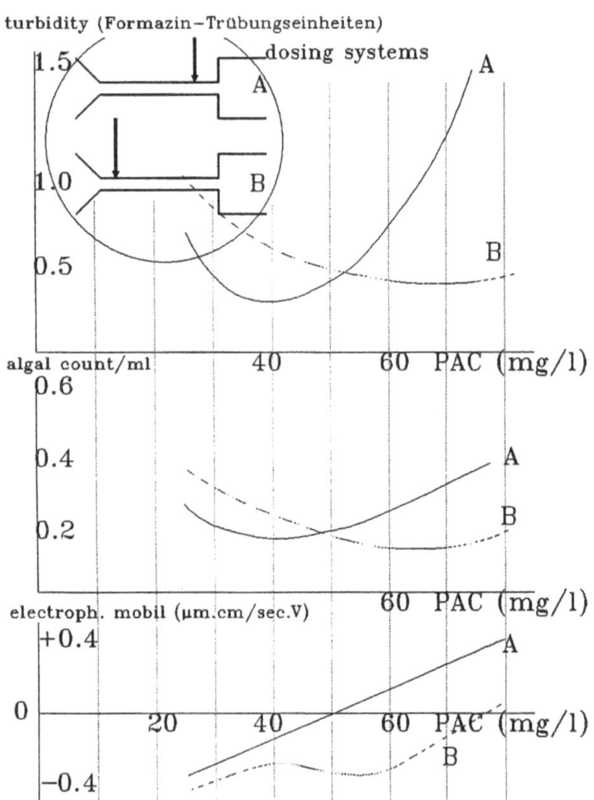



**Fig. 4** The effectivity of two slightly different designs of chemical dosing points expressed in terms of resulting electrophoretic mobility and the rate of the coagulation and separation process (measured by reduction in algae and turbidity after [4])

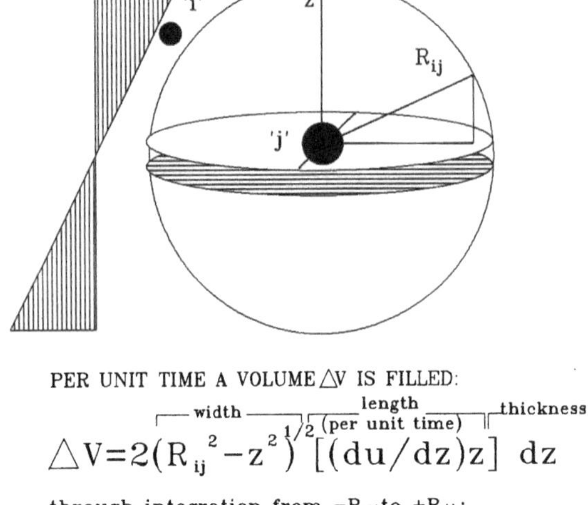



PER UNIT TIME A VOLUME $\triangle V$ IS FILLED:

$$\triangle V = 2(R_{ij}^2 - z^2)^{1/2} \, [(du/dz)z] \; dz$$

through the width — length (per unit time) — thickness

through integration from $-R_{ij}$ to $+R_{ij}$:
(with a concentration of $n_j$ particles per unit volume)

COLLISIONS PER UNIT TIME WITH CENTRAL PARTICLE

$$J = (4/3)n_i R_{ij}^3 (du/dz)$$

**Fig. 5** Smoluchowski's model for the collision between colloids in shear flow systems (after [1])

or chlorophyll alpha in the case of algal matter). If fixture "B" is used then for the same sol and the same chemical a (significantly) larger dosage is needed to attain values of the electrophoretic mobility that signal sufficient destabilization. Turbidity measurements and chlorophyl alpha recordings again support these electrophoresis data.

Clearly, the mixing situation with apparatus "A" leads to a more homogeneous or more efficient or faster mixing than the apparatus "B". And since the surface reaction is not readily reversible within the given detention time, this leads to satisfactory destabilization in one case and unsatisfactory destabilization in another.

## Physical effects upon the collision of chemically completely destabilized colloids

Collision of colloids can be effected by Brownian motion, by velocity gradients resulting from laminar and turbulent flow and also by differential movement in the sedimentation of such particulates. If coagulation is used as technical process for the improvement of solid separation through aggregation, then velocity gradient induced collisions predominate.

Von Smoluchowski [1] has developed a readily understandable and practically useful mathematical model for the rate of collisions in a locally linear shear flow system (cf. Fig. 5). He assumes that all particulates collide if they move within a sphere of a diameter that corresponds to the sum of the diameters of the potentially colliding solids. The calculation of the flux through such a sphere of influence leads to the (potential) number of collisions of all particles "i" transported into that sphere with a particle "j" located in the center.

This collision number (see Fig. 5) describes the rate of collisions as directly dependent on the number $n_i$ of particles "i", i.e., the ones that are moving into the sphere of collision. The total number of collisions is obtained by multiplying it with the concentration $n_j$ of the central particle "j". Furthermore, the collision rate is directly proportional to the linear velocity gradient. In most practical applications such linear velocity gradients will exist only locally. Camp et al. [5] have shown the relationship between such conceptual linear velocity gradients and the energy input per unit volume of reactor space. This parameter is used widely in the description of collision rates in practical applications. The expression for the collision rate can be rewritten as a rate of particle aggregation or as a rate of disapperance of primary or total particles (see for instance [6]). The resulting rate law for coagulation still contains the linear dependence on the volume related energy input. This has been the basis of most quantitative assessments of coagulation processes: it is postulated that the energy input or the torque at the shaft of the stirrer in the coagulation chamber is the physical variable that exclusively describes the control of the process rate. The type of stirrer, for instance, should be of no effect.

One may assume, however, that the energy dissipation in stirred real reactors leads to a non-homogeneous distribution in terms of the absolute size of the (locally linear) velocity gradient. There will therefore be locally differing aggregation conditions. This will be superimposed by the characteristics of the coagulating chemicals, i.e., the ability to form stronger or less strong flocs. The possible consequences of such non-uniform boundary conditions for the aggregation process are shown in Fig. 6 (where the first column depicts schematically the various alternatives for particles to aggregate, the second column indicates the effect of such aggregation for the selection of a separation process, and the third column indentifies quantitative measures to assess this specific reaction progress). The indications given in Fig. 6 as to the effects of these aggregate properties upon the liquid-solid-separation process and how to assess these characteristics are significant for the operator. Measuring, for instance, the $d_{60}/d_{10}$, i.e., the diameter of the 60-percentile in the particle distribution curve, relative to the diameter of the 10-percentile, furnishes quantitative information on the heterodispersity of the system. The products of the coagulation reaction will differ in terms of:

1) absolute size of the aggregates formed;
2) the degree of heterodispersity of the aggregated sol;
3) the shape of the agglomerates;
4) the porosity of the aggregates;
5) the destruction of the aggregates formed under changed shear conditions if the adherence is a weak one.

These aggregate characteristics have been investigated to a lesser degree, yet, they are of great significance for the practical or technical application of the process.

Such deviation of the reaction progress from the course described by the collision term developed by [1] can be observed if particle counting devices are used to follow the reaction. Figure 7 (after [7]) describes such reactions in stirred cylindrical batch reactors where for identical stirrer speed the reaction progress should be the same in all systems. The observed reaction progress, however, is different in each system. One would conclude that the so-called turbine-type stirrer is the most effective in generating a reaction-favorable environment, while the

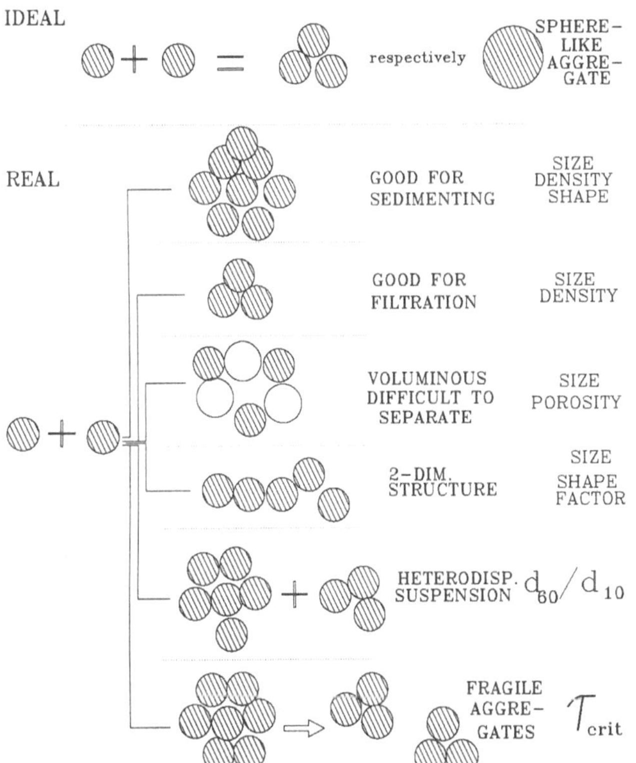



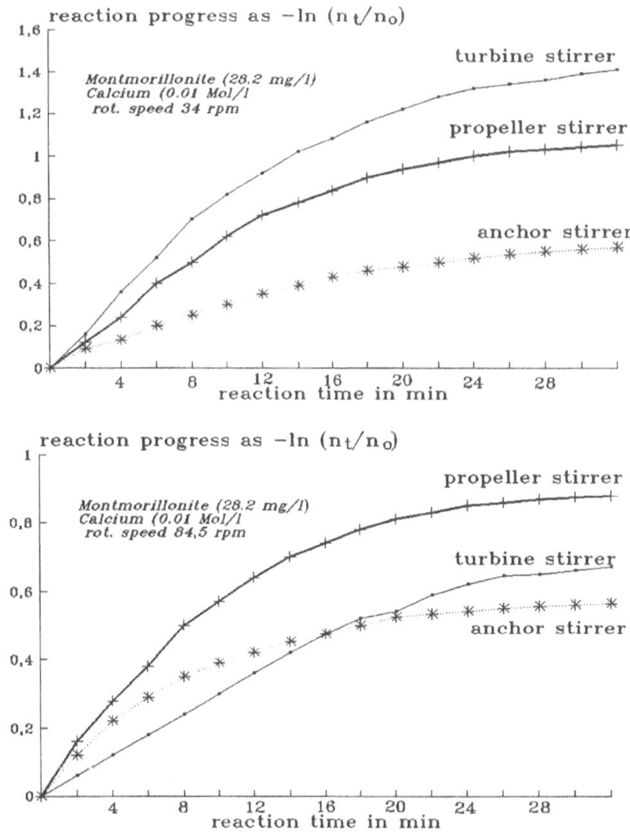



**Fig. 6** Schematic depiction of the consequences for colloid aggregation if energy dissipation (and chemicals distribution) is non-homogeneous

**Fig. 7** Observations on the different rate of coagulation in stirred systems under identical conditions of energy input for differently designed stirrers (after [7])

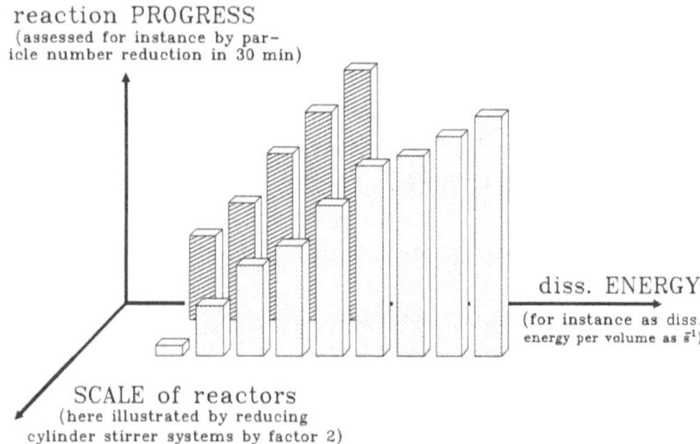



**Fig. 8** The effect of heterogeneous energy dissipation on the progress of the coagulation reaction: in geometrically similar systems of different scale the reaction progress is different even if the (microscopically determined) overall energy dissipation is identical (after [8]). There are no numbers given since they depend exclusively on the experimental boundary conditions

propeller stirrer is less effective. For higher rotational speed the stirrer characteristics appear to change, the propeller stirrer generates at higher rotational speeds the most favorable flow regime while the turbine stirrer is less favorable. In all instances the ancho-type of stirrer is least effective for this type of destabilized suspension. If other chemicals are used, for instance, organic polymers instead of $Ca^{2+}$ then the sequence of effectivity of stirrers is a different one again (compare [7]).

These observations suggest that there are inhomogeneous patterns of energy dissipation in stirred coagulating systems. This might lead to a reaction progress of differing magnitude if different stirrers are employed. More recent investigations on the fluid structure in stirred coagulation chambers [8] and its effects upon the coagulation rate confirm this notion. In these studies the fluid regime was analyzed by Laser-Doppler-Anemometry techniques, permitting an instantaneous recording of all three components of the convective and turbulent fluid motion. The reaction progress itself was also analyzed in a microscopic and close to instantaneous way by using a most detailed particle-counting technique, based on the blockage of a rotating laser beam (see [9]). Figure 8 shows first results in a generalized form: the dimensionless plotted reaction progress grows nearly linearly with increasing energy dissipation (calculated in this instance with the help of the observed and detailed fluid motion parameters). Yet, if the reactor-stirrer system is enlarged, i.e., only the scale changed, then the same observed overall energy dissipation (however, with locally differing flow patterns) will yield a different overall reaction progress.

## Interaction of chemical and physical parameters in the removal of aggregates in specific (technical) separation units

As indicated earlier, coagulation is frequently used in a technical context for the improvement of the liquid-solid separation. Separation techniques employed today are sedimentation, flotation, and filtration. Sedimentation and flotation, which might be looked at as analogous unit processes, are considered separation methods with a large capacity in terms of solids loading. They also appear as highly economical due to their favorable loading characteristics. The following discussion will therefore focus on sedimentation, or flotation, respectively, as a separation step following the coagulation process.

The notion of technically successful sedimentation describes the course of a sedimenting particle or aggregate through the sedimentation reactor in the way depicted schematically in Fig. 9. It is postulated that the particle which reaches the bottom of the tank before it enters the outlet zone is removed permanently. By the same arguments, a particle whose travel time for horizontal movement through the basin is shorter than its sedimentation time is not removed at all. This concept of the sedimentation process, which can be adjusted for different reactor geometries, has been confirmed by many observations from large-scale technical systems. The most significant conclusion from this concept is that only the size and density, i.e., the Stokes' velocity of spheres controls the removal rate. In an analogous way it is formulated and proven useful for flotation.

If a suspension is destabilized by a given amount of metal ions then the coagulation reaction described as destabilization ratio — or for defined energy input as reaction rate constant — and the result in terms of aggregate size and density should be identical. And then the removal rate should also be identical. However, it is frequently observed that some destabilizing reagents produce better sedimenting aggregates than others with very

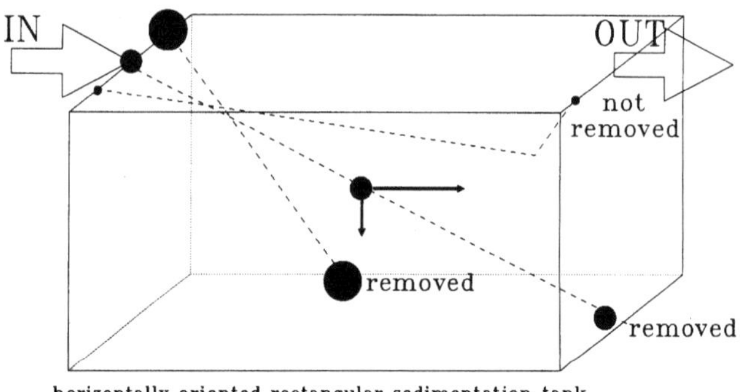



**Fig. 9** A simplified model for the removal of solid matter in a sedimentation tank

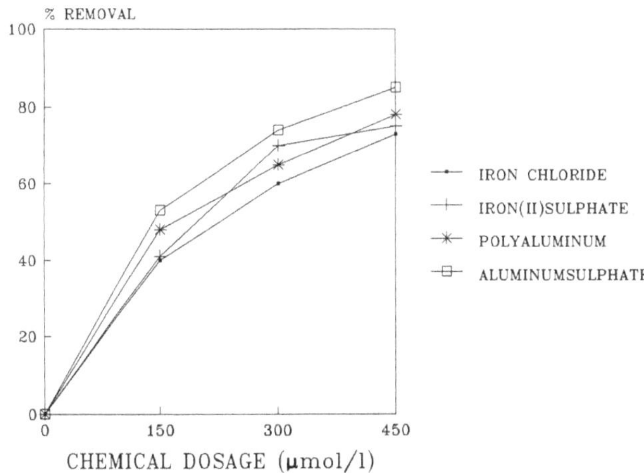



**Fig. 10** Observations of the different removal efficiency of suspended solids coagulated with different chemicals at identical dosages in a sedimentation tank (after [10])

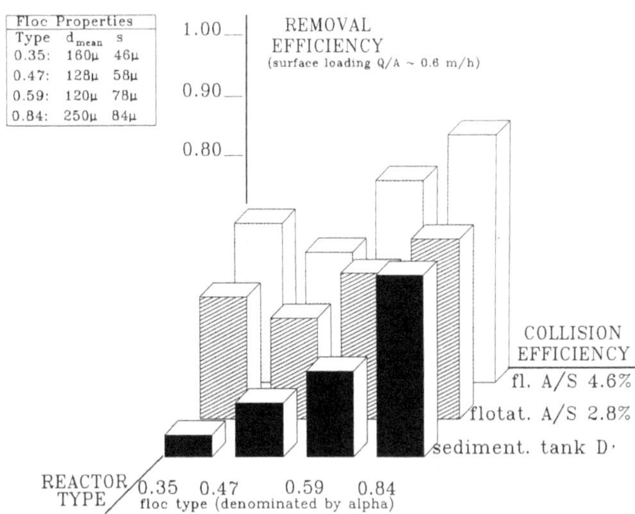



**Fig. 11** Removal efficiency of aggregates in different separation reactors showing the interdependence of the coagulation step with the separation step. The X-axis describes the different reactor types investigated, the Y-axis the various aggregate types generated and separated, and the Z-axis the observed removal efficiency

similar chemical characteristics applied at comparable concentrations (see Fig. 10 drawn after [10]). The data indicate that for these boundary conditions aluminum sulphate produces the most effectively settled flocs, while iron chloride generates flocs which do not all reach the bottom of the tank in a sufficiently short time to be removed completely. Other investigations have shown that $Al^{3+}$ coagulated flocs sediment less readily than iron coagulated flocs at identical concentrations [11].

Such deviation from the simple concept of sedimentation — the analogy holds for flotation as well with only the direction of motion changed — can be explained by the interaction of aggregate properties with the flow structure in the separation unit: if there are regions with higher shear stress in a separation apparatus then the more stable floc will be separated more effectively than the shear-endangered one. Or, similarly, the less heterodisperse suspension will be separated more effectively in sedimentation tanks than suspensions with a broader spectrum of aggregate sizes.

This concept of interdependence of the coagulation step and the separation step is confirmed by data from systematic investigations on the separation effectivity of geometrically different units for systematically varied boundary conditions of the coagulation process. Figure 11 (after [12]) illustrates some results of these studies. Suspensions were coagulated quite differently using aluminum or iron or a combination of iron and polymeric flocculants. This is described quantitatively by the observed collision efficiency factor "alpha". Increasing values of "alpha" indicate more rapidly coagulating systems. The resulting aggregates are described in more detail in terms of first and second moment of the statistical distribution of the aggregate diameters in the insert in Fig. 11. The removal of such aggregates has been studied in reactangular sedimentation tanks and in a

flotation unit of a geometry similar to one of the sedimentation units. It has been recorded as $(C_{in}-C_{out})/C_{in}$, i.e., as removal efficiency. We conclude from these data that there is a very clear effect of the floc type upon the separation efficiency, even if the chemical conditions are kept constant, i.e., for unchanged "alpha" values. More rapidly aggregating systems, characterized by a higher collision efficiency factor "alpha" are sedimenting better than a slowly coagulating one. This is not the same for flotation as a separation process as is shown by the data (the geometry of the flotation tank is the same as that of the sedimentation tank). For practical purposes of designing and operating a liquid-solid separation step, one can deduce that

1) flotation is more effective than sedimentation, under conditions of geometrically similar reaction units;
2) flocs sedimenting relatively effectively cannot in all instances be separated with good results by flotation (the difference results from the type of aggregation chemical), and
3) very dense and/or heavy flocs are sedimented and floated equally well.

It has been described elsewhere [12] that the flow pattern in the tanks investigated and described by the dimensionless dispersion number (DN = $D/\{u.L\}$, where $D$ is the turbulent dispersion coefficient, $u$ the convective transport, and $L$ a characteristic length) has, in conjunction with the floc properties, a more pronounced effect upon the separation effectivity than can be concluded from the coagulation rate alone.

## Conclusions/Summary

The coagulation reaction, as it occurs in the laboratory, can be described as the result of two mostly independent reaction steps, a more chemically controlled destabilization and a transport step that is more controlled by physical parameters, i.e., fluid dynamics of the system. In large-scale or technical realization of this process there is at least one more reaction step to the considered, the addition and homogeneous mixing of coagulating chemicals. Furthermore, the conceptually very clear separation between the chemically controlled destabilization and the physically controlled transport fails. There are numerous observations on the interaction of physical, mostly fluid dynamics parameters with chemical reactions and vice versa.

Interactions manifest themselves in the irreversible and incomplete destabilization of colloids in inefficiently mixed systems or in the heterodispersity of sols formed under conditions of non-homogeneous energy dissipation or through the differing effectivity of liquid-solid-separation for coagulated systems.

In the past, these phenomena have been studied to a lesser degree since existing analytical tools did not allow microscopic description of the coagulation and the separation process. And furthermore, such reductions in effectivity were frequently compensated by the addition of larger amounts of destabilizing reagents. Today the user of coagulation processes for technical systems attempts an optimization of operating conditions such that with a minimum amount of costly chemicals a maximum effect in aggregation and liquid-solid-separation is obtained.

## References


1. von Smoluchowski M (1917) Z Phys Chem 92:129—136
2. Masides J, Mata J, Soley J (1968) Proceedings of the Second International Gothenburg Symposium, ISSW Reihe Universität Karlsruhe 45, pp 333—349
3. Dittmann W (1990) VDI Reihe 3: Verfahrenstechnik Düsseldorf: pp 71—80
4. Kleinschmidt A, cited in R Klute (1985) Proceedings of the First International Gothenburg Symposium, G Fischer Verlag Stuttgart, pp 53—67
5. Camp TR, Stein PC (1943) J Boston Soc Civ Engrs 30:219—228
6. Friedlander SK, Wang CS (1966) J Colloid Interface Sci 22:126—138
7. Klute R, Hahn HH (1975) Vom Wasser 43:215—235
8. Hoffmann E (1992) Berichtsband des SFB 210 zum Kolloquium 1990: 153—154
9. Mihopoulos J, Hahn HH (1992) Proceedings of the Fifth International Gothemburg Symposium, Springer Verlag Berlin, Heidelberg, New York, pp 81—96
10. Gillberg L, Eger L, Jepsen SE (1990) Proceedings of the Fourth International Gothenburg Symposium Springer Verlag Berlin, Heidelberg, New York, pp 243—256
11. Rosén B (1991) Personal communication
12. Mihopoulos J, Dissertationsschrift, Fakultät für Bauingenieur- und Vermessungswesen, Universität Karlsruhe, (eingereicht 1994)

Progr Colloid & Polym Sci (1994) 95:161—167
© Steinkopff Verlag 1994

A. Bartelt
D. Horn
W. Geiger
G. Kern

# Control and optimization of flocculation processes in the laboratory and in plant

Dr. A. Bartelt (✉) · Dr. D. Horn
Polymer Research Division
Polymer and Solid-State Physics, ZKM

D. I. W. Geiger · D. I. G. Kern
Waste Water Treatment Plant
DUR/WK
BASF AG
67056 Ludwigshafen, FRG

**Abstract** Flocculation processes are of great importance for numerous industrial processes, such as paper manufacture, mining, and the treatment of industrial and municipal waste waters. Polymeric flocculants having molecular weights up to $1 \cdot 10^7$ g/mol are generally used for this purpose. The flocculation mechanism (bridging, charge-mosaic) is determined by the chain length, the charge density and the molecular weight of the polymeric flocculant. There is a closely defined optimum for the dosing rates of flocculants, from both an ecological and an economic point of view. The specific choice of flocculants therefore requires a reliable measuring method with which tests of effectiveness can be carried out, ideally under service conditions. This paper describes a fiber-optic flocculation sensor (FOFS) for measuring the flocculation state in flowing systems, which has proved useful in practice for both laboratory flocculation tests and as a measuring component of control systems for dosing equipment. The focus of this contribution is the use of the FOFS for assessing the effectiveness of polymeric flocculants in relation to waste water treatment.

**Key words** Flocculation — polymeric flocculants — fiber-optic flocculation sensor — waste water treatment

## Introduction

Flocculation processes with following solid/liquid separation are the basis of numerous industrial processes. Significant sectors, such as the paper industry, textile finishing, the detergent industry and the ecologically important sector of water treatment and protection, owe their technological progress in large measure to the use of organic flocculants. These are polymeric, water-soluble materials of different charge density having molecular weights up to $1 \cdot 10^7$ g/mol [1]. The most important product group for the waste water sector is that of cationically and anionically modified polyacrylamides.

Flocculation processes involve a complex interaction of individual, elementary processes (Fig. 1) determined by aspects of colloid and surface chemistry:

— production of a homogeneous mixture of the polymeric flocculation aid and the substrate to be flocculated;
— adsorption of the polymer onto the particle surface which thus becomes destabilized after a particular degree of polymer coating;
— rearrangement of the adsorbed polymer on the particle surface to achieve an equilibrium conformation;
— collision of destabilized particles (before or after rearrangement) which initiates aggregation or flocculation;
— floc growth;
— breakup of flocs by shear stress.

The nature of the polymers determines their mode of action, which is explained by structural differences in

## Adsorption and Flocculation Kinetics

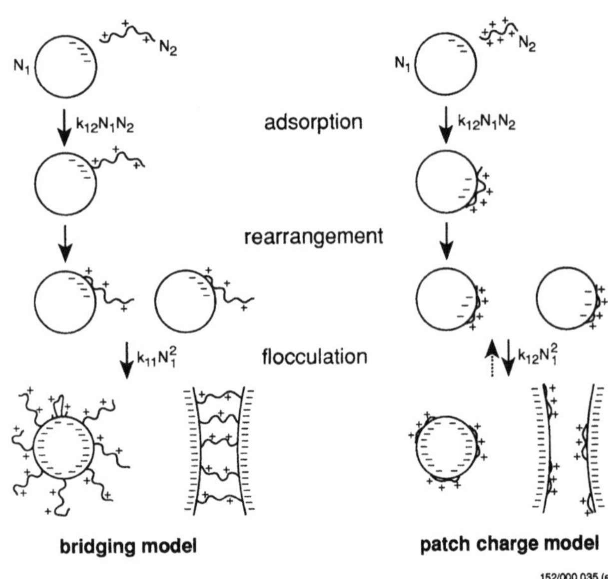


**Fig. 1** Interaction of adsorption and flocculation kinetics

the adsorbed layers [3—5]. The flocculation mechanism of bridging is enhanced by high molecular weights and a low affinity between the flocculation aid and the surface. Destabilization and flocculation is achieved when a voluminous far-reaching polymer conformation is present and the surface is about half-coated. Polyelectrolytes with a high charge density tend to give a flat surface adsorption; the resulting mosaic structure at a particular degree of polymer coating is therefore characterized by regions having positive and negative charges (patch charge model). A correlation between flocculation effect and surface charge can only be derived for the charge-mosaic mechanism.

Differentiation of the two flocculation mechanisms is often possible by observing the shear stability. Polymer bridges are relatively shear-stable, the destruction of flocs by increased shear stress generally being irreversible. Flocs formed by the charge-mosaic mechanism, on the other hand, are substantially more compact and shear-labile; however, destroyed flocs return reversibly to the original state after the stress is stopped [6, 7].

For those flocculation processes which are initiated by polymers, the depletion flocculation mechanism is also being discussed in addition to the mechanisms based on polymer adsorption. In the depletion mechanism, the osmotic pressure causes the displacement of polymer molecules from between neighboring particles as soon as these are closer than a particular minimum distance [8—10]. Theoretically, this mechanism is quite feasible, although it has as yet not been experimentally verified, in particular for real systems.

Besides the polyelectrolyte structure, the dosing and mixing conditions play a decisive role in the flocculation

result [2, 16]. Both ecological and economic factors are important in driving the targeted development of flocculants, their specific selection and their optimal dosing. This requires effective measurement methods which allow assessment of the prevailing flocculation state under defined conditions which are close to those in practice. This paper describes a fiber-optic flocculation sensor (FOFS) for measuring the flocculation state in flowing systems, which has proved its value both for basic studies in the laboratory and as the measuring component of control systems for actual dosing equipment [6, 11—13, 16]. The focus of this contribution is the use of the FOFS for assessing the effectiveness of polymeric flocculants in the treatment of industrial and municipal waste water.

## Fiber-optic flocculation sensor (FOFS)

The measuring device is made up of three integrated units:

— an arrangement for mixing in the flocculation aid and a region for floc formation,
— a fiber-optic sensor fitted in an enveloped-flow cell,
— detector electronics connected to a computer for combined data analysis and control of dosing.

The principle of the experimental arrangement of the FOFS is shown schematically in Fig. 2. The medium to be flocculated is turbulently mixed with the solution of the flocculation aid in a mixing chamber, which achieves rapid homogeneous distribution of both the components. Flocculation takes place slowly under defined shear conditions in a region of laminar flow. The length of the laminar-flow tube can be matched to the applicable practical conditions.

After passing through the laminar-flow region, the flocculated material flows through the fiber-optical measuring cell (Fig. 3). The transmitted light of a laser beam passing through the sample is detected (IR laser

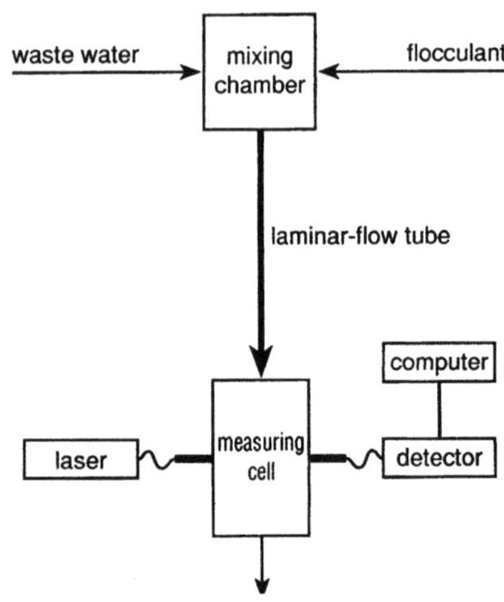


**Fig. 2** Principle of the flocculation sensor

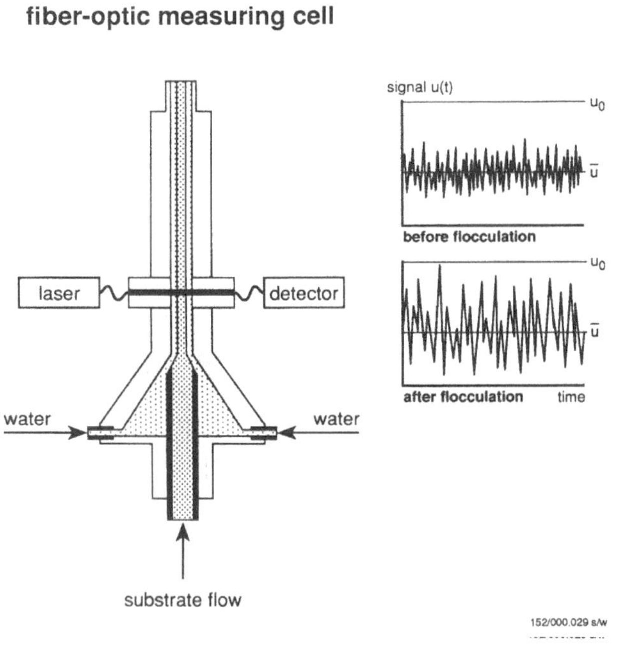

**fiber-optic measuring cell**

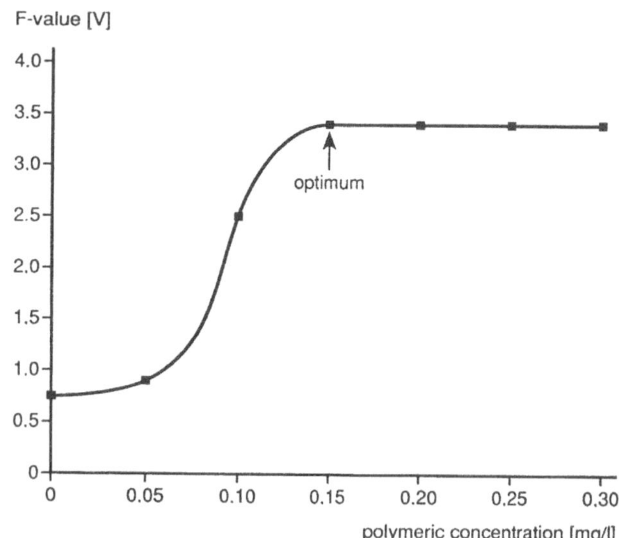


**Fig. 3** Laser-optic flocculation monitoring technique

beam 835 nm; diameter in the suspension 1 mm). The intensity fluctuations of the transmitted laser light resulting from particle density variations very sensitively mirror the decrease in particle number caused by flocculation and the corresponding increase in particle size. The electrical signals of the detector are converted by an integrated electronic component into the flocculation value $F$ (in volts) by means of a root-mean-square analysis [14], this value serving as the assessment criterion for the effectiveness of flocculation.

It is of decisive importance for the optical measurement and for the use of the sensor that contamination of the measuring cell is avoided. This is achieved by means of an enveloping stream of deionized water which surrounds the substrate stream and so prevents the contact of the latter with the wall of the measuring cell (Fig. 3). Furthermore, the optical path length can be adjusted for an optimal signal-to-noise ratio by varying the ratio of the sample stream to the enveloping stream, thus enabling the measurement of typical samples of practice, such as sewage sludge, paper pulp, coal suspensions, etc., in their original concentration without prior dilution.

The measurement and evaluation of flocculation state is typically carried out over 10 min by means of programmed dosing, in which the flocculation value $F$ is recorded during a stepwise increase in flocculant dosing. Figure 4 shows an example of a flocculation curve. The amount of flocculant required for optimal flocculation is that where $F$ reaches an asymptotic limit (bridging mechanism) or a maximum (patch charge mechanism).

The degree of flocculation $F$ derived from intensity fluctuations correlates with the floc size, which can be

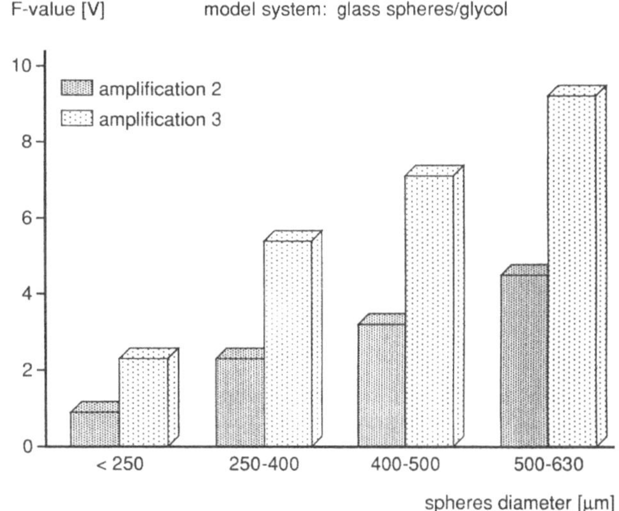


**Fig. 4** Flocculation curve: dependence on the degree of flocculation $F$ on flocculant dosing



**Fig. 5** Dependence of the degree of flocculation $F$ on floc size for the model system of glass spheres/glycol

shown both theoretically [15] and experimentally [13]. Figure 5 shows the results of model experiments with glass spheres of differing size with glycol as the continuous phase. A linear relationship between the degree of flocculation $F$ and the sphere size is obtained, regardless of the amplification of the detector.

## Screening method for assessing the effectiveness of flocculation aids for waste water treatment

In waste water treatment, flocculation (coagulation) of the solid particles present in the water or sludge is induced by polymeric flocculants. Voluminous, stable flocs are formed, which speeds up the process of sedimentation and makes dewatering of the sludge possible. Depending on the nature of the water or sludge, it is necessary to select polymeric flocculants which have optimal properties with respect to effectiveness and economics. The fiber-optic flocculation sensor has the particular advantage, when compared with conventional laboratory tests such as sedimentation experiments (jar test), that the monitoring of the flocculation process and therefore the assessment of the flocculant is possible within a few minutes.

## Flocculation of excess sludge

Excess sludge occurs in the treatment of industrial effluents in concentrations of 10—15 g/l. Figure 6 shows flocculation curves for excess sludge, in which anionically modified polyacrylamides from the Sedipur (registered trademark of BASF) products were used (A—D) which are different concerning their molecular weight and their charge density. Assessment criteria are the amount of flocculant needed for optimal flocculation and also, as a measure of the floc size, the derived flocculation parameter $F$. The differing effectiveness of the products is clear from a comparison of the results (Fig. 6).

The flocculation curve of flocculant $A$ is almost ideal: after a short initiation or response time, during which the amount of polymer is still too low for sufficient destabilization of the particles, there is a significant increase in the degree of flocculation at higher polymer concentrations. The flocculation curve is linear over a wide range and finally approaches an asymptote at even higher polymer concentrations. The flocculation curves

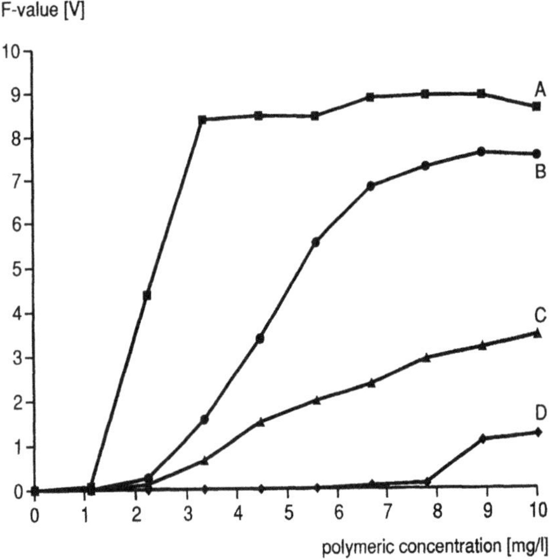


**Fig. 6** Flocculation curves for excess sludge with Sedipur products (A, B, C, D), which are different concerning molecular weight and charge density

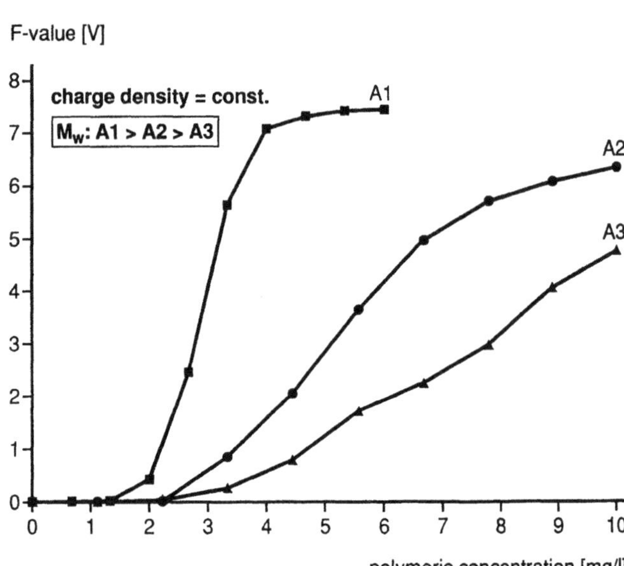


**Fig. 7** Flocculation curves for excess sludge with Sedipur products (A1, A2, A3); variation of molecular weight of the flocculants

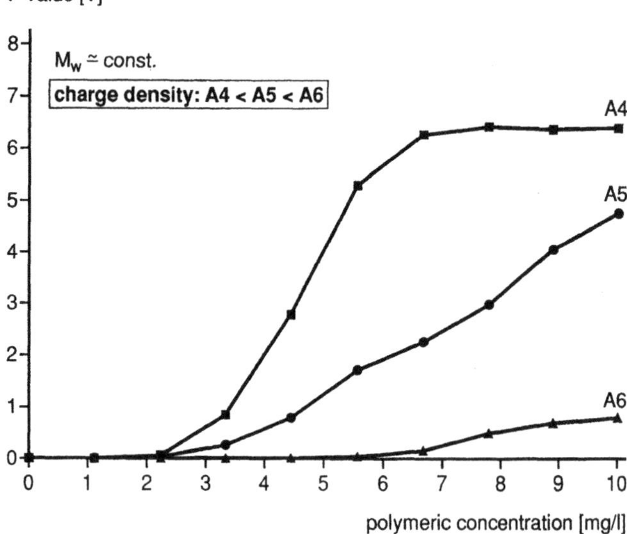


**Fig. 8** Flocculation curves for excess sludge with Sedipur products; variation of charge density: A4: 2.0 meq/g, A5: 4.2 meq/g, A6: 8.4 meq/g

of the polymers $B$, $C$ and $D$ are flatter, which implies a higher usage of flocculant and a poorer flocculation effectiveness.

Systematic tests with selected anionic products have shown high molecular weight flocculants with low to medium charge densities (—2 to —4 meq/g) to be optimal for flocculating excess sludge (Figs. 7, 8). In these flocculation processes calcium ions serve as links between the anionic polymers and the negative particle surfaces. Because the bridging mechanism is responsible for the destabilization, measuring methods based on a charge characterization of the systems for assessing flocculation (e.g. electrokinetic methods) are not appropriate.

## Flocculation of digested sludge from a municipal waste water treatment plant

For more highly concentrated grades of sludge, too, the effectiveness of flocculants can be assessed without difficulty by means of the fiber-optic flocculation sensor. Figure 9 shows this for the example of a digested sludge (30 g/l) from the waste water treatment plant of Obererlenbach, which was flocculated with cationically modified polyacrylamides (K1, K2).

For the examination of highly concentrated systems (30 to 60 g/l), such as thickener sludge, a modification to the mixing chamber is required. The usual mixing chamber operates as a static mixer, i.e., the mixing energy is derived only from the volume flows of the two components. For concentrations >30 g/l, this arrangement can no longer ensure turbulent mixing. The use of a mixing chamber to which additional mixing energy can be provided (Ultra Turrax, blade stirrer) is therefore required.

## Use of the fiber-optic flocculation sensor as on-line control device in the Sedipur dosing facility of the BASF waste water treatment plant

The individual steps of mechanical and biological waste water treatment in the BASF plant (Ludwigshafen, FRG) [17, 18] are shown in Fig. 10. After neutralization of the

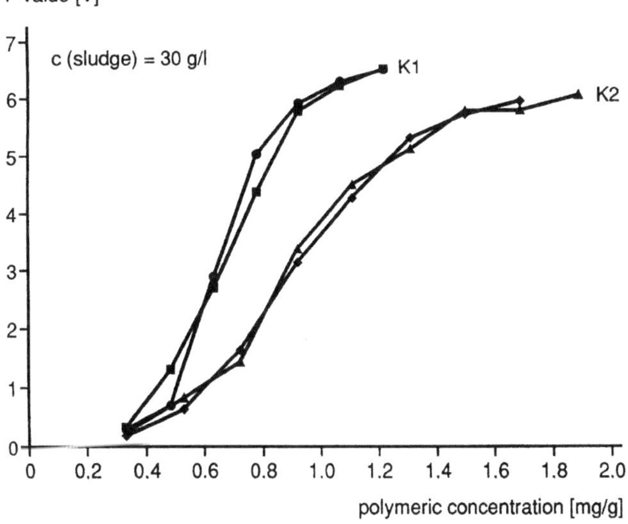


**Fig. 9** Flocculation of municipal sewage sludge with Sedipur products (K1, K2)

**BASF Waste Water Treatment Plant**

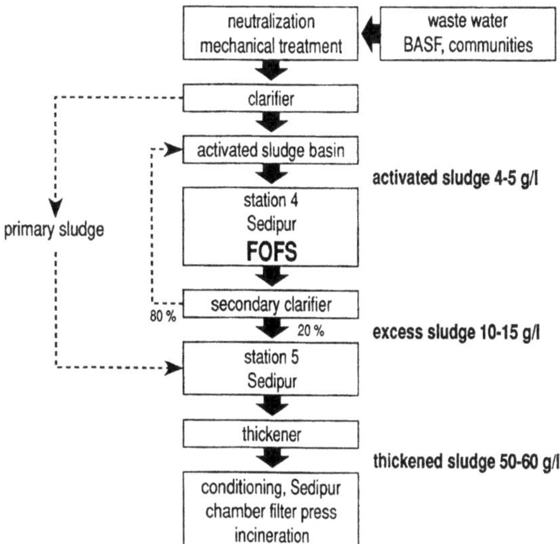


**Fig. 10** Flow diagram of the BASF waste water treatment plant

waste water, mechanical treatment is carried out in a screening plant and in clarifier. This is followed by biological treatment in the activated sludge basins using bacterial sludge. The bacteria degrade dissolved organic substances with the aid of oxygen from the aeration, forming water, carbon dioxide, and a new biomass, the excess sludge mentioned earlier (70% protein). The sedimentation of the activated sludge in the secondary clarifier (4—5 g/l) and the concentration of the excess sludge in the thickeners (10—15 g/l) is aided by anionic Sedipur products. After addition of conditioning agents (ash and carbon), the thickened sludge (50—60 g/l) is flocculated with a cationic Sedipur product. After dewatering, the filter cake is finally incinerated, the ash being largely reused in the mining industry.

The flocculant usage of the waste water treatment plant is about 8 tons per month. At present, flocculation dosing occurs on the basis of empirical data. With the aim of achieving more effective and more economical dosing which will, in particular, also allow the flocculant usage to match variations in the nature of the sludge, it is planned to install the fiber-optic flocculation sensor as the control unit in the Sedipur dosing facility.

The FOFS is at present installed in the distributor 4 of the waste water treatment plant and is fed with a representative bleed stream which is taken from the unflocculated main stream of an aeration basin via a bypass. The optimal amount of flocculant is determined for this bleed stream every 15 min. In the long term, it is planned to connect the flocculation sensor to the process control system of the waste water treatment plant. After a proportional transmission control factor has been determined for the process and plant, it will be possible to control the

## Original Record of the FOFS in the Waste Water Treatment Plant

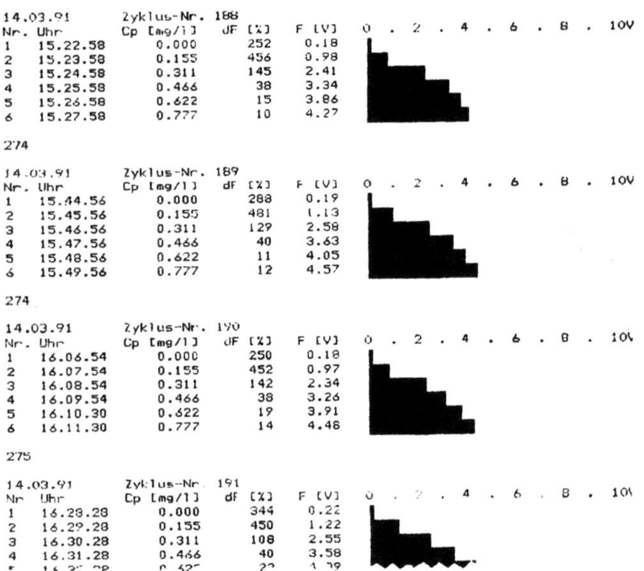


**Fig. 11** Extract of an original record of the FOFS in the waste water treatment plant

dosing in the main stream with the aid of the optimal flocculant amount in the bypass.

Changes to the laboratory version of the FOFS relate essentially to data processing. Instead of the computer used in the laboratory, the control program will be installed on a programmable interface (Multi Data), enabling connection to the process control system of the waste water treatment plant. A printer coupled to the interface logs the results continuously. Figure 11 shows an extract of an original record.

The time of day, the actual flocculation aid concentration, the change in the flocculation value in comparison with the previous concentration step (dF), and the absolute flocculation value (F) are given; the bar chart corresponds to the flocculation curve shown before. The optimal flocculant concentration is defined as that at which increasing of the Sedipur concentration twice no longer increases the flocculation value by more than 20%. On the basis of this Sedipur concentration, which is transmitted to the process control system, the optimal dosing for the main stream is determined, taking the varying conditions (flow rates, concentrations, dimensions, etc.) into account.

As long-term experiments, the optimal flocculant concentrations together with the corresponding F-values were recorded over several months. Evaluation of these results has made it clear that the activated sludge flocculation is being carried out with considerable over- or under-dosing of flocculant over extensive periods of time. Controlled dosing matched to the sludge properties would save flocculant and give improved flocculations.

The FOFS in distributor 4 of the waste water treatment plant is still in the test phase, although connection to the process control system is approaching.

## Conclusions

The fiber-optic flocculation sensor has proved its utility as a robust measuring device with short response times and high resolution, both in research investigations and in field tests.

Systematic investigations with model suspensions and with real systems lead to clarification of flocculation mechanisms. Knowledge of the applicable flocculation mechanisms is a fundamental requirement for understanding and optimizing flocculation processes. The fiber-optic flocculation sensor is, furthermore, useful as a screening method which is close to practice for developing and for assessing the effectiveness of flocculants.

Use of the flocculation sensor as an on-line control device in real dosing facilities is possible, which could in the future be of particularly great value for water treatment in industrial and municipal waste water treatment plants. With regard to the economics and effectiveness of process operation, the advantages of optimized flocculant dosing have already been shown in activated sludge treatment, and will certainly become even clearer with highly concentrated types of sludge (excess, thickener, digested sludge).

**Acknowledgement** We thank Mrs. H. Debus for her diligence in carrying out the measurements and for her assistance in the implementation of all these projects.

## References

1. Eisenlauer J, Horn D (1985) DVGW-Schriftenreihe Wasser Nr. 42:59—72
2. Gregory J, Guibai L (1991) Chem Eng Comm 108:3; Gregory J (1988) Colloids and Surfaces 31:231
3. Gregory J (1989) Critical Reviews in Environmental Control 19:185
4. Dickinson E, Eriksson L (1991) Advances in Colloid and Interface Science 34:1
5. Gill RIS, Herrington TM (1987) Colloids and Surfaces 28:41
6. Ditter W, Eisenlauer J, Horn D (1982) In: Tadros (ed) The Effect of Polymers on Dispersion Properties. Academic Press, London, pp 323—342
7. Leif Eriksson, Barbro Alm, 5th International Gothenburg Symposium on Chemical Treatment of Water and Wastewater, Nice, Sept. 28—30, 1992. "Model system studies of formation and

properties of flocs obtained with cationic polyelectrolytes."

8. Sperry PR et al. (1981) J Coll Interf Sci 82:62
9. Dickinson E (1989) J Coll Interf Sci 132:274
10. Dobias B (ed) (1993) Coagulation and flocculation — Theory and Application. Surfactant science series vol 47. Marcel Dekker, Inc, New York, Basel, Hong Kong
11. Eisenlauer J, Horn D (1983) Z Wasser Abwasser Forsch 16:9
12. Eisenlauer J, Horn D, Linhart F, Hemel R (1987) Nordic Pulp & Paper Research Journal 2:132
13. Eisenlauer J, Horn D (1987) Colloids and Surfaces 25:111
14. Gregory J, Nelson DW (1984) In: Gregory J (ed) Solid-Liquid Separation. Society of Chemical Industry/Ellis Horwood, London, pp 172—182
15. Gregory J (1985) J Colloid Interface Sci 105:357
16. Eisenlauer J, Horn D (1985) Colloids and Surfaces 14:121
17. Engelhardt H, Neuwirth M, Weisbrodt W (1985) Korrespondenz Abwasser 32:940
18. Wilms HE, Wolf G, Ahrens K (1992) Abwassertechnik Heft 5:60

Progr Colloid & Polym Sci (1994) 95:168—174
© Steinkopff Verlag 1994

M. Liphard
W. von Rybinski
B. Schreck

# The role of surfactants and polymers in the filler flotation from waste paper

**Abstract** High contents of fillers such as kaolin or calcium carbonate limit the use of waste paper, especially in tissue paper production. In order to determine the effect of flotation reagents on the removal of fillers, adsorption, zeta potential, and particle size measurements, as well as flotation experiments using model dispersions of calcium carbonate, kaolin, and cellulose fibers were carried out. The adsorption of the cationic polymer starts at low initial concentrations on the negatively charged filler surfaces and cellulose fibers. However, due to the steeper slope of the adsorption isotherm on the fillers, the polymer is preferentially adsorbed on the fillers. Furthermore, the adsorption of the polymer causes an increase in the particle size of the fillers. Anionic surfactants are generally better suited for waste paper systems containing calcium carbonate than for those with kaolin. This is due to the fact that the adsorption onto calcium carbonate occurs at lower concentrations than that onto kaolin. Calcium ions dissolved in the pulp improve the adsorption of anionic surfactant onto kaolin and are necessary for a sufficiently high recovery of the fillers.

**Key words** Fillers — flotation — adsorption — anionic surfactant — cationic polyelectrolyte

Maria Liphard (✉) · W. von Rybinski · B. Schreck
Henkel KGaA
Forschung/Physikalische Chemie
D-40191 Düsseldorf

## Introduction

Waste paper has become one of the most important raw materials for the production of paper and cardboard [1]. Besides simple repulping, deinking has become the best known technique for waste paper recycling. Deinked stock is the predominant raw material in the manufacture of newsprint, recycled graphic paper and sanitary paper in Central Europe, and its use in wood-containing paper for magazines is likely to become possible within the near future [2].

However, not only the removal of printing inks but also the control of the filler content is necessary for their production. In the Central European paper industry, kaolin and calcium carbonate are the most important fillers and coating pigments. They not only improve paper quality, but also reduce the production costs by economizing on fiber utilization. Therefore, there has been a trend towards utilization of more white pigments in paper production. Together with the high recycling rate, this led to a high filler content in the recycled stock, limiting its use in special recycled papers.

In high quality sanitary papers, for instance, the ash content should be as low as possible (at about 2%). At present, the filler content is usually reduced by a washing process and cannot be decreased by flotation, the most important deinking process in Europe. As flotation has proved to be superior to washing in terms of economics and minimal ecological impact, specific reagents have been developed to improve filler removal by flotation [3—5]. In a previous paper, we investigated the influence of the type of filler, its particle size and the presence of dispersants on the flotation recovery [6]. This paper deals with basic aspects of the selectivity of special surfactants and polymers in this process.

## Experimental

### Flotation tests

Flotation tests were carried out in a Denver-type, 2-l laboratory cell. The concentrations of the reagents were specified as solution concentrations. The procedure of waste paper flotation is described elsewhere [5]. For the filler flotation an aqueous dispersion with 0.2% filler in tap water (pH 8.5) was stirred with the respective collector for 10 min and then floated for 5 min. The floated fillers were subsequently filtered; the filtrate was dried at 150°—180°C and weighed to determine the recovery.

### Particle size mesurement

The particle size distributions were determined by means of laser diffraction (Particle Size Analyzer, Sympatek, model Helos KA). The filler suspensions were prepared under the same conditions as the corresponding flotation tests (pH, water hardness, collector concentrations, stirring conditions). The solid material concentration was between 40 and 100 mg/l.

### Zeta potential measurement

The electrophoretic mobilities of the minerals were measured with a Zeta-Meter (Pen-Kem Lazar Zee Meter, type 501, USA), the mobilities being converted into zeta potentials by means of the Helmholtz-Smoluchowski equation. The filler concentration of the suspension was 20 mg/l. The ionic strength was kept constant by adding $1 \times 10^{-3}$ mol/l KCl. Measurements were carried out after a conditioning and pH regulating time of 30 min in a nitrogen atmosphere at pH 8.5.

### Adsorption measurements

The amounts of alkyl sulfosuccinate and of the cationic polymer adsorbed on the fillers and fibers were determined by the depletion method. 1 g of the filler or 0.5 g of the fibers were shaken with 100 ml of surfactant or polymer solution of known initial concentration at pH 8.5 in a glass tube for 60 min. The suspension was subsequently separated by centrifugation (labcentrifuge UJ II E Heraeus-Christ, FRG) at 4000 rpm for 60 min. Control experiments without the addition of solids did not indicate any change in the polymer concentration during centrifugation.

The concentration of the surfactant were determined by titration with cetylbenzyldimethylammonium chloride, which forms water-insoluble complexes with the anionic surfactant. Potentiometric titrations were carried out with a Titroprocessor 672 (Metrohm, Switzerland) using a surfactant-sensitive electrode [7]. The cationic polymer was analyzed by polyelectrolyte titration with poly-(ethylensulfonate) (PES). The equivalence point of the polyelectrolyte complexation reaction was indicated either by streaming potential (PCD 2, Mütek GmbH, FRG) [8] or, for lower concentrations of the cationic polymer, by a color change due to the cooperative binding of a metachromic dye on the excess chromotrope titrant [9, 10]. Brilliant Yellow was used as dye indicator.

## Substances

### Fillers

The following commercially available fillers were used: Kaolin Spex IF, calcium carbonates of the type Hydrocarb 90 OG (powder and slurry). The specific surface area of the fillers used for adsorption was determined by the BET method (Sorptomatik 1800, Carlo Erba Strumentazione, Italy). It was 7.5 m²/g for kaolin and 10.3 m²/g for Hydrocarb 90 OG. Fibers: Sulfate-bleached cellulose from soft wood was used for the adsorption experiments. The cellulose fibers were disintegrated and left to swell after several treatments with distilled water. The water-soluble impurities were removed from the material by hot water extraction for 1 h, which was repeated two times.

### Reagents

A commercially available fatty acid (Olinor 4020, mainly C18, Henkel KGaA, FRG) was used for deinking. The filler collectors are laboratory products: two different anionic surfactants — di-sodium alkyl sulfosuccinate (alkyl = tallow, type A) or sulfonated oleic acid (type B) — and a cationic polymer (all from Henkel KGaA, FRG).

### Fundamentals

The flotability of particles, i.e., the formation of bubble particle aggregates and the enrichment of these aggregates in the foam layer, is given by a complex relationship of interfacial properties, particle size and hydrodynamic conditions [11]. The two parameters, surface properties and particle size, can be controlled with the aid of surface-active chemicals, such as surfactants or polymers.

A certain hydrophobicity, which is a prerequisite for the flotation of a solid particle, can be achieved by the adsorption of surfactants or polymers ("collectors") on the solid surface. In the case of filler flotation, selective separation means that the amount of collector adsorbed on the fillers as well as the degree of hydrophobicity have to exceed the adsorbed amount on the fibers in the concentration range generally used for flotation.

The relationship between flotation rate and particle size is complex [12]. The flotation efficiency decreases distinctly below a certain particle size (often below 1 μm). On the other hand, paper manufacturers prefer fine pigments, because many properties of the paper, such as degree of brightness, opacity and smoothness, are improved by decreasing particle size of the fillers. Therefore, a large amount of fillers with a particle size below 1 μm is expected in the waste paper.

The adsorption mechanism of collectors strongly depends upon the surface charge and characteristics of the solids involved. With the exception of the ink particles, the electrokinetic behavior of all solid materials which are present during waste paper recycling has been investigated thoroughly [13—20], and the complex surface models of kaolin and calcium carbonate are not dealt with here. Under flotation conditions, i.e., in the pH range from 8 to 10, the commercially available fillers used for the experiments as well as the fibers are negatively charged. The negative charge is improved by the adsorption of anionic dispersants, which are used for the stabilization of filler slurries [4].

Different mechanisms have been suggested for the adsorption of anionic surfactants on negatively charged surfaces [21]. Multivalent cations in solution such as calcium ions can improve the adsorption of anionic surfactants when the cations adsorb on the surface and do not form insoluble metal soaps with the surfactants. Sulfonate and phosphate group containing surfactants have proven to be particularly suitable [22].

In general, polymer molecules are adsorbed on a solid surface from solution more or less as random coils. Some of the polymer segments are anchored on the surface and the rest of the segments form loops or tails. The extent to which the segments are anchored to the surface is supposed to depend on the affinity between polymer segments and solid surface. The adsorption of cationic polymers is based either on the electrostatic attraction with anionic surface sites or on an ion exchange in the Stern plane. A similar mechanism has been suggested for the adsorption of polymers onto cellulose [23]. In the interaction of cationic polymers with cellulose fibers, the main factors are supposed to be the cationic charge of the polymer and the negative charge of the cellulose fibers [17]. Most of the cationic polymers, such as poly-(ethylenimines), which are used as retention aids to control drainage and formation properties during paper manufacturing, are strongly adsorbed on cellulose fibers.

## Cationic polymers

The adsorption isotherms of the cationic polymer with both fillers — kaolin and calcium carbonate — are il-

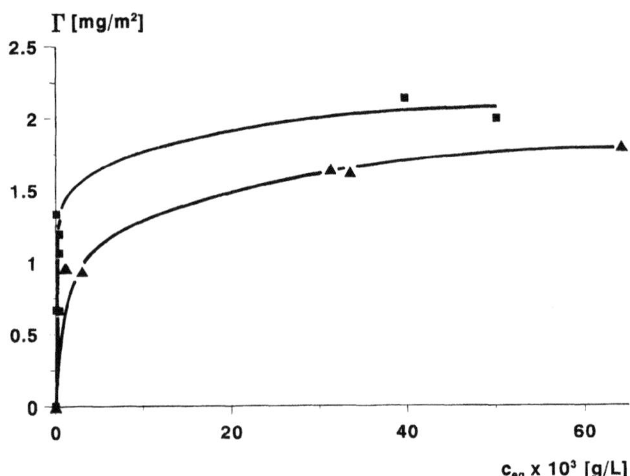

**Fig. 1** Adsorption isotherm of the cationic polymer on kaolin (■) and calcium carbonate (▲) at pH 8.5

lustrated in Fig. 1. The adsorption experiments were conducted at a constant pH value of 8.5 in distilled water. There is a steep increase of the adsorption isotherms at low equilibrium concentrations. At higher concentrations the polymer adsorption reaches a point of saturation. For kaolin, this point of saturation is slightly higher than for calcium carbonate. The steep slope of the isotherms indicates a high affinity between the cationic polymer and the negatively charged filler surfaces.

Futher information on the adsorption of polyelectrolytes can be obtained from zeta potential studies. Figure 2 shows the zeta potential of both pigments as a function of the polymer concentration in a semi logarithmic plot. In the absence of polymers at pH 8.5 the zeta potentials of kaolin and calcium carbonate are

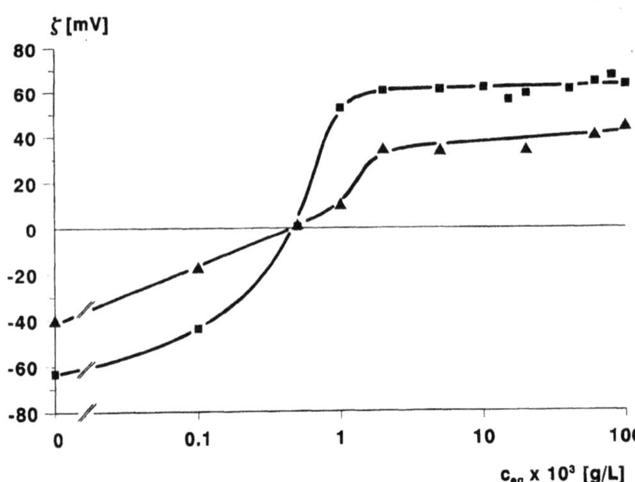

**Fig. 2** Zeta potential of kaolin (■) and calcium carbonate (▲) as a function of the cationic polymer concentration at pH 8.5 in deionized water

negative and lie in the range of $-60$ mV and $-40$ mV, respectively. With increasing polymer concentration the surface charge of the pigments becomes less and less negative, and at a concentration of $5 \times 10^{-4}$ g/l the sign of the charge is reversed.

For the zeta potential measurements a solid/liquid ratio of 20 mg/l was chosen. Under these conditions the decrease in polymer concentration due to adsorption can be neglected and the equilibrium concentration is approximately equal to the initial concentration. Thus, a direct comparison of the adsorption isotherms and the zeta potentials as a function of the polymer is possible. For both pigments the information correlates: The change in the negative charge of the pigment at low concentration is due to the polymer adsorption. Further adsorption reverses the sign of the zeta potential and at the plateau value of the adsorption isotherm the zeta potential remains constant. The corresponding zeta potential value is higher for kaolin than for calcium carbonate, correlating with the different adsorbed amounts of the cationic polymer. Thus, the positive charge of the polymers superimposes the negative surface charge, and the positive charge of the polymer adsorption layer determines the electrokinetic behavior of the pigments.

In practice, multivalent cations in the flotation pulp strongly affect the recovery of printing inks and fillers. The formation of calcium soaps is supposed to be the basic mechanism for the deinking with fatty acids as collector [24]. The specific adsorption of calcium ions on pigment surfaces also modifies the electrokinetic behavior of the fillers. The zeta potential of pure filler pigments can change its sign in the presence of calcium and aluminium ions [16]. In the case of the commercially available fillers investigated in this study, the negative zeta potential of calcium carbonate and kaolin decreases with an increasing amount of calcium ions, but does not change its sign [25].

Due to the still negatively charged surfaces, the adsorption of the cationic polymer is only slightly affected by calcium ions in solution. Thus, the zeta potential of calcium carbonate and kaolin at high polymer concentrations is nearly the same in tap water with 286 ppm calcium ions as in distilled water (Fig. 3). Due to the calcium adsorption on the filler surface and the decreasing number of negative surface sites, the charge reversal occurs at five times lower concentrations than in calcium-free solution. The recovery of the fillers is not influenced by the water hardness, tested from 0 to 1780 ppm calcium ions as $CaCO_3$ (not shown here).

The data of the polymer adsorption on cellulose fibers are represented in Fig. 4. They show a Langmuir-type adsorption isotherm. The plateau value obtained from Fig. 4 is more than 25 mg/g. Somewhat lower values have been reported for the adsorption of other cationic polymers such as poly-DADMAC or poly(ethylenimine) on various cellulosic materials [26, 27]. The cationic

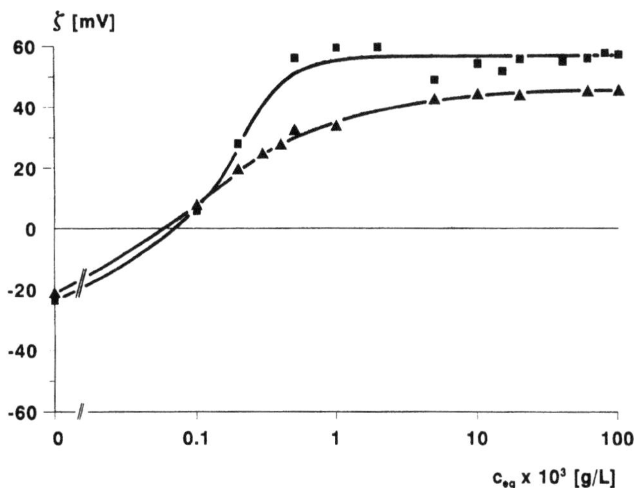



**Fig. 3** Zeta potential of kaolin (■) and calcium carbonate (▲) as a function of the cationic polymer concentration in tap water (286 ppm as $CaCO_3$) at pH 8.5

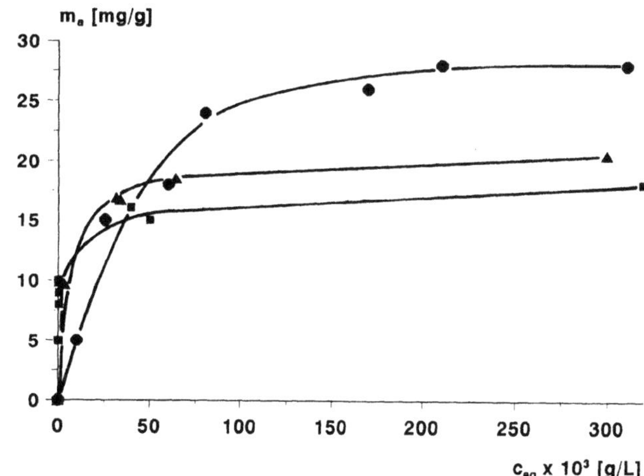



**Fig. 4** Adsorbed amounts of the cationic polymer on kaolin (■), calcium carbonate (▲) and cellulose (●) as a function of the equilibrium concentration at pH 8.5

polymer not only contains monomer units with cationic groups like poly-DADMAC, but also hydrophobic elements. The interfacial properties of the polymer, i.e., their ability to attach to and hydrophobize the pigment surface, result from the systematic optimization of the hydrophobic/hydrophilic ratio of the monomers. Due to the hydrophobic parts it·seems reasonable that the polymer does not adsorb on surfaces in a flat conformation, but rather in the form of loops or tails. This conformation of the adsorbed polymers causes higher adsorbed amounts than in the case of poly-DADMAC. A flat adsorption of poly-DADMAC has been assumed for different types of cellulose fibers [28].

The onset of the isotherm starts at low equilibrium concentrations, but the slope of the isotherm is steeper in the case of the fillers (Fig. 4). The specific surface area

M. Liphard et al.
Filler flotation from waste paper

of swollen cellulose fibers is reported to be between 50 and 200 m²/g, depending on the size fraction and pretreatment [20]. Corresponding to these values, the adsorption of the collector per unit area is considerably lower than in the case of the fillers (Fig. 1). However, the hydrodynamically accessible surface area for the polymer adsorption lies in the range from 1.5 to 15 m²/g, depending on the fine fraction and the molecular weight of the polymer [29].

Nevertheless, for waste paper recycling the plateau values of the isotherm are not decisive. With 0.2% dosage of the cationic polymer and 1% stock consistency, the initial concentration is only 20 mg/l. Due to the steeper slope of the adsorption isotherm on the fillers, the polymer should preferentially be adsorbed on the fillers. Figure 5 shows the flotation recovery of all three solid materials — calcium carbonate, kaolin, and cellulose. These flotation tests were performed with pure systems. The solid content was the same as in waste paper recycling: 1 and 0.2% for fibers and fillers, respectively. At a concentration of 20 mg/l, the recovery of kaolin is higher than 90%, while 70% of calcium carbonate are floated. However, with the same initial concentration the recovery of the cellulose fibers is below 20%. In comparison with the adsorption isotherms it can be concluded that the adsorbed amount of the cationic polymer is too low in order to hydrophobize the cellulose surface. Thus, the cationic polymer is a selective collector with respect to filler pigments. This fact has been confirmed by flotation tests with synthetic mixtures of cellulose and fillers [30].

Anionic collectors

Since the adsorption of anionic surfactants on calcite has already been thoroughly studied, adsorption experiments

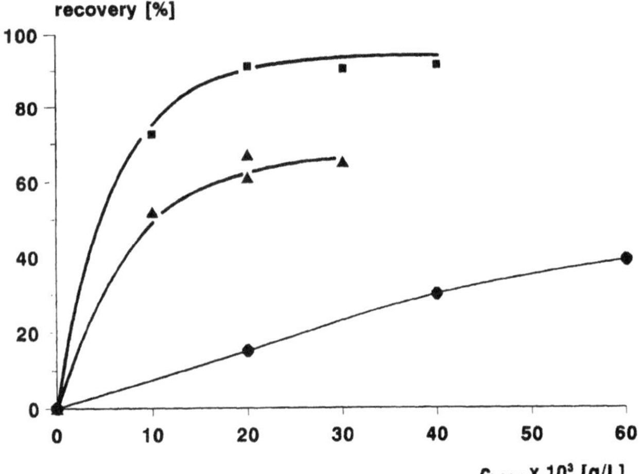

**Fig. 5** Flotation recovery of kaolin (■), calcium carbonate (▲) and cellulose [30] (●) as a function of the initial polymer concentration at pH 8.5

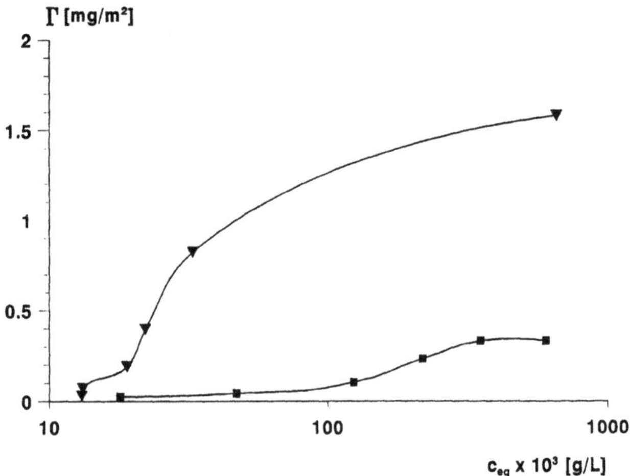

**Fig. 6** Adsorption isotherm of sodium alkyl sulfosuccinate on kaolin at pH 8.5 with 0 (■) and 286 ppm (▼) CaCO₃

have been carried out on kaolin fillers. For these experiments a well-characterized sodium alkyl sulfosuccinate was used. In Fig. 6 the adsorption isotherms are shown both with and without 286 ppm calcium hardness. In distilled water the amount of adsorbed surfactant is very low, reaching a plateau value of about 0.35 mg/m² at higher equilibrium concentrations. From this amount an average surface area of 2.2 nm² per molecule can be calculated. Taking 0.53 nm² as the basis for the cross-sectional area of a surfactant molecule [31], this corresponds to 20—25% monolayer coverage. As a layered silicate, kaolin has different chemical reactivities of the basal plane and the edge surfaces. It is quite possible that the anionic surfactant only adsorbs at the edges of kaolin which possess a higher number of cationic surface sites [14]. A plateau coverage between 10 and 25% of a monolayer was reported with other sulfates and sulfonates above pH 5 [32, 33]. The adsorption of the alkyl sulfosuccinate at the edges of kaolin is not reflected in a noticeable change of the zeta potential (Fig. 7). This is partly due to the fact that the electrophoretic mobility of kaolin is not primarily determined through the edges, but through the charge of the basal planes. On the contrary, the zeta potential of calcium carbonate decreases by 15 mV in the concentration range from 0 to 20 mg/l alkyl sulfosuccinate; this suggests an adsorption of the alkyl sulfosuccinate on filler calcium carbonate already at low concentrations and confirms direct adsorption experiments on pure calcite [25].

In the presence of 286 ppm CaCO₃ the onset of the adsorption isotherm starts at much lower equilibrium concentrations, reaching a fivefold higher plateau value (Fig. 6). As a result, calcium ions improve the adsorption of anionic surfactant molecules on the kaolin surface. However, under these circumstances the precipitation of Ca-surfactant complexes is noticeable. The adsorption

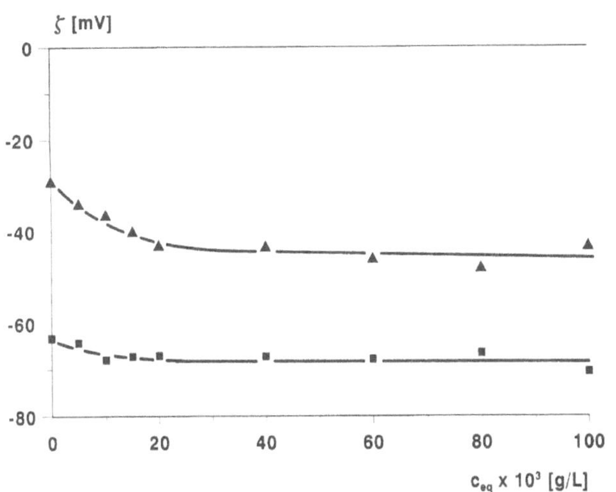

Fig. 7 Zeta potential of kaolin (■) and calcium carbonate (▲) as a function of the anionic surfactant concentration

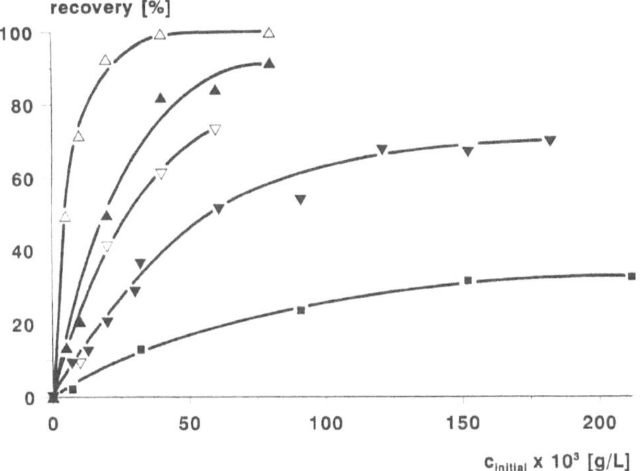

Fig. 8 Flotation recovery of kaolin as a function of the concentration of sodium alkyl sulfosuccinate with 0 (■) and 286 (▼) ppm $CaCO_3$; recovery of kaolin (▽), calcium carbonate Hydrocarb 90 OG (▲) and calcium carbonate Hydrocarb OM 50 (△) with sulfonated fatty acid in tap water

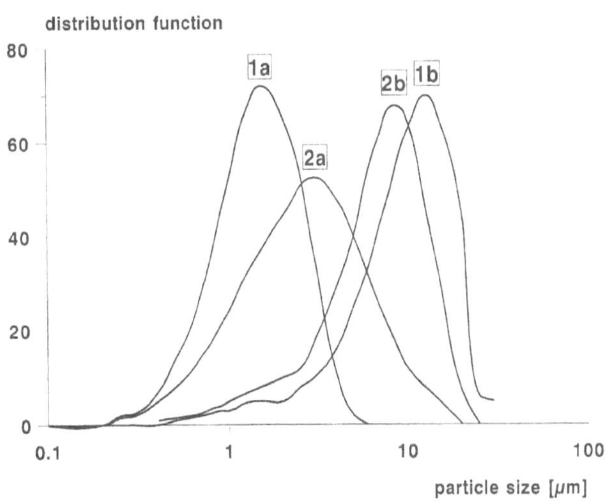

Fig. 9 Particle size distribution of two different types of calcium carbonate with and without the addition of 20 mg/l of the cationic polymer:
1a. Hydrocarb 90 OG
1b. Hydrocarb 90 OG + 20 mg/l cationic polymer
2a. Hydrocarb OM 50
2b. Hydrocarb OM 50 + 20 mg/l cationic polymer

isotherm includes both adsorption and precipitation of the surfactants in solution.

The flotation behavior of kaolin in distilled water is poor. Even with high initial concentrations, only 30 to 40% of the kaolin are recovered with the foam (Fig. 8). On the other hand, in hard water the recovery increases more strongly even at low surfactant concentrations, reaching a recovery of 75% in the plateau region. The high recovery in hard water shows that the adsorption of the surfactant on the surface predominates the precipitation in solution. A comparison of the recovery of fillers with a technical collector in hard water (Fig. 8) leads to the conclusion that it is distinctly easier to flotate calcium

carbonate than kaolin. However, the flotation capability is highly dependent on the respective calcium carbonate type.

The main differences in the behavior of the calcium carbonate types have been attributed to their different particle size. Figure 9 shows particle size distributions of the used fillers. For these measurements the filler suspensions were produced with the same concentrations of collectors and water hardness and under corresponding stirring conditions as in the flotation process. This procedure is used to register the particles present under flotation conditions. It includes agglomerates which determine the flotation behavior and not just the particle sizes of the filler particles present under optimal dispersion conditions (complete stabilization in the aqueous medium of the particle size produced by grinding). The mean particle size decreases in the following order: OM 50 > Hydrocarb 90 OG. As was expected, the filler recovery during flotation diminishes in the same order (Fig. 8).

Figure 9 also displays the particle size distribution of both calcium carbonate types under flotation conditions with the cationic polymer. It is obvious that the particle diameters have considerably increased for both types. The adsorption of the polymers causes agglomeration of the fillers, i.e., an increase in particle size. There are hints that in agglomeration of colloidal systems, e.g., water-based printing inks, the polymer acts via electrostatic effects rather than bridging mechanisms [22]. Due to this agglomeration effect, the recovery of all calcium carbonate types is high, independent of their original particle size.

## Filler flotation from waste paper

Based on these fundamental results the filler flotation from different types of waste paper was investigated. The flotation experiments were conducted in tap water (286 ppm $CaCO_3$) using the standard deinking formulation. Besides alkali this formulation contain hydrogen peroxide and sodium silicate, which increases the selectivity and has been proven to be a dispersant for fillers [4]. 0.2% of the anionic surfactant or of the cationic polymer was applied. The results are summarized in Fig. 10 and are given as percent ash (filler) content in the flotated pulp. A low ash content means a high recovery of the fillers.

The cationic polymer has a comparably good effect for kaolin and calcium carbonate (Fig. 10). The ash content of paper mixture 1 (kaolin containing) and of mixture 2 (calcium carbonate containing) is reduced from 18 to 2% or from 21 to 3%, respectively. The filler recovery exceeds 80% for both mixtures. In accordance with the results from the model system, the cationic polymer is suitable as a collector for both types of fillers, calcium carbonate, and kaolin.

The ash content obtained with the anionic surfactant was considerably lower in waste paper mixtures with a high calcium carbonate content than in mixtures containing only kaolin. 0.2% of the surfactant reduces the ash content of paper mixture 1 (kaolin) from 18 to 10%, but in the case of paper mixture 2 (calcium carbonate) from 21 to 5%. Thus, the anionic surfactants are generally bet-

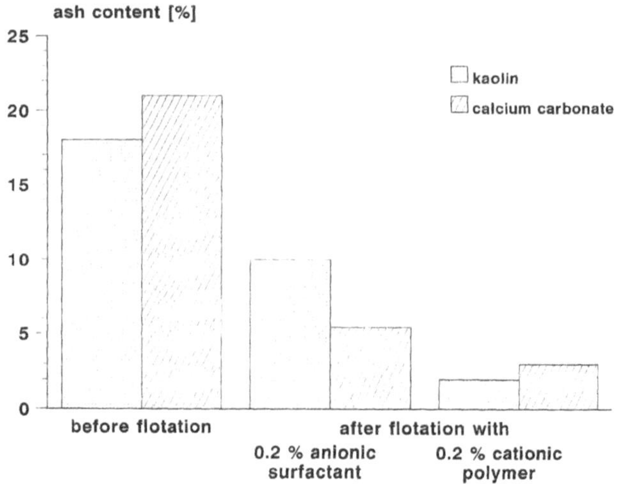


**Fig. 10** Ash content of two different mixtures of waste paper with the cationic polymer and the anionic surfactant

ter suited for waste paper systems containing calcium carbonate than for those with kaolin. This is due to the fact that the adsorption onto the calcium carbonate surface occurs at lower concentrations than that onto the kaolin surface.

**Acknowledgement** Financial support from the German Federal Ministry of Research and Technology (BMFT) is gratefully acknowledged.

## References

1. von Raven A, Uutela E (1992) Proceedings of the 5th PTS-PTI-Deinking-Symposium 1992, p 3
2. Holzey G (1992) Proceedings of the 5th PTS-PTI-Deinking-Symposium 1992, p 1
3. Hornfeck K, Liphard M, Schreck B (1990) Tappi Proceedings, Pulping Conference, Toronto, Kanada, pp 965—973
4. Liphard M, Schreck B, Hornfeck K (1993) 1st Research Forum on Recycling, Toronto, Kanada, to be published in Pulp and Paper Canada
5. Hornfeck K, Liphard M, Schreck B, 5th PTS-PTI-Deinking-Symposium, München, to be published in Wochenblatt für Papierfabrikation
6. Hornfeck K, Liphard M, Schreck B (1991) Tappi Proceedings, 1991 Pulping Conference, Orlando, Florida, p 1031—1038
7. Kurzendörfer CP, Schlag M (1986) In: Dechema Monographien 102, VCH Verlagsgesellschaft, Weinheim
8. Schempp W, Vortrag 5, Zeta-Potential Symposium, München 1988
9. Horn D (1978) Progr Colloid Polym Sci 65:251—264
10. Schröder U, Horn D, Waßmer KH (1991) Seifen — Öle — Wachse 117:311—314
11. Schulze HJ (1984) Physicochemical elementary processes in Flotation. Elsevier, Amsterdam
12. Sivamovahan R (1990) Inter J Miner Process 28:935
13. Andersen JB, El-Mofty SE, Somasundaran P (1991) Colloids and Surfaces 55:365—368
14. Smith RW, Narimatsu Y (1993) Minerals Engineering 6:753—763
15. Weigl J, Huggenberger L (1974) Wochenblatt für Papierfabrikation 23:886—895
16. Siffert B, Fimbel P (1984) Colloids Surfaces 11:377—389
17. Pierre A, Lamarche JM, Mercier R, Foissy A, Persello J (1990) J Dispersion Sci Technology 11:611—635
18. Doering H (1956) Das Papier 10:140—141
19. Sjöström E, Haglund P (1961) Sven. Papperstidn 64:438—446
20. Budd J, Herrington ThM (1989) Colloids Surfaces 36:273—288
21. Somasundaran P, Agar GE (1967) J Colloid Interface Sci 24:433—440
22. Hornfeck K, Liphard M, Schreck B (1990) Wochenblatt Papierfabrikation 118:935
23. van de Steeg HGM, de Keizer A, Cohen Stuart MA, Bijsterbosch BH (1993) Colloids Surfaces A 70:77
24. Larsson A, Stenius P, Ödberg L (1984) Svensk Papperstidning 18:R158
25. von Rybinski W, Schwuger MJ (1985) Aufbereitungs-Technik 26:632—639
26. Kindler WA, Swanson JW (1971) J Polym Sci A-2 9:853
27. Nedelcheva MP, Stoilkov GV (1976) J Appl Polym Sci 20:2131
28. Onabe F (1978) J Appl Polym Sci 22:3495—3510
29. Laatikainen M (1989) J Colloid Interface Sci 132:451—461
30. Baldauf H, unpublished results
31. Rosen MJ (1989) In: Surfactants and Interfacial Phenomena. Wiley, New York, p 70
32. Xu Q, Vasudevan TV, Somasundaran P (1991) J Colloid Interface Sci 142:528—534
33. Köster R, Schreck B, von Rybinski W, Dobiás B (1992) Minerals Engineering 5:445—456

Progr Colloid & Polym Sci (1994) 95:175—180
© Steinkopff Verlag 1994

N. Buske

# Application of magnetite sols
# in environmental technology

**Abstract** Theoretically and experimentally, it could be shown that 10 nm magnetite particles can be stabilized in aqueous carriers by electrostatic and/or Born repulsion, also under action of external magnetic fields.

The bare or monolayer and bilayer covered particles can be used in environmental protection to eliminate dangerous components from waste water, polluted air and soil by adsorption, and extraction or heterocoagulation in combination with magnetophoresis of the modified magnetite particles in presence of an external magnetic field.

A mixture of water and the ammonium salt of a perfluoropolyether (PFPE)-acid was used to determine the adsorption capacity of stable dispersed, positively charged magnetite particles. Under the condition of monolayer formation the PFPE-salt can be nearly totally eliminated from the water carrier. At bilayer formation the adsorption capacity is increased two fold, but there is an equilibrium between the adsorbed and in the carrier solved PFPE salt.

**Key words** Magnetite sol — ferrofluid — adsorption isotherm — filtration — interparticle interaction — perfluoropolyether

Dr. N. Buske
Fraunhofer-Institute for Production
Systems and Design Technology
Branch
House 1.1
Rudower Chaussee 5
12484 Berlin, FRG

## Introduction

The preparation of stable, concentrated hydrosols containing nanometer particles is well known. An aqueous iron-oxidhydratsol (Graham sol) is a typical example.

If the nanometer particles consist of ferri or ferromagnetic nm-particles such as magnetite or iron, the prepared magnetic sols show — besides the high colloidal stability — a superparamagnetic behavior: no hysteresis in the magnetization curve.

Surprisingly, the sols are even sedimentation stable in strong magnetic field gradients: Therefore, the whole fluid volume can be moved towards a magnetic pole and can be fixed on the pole. In literature, these stable sols are known as magnetic or ferrofluids [1]. The magnetic fluids are mostly prepared with magnetite as the magnetic component and with dispersion media such as water, hydrocarbons, hydrocarbon esters, silicones, and perfluoropolyethers. The main problem for the preparation of stable sols is to find the right surfactant to cover the particles. There is current aim to use the volume properties of the magnetic fluids in environmental protection: On the one hand, ferrofluidic exclusion seals [2] prevent dangerous gas, oil-drops, and dust leakage of chemical and biotechnological reactors, and on the other hand, the float-sink technology [3] in magnetic fluids can be used for separation of non-ferrous metals of automobile and electronic scrap.

In this paper the dispersed 10 nm magnetite particles were used for separation processes in environmental protection, such as was already proposed for the magnetic separation of oil from the water surface or from oil-in-water emulsions after mixing with magnetic fluids [4—6], the decontamination of radioactive waste water [7], the selective deposition of fine particles such as Sn of Sn-hydrosols [8] or minerals [9] onto magnetite particles, and the elimination of biological components [10].

The general conditions for obtaining stable, water-based magnetite sols are described by superposition of the individual interparticle interaction energies (DLVO theory) of dispersed magnetite particles: the electrostatic repulsion, the Van der Waals attraction, the Born repulsion, and the induced magnetic attraction. Stable aqueous magnetite sols can be most effectively used for elimination purposes. The high adsorption capacity of the magnetite particles will be demonstrated by determining the adsorption isotherm of a perfluoropolyether surfactant.

## Stabilization mechanisms

According to DLVO-theory [11], the interparticle interaction of dispersed particles can be described by superposition of their individual interparticle interaction energies as function of their particle distance (d).

The energies can be attractive or repulsive: generally, the electrostatic repulsion energy $(V_{el})$ [12] and Van der Waals attraction energy $(V_d)$ [13] act between non magnetic particles dispersed in aqueous carriers.

In an external homogeneous or inhomogeneous magnetic field an induced interparticle magnetic attraction energy $(V_m)$ [14] between subdomain magnetic particles must be considered.

Particles with anchored surfactant layers cannot come into direct contact. The distance of closest approach can be, at most, two time the adsorption layer thickness. Using nm particles the $V_d$-energy of particle core no longer acts at closest approach, therefore the Van der Waals attraction energy of the adsorbed layer $(V_l)$ [15], the Born repulsive energy $(V_b)$ [16], and the $V_{el}$ of the external ionic adsorption layer determine the colloidal stability of the dispersed particles.

For the interparticle interchange energy calculations the experimentally determined properties of the particles and the following Hamaker constants were used:

the Hamaker constant of magnetite: $2.3 \cdot 10^{-19}$ J,
the Hamaker constant of water: $3.5 \cdot 10^{-20}$ J;
and the Hamaker constant of the adsorption bilayer should be very similar to the Hamaker constant of the carrier: $3.4 \cdot 10^{-20}$ J.

Further parameters are given in the figure legends.

The stabilization mechanisms of diversely stabilized magnetite particles in aqueous carriers are now described in detail. For the calculation a special computer software [17] was used: The dimensionless individual energies $V/kT$ and the superposition of these energies were calculated as functions of particle distance d.

## Electrostatically stabilized magnetite particles

The calculated individual interaction energies and their superposition $(V)$ curves of 10 nm magnetite particles are shown in Fig. 1, but in the presence of a magnetic field in Fig. 2, respectively. For a 1 : 1 electrolyte with a con-

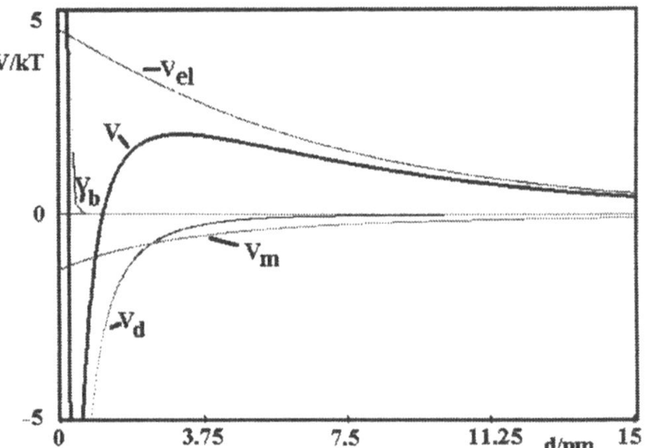

**Fig. 1** Dimensionless interparticle interaction energies $V_{el}$, $V_d$, $V_b$ and their superposition $V$-distance curves for electrostatically stabilized 10 nm spherical magnetite particles. Hammaker constant of the magnetite $= 2.3 \cdot 10^{-19}$ and of the carrier $= 3.5 \cdot 10^{-20}$ J, an 1 : 1-electrolyte concentration of 0.003 mol/l, and a zetapotential of 35 mV

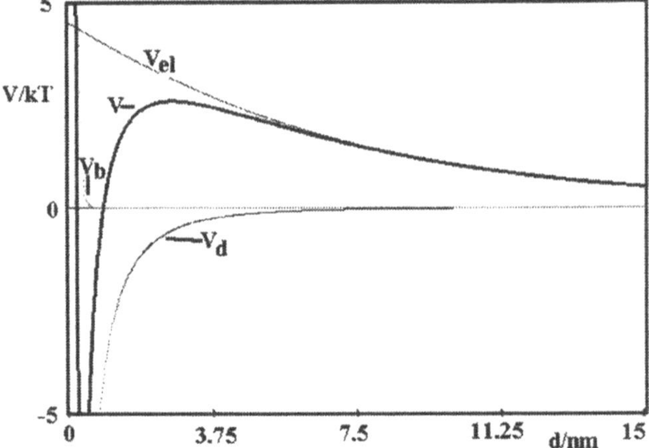

**Fig. 2** Dimensionless interparticle interaction energies $V_{el}$, $V_d$, $V_b$, $V_m$ and their superposition $V$-distance curves of electrostatically stabilized magnetite particles in an external magnetic field; same constants as in Fig. 1

centration of $1 \cdot 10^{-3}$ mol/l and an assumed zetapotential of 35 mV, both $V$-curves have an energy barrier of $V/kT > 2$. That means the sols should be stable against coagulation, even in a magnetic field.

An increase of the electrolyte concentration or a decrease of the zetapotential or a change of the pH value would decrease the energy barrier shown in Figs. 1 and 2, and the particles would flocculate.

## Sterically stabilized magnetite particles

10 nm magnetite particles can also be stabilized in water by a bilayer of surfactant. If the external layer is a nonionic one, the particle charge is nearly zero. These particles should be sterically stabilized.

Figures 3 and 4 show the interaction energy curves outside of and inside an external magnetic field.

The figures show that Born repulsion creates the high particle stability against coagulation.

Under action of additional external forces such as centrifugal or magnetic ones, the particles come into more or less temporary contact at the distance of closest approach. $V$ becomes negative, so that small attraction energies are created (Fig. 4). But the energies are much smaller than $1 \, kT$. That means the particles can be redispersed by their own kinetic energy $kT$, if additional forces do not act any longer.

The sols should be extremely stable: The pH-value and the electrolyte concentration can be changed in a wide range without any particle flocculation.

To accomplish coagulation the Hamaker constant of the carrier or of the bilayer must be changed, i.e., by addition of less-polar solvents.

## Sterically plus electrostatically stabilized magnetite particles

10 nm magnetite particles covered by a double layer can be stabilized sterically and electrostatically, if the external layer has a positive or negative charge. This can be realized by using cationics and anionics as external layers.

The particles should be very stable against coagulation, but a destabilization is possible, i.e., by changing the pH or ionic strength of the carrier. Figures 5 and 6 show the interparticle energies curves outside of and inside a magnetic field.

At the distance of closest approach the stabilization occurs by Born repulsion alone, at larger distances by a superposition of Born plus electrostatical repulsion.

If the external adsorption layer is a salt of an organic acid the sols can be reversibly flocculated by changing the pH value of the carrier.

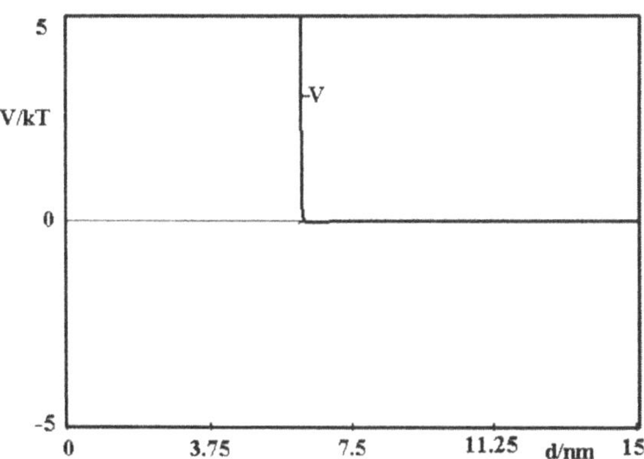


**Fig. 3** Dimensionless interparticle interaction energy $V$-distance curve of 10-magnetite particles, modified by a bilayer: the external layer is a nonionic one. Adsorption layer thickness: 3 nm, Hamaker constant of the magnetite: $2.3 \cdot 10^{-19}$ J, of the carrier: $3.5 \cdot 10^{-20}$ J, of the surfactant layer: $3.4 \cdot 10^{-20}$ J. The $V_b$ and $V$ curve are identical, $V_{el}$ and $V_l$ can be neglected and not shown

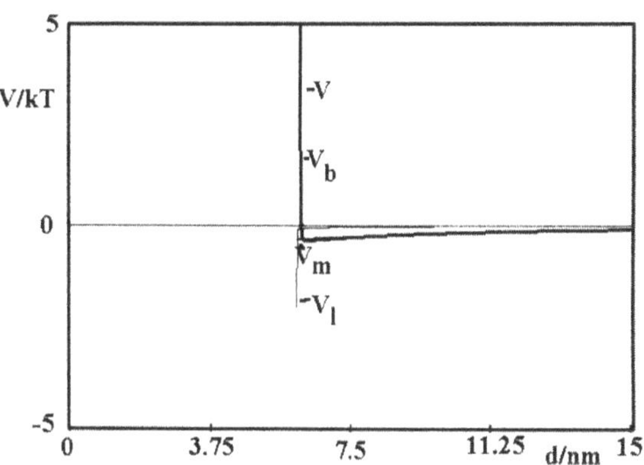


**Fig. 4** Dimensionless interparticle interaction energies $V_b$, $V_{el}$, $V_l$, $V_m$-distance-curves in presence of an external magnetic field; same constants as in Fig. 3

## Magnetophoresis

Magnetic particles can move in an external magnetic field gradient by interaction of the particles with the external magnetic pole, called magnetophoresis.

The velocity $v/m \, s^{-1}$ of pure magnetite particles with their magnetic polarization $I/T$ in a magnetic field gradient grad $H/A \, m^{-2}$ is given in a first approximation [18]:

$v = 2 \, I a^2 \, \text{grad} \, H/9 \, \eta$
$a = $ particle radius/$m$
$\eta = $ viscosity of the carrier, for water: $10^{-3}$ Pa s.

The migration velocity $v$ was calculated for different particle aggregate sizes. A technically realizable grad $H$ of $10^7$ A $m^{-2}$ was assumed. The results are:

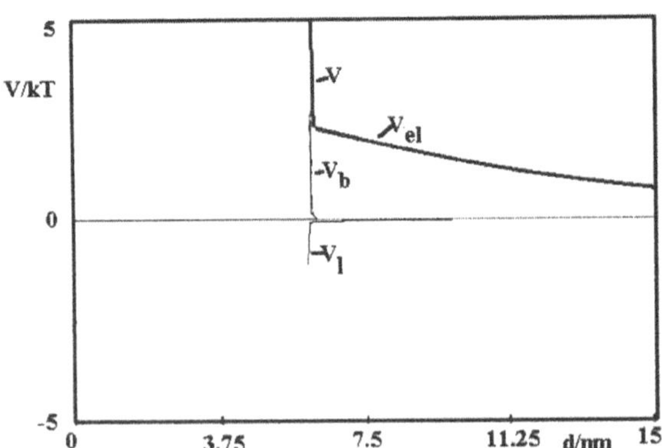

**Fig. 5** Dimensionless interparticle interaction energies $V_b$, $V_{el}$, $V_l$, — distance-curves of 10 nm magnetite particles modified by a bilayer. Adsorption layer thickness: 3 nm, zetapotential of the external adsorption layer: 20 mV, 1.1 electrolyte concentration = 0.002 mol/l, Hamaker constant of the magnetite: $2.3 \cdot 10^{-19}$ J, of the carrier: $3.5 \cdot 10^{-20}$ J, of the surfactant layer: $3.4 \cdot 10^{-20}$ J

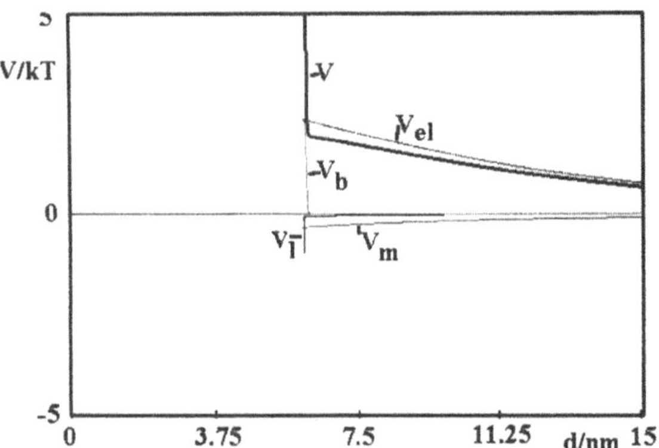

**Fig. 6** Dimensionless interparticle interaction energy $V_b$, $V_{el}$, $V_l$, $V_m$ — distance curves of magnetite particles modified by a bilayer, in presence of an external magnetic field; the external layer is an ionic one; same constants as in Fig. 5

| Size of particle aggregate | $v/\text{mm s}^{-1}$ |
| --- | --- |
| 1 magnetite particle (mp) | $6 \cdot 10^{-5}$ |
| 10 mp | $4 \cdot 10^{-4}$ |
| 1000 mp | $8 \cdot 10^{-3}$ |

The calculations indicate that particle aggregates of more than 1000 individual particles can be effectively separated by a simple laboratory magnet.

---

## Experimental results

### Preparation and properties of the magnetite particles

The magnetite ($Fe_3O_4$) particles were prepared in the water carrier by coprecipitation of concentrated iron(II) and iron(III) salt solutions with ammonium hydroxide.

The particles have the following physical properties:
Medium particle core diamter: 10 +/−1 nm
Specific area (BET-method): 130 m²/g
Saturation polarization: about 400 mT,
No hysteresis in the magnetization curve.

### Preparation of magnetite sols

Magnetite sols (ferrofluids) have been well known for about 30 years. Our preparation methods are based on Khalafallas' [19, 20] and Massarts' [21] suggestions.

Using the above-mentioned nm magnetite particles the following water based magnetic fluids were prepared:

a) Magnetic fluids in surfactant free acidic aqueous carriers at electrolyte concentrations of 1:1-electrolytes below an electrical conductivity of $10^{-3}$ $\Omega^{-1}$ cm$^{-1}$, corresponding to electrostatically stabilized magnetite sols.

b) Magnetic fluids in alkaline carriers stabilized by a double layer of carbonic acids [22] or perfluoropolyether-mono carbonic acids, corresponding to sterically plus electrostatically stabilized magnetite sols.

c) Magnetic fluids in neutral, acidic or alkaline carriers stabilized by two different layers: the first one consists of a carbonic acid, the second one of nonionics [23], corresponding to sterically stabilized magnetite sols.

These fluids are environmentally friendly and therefore suitable for the separation of non-ferrous metals using the magneto-levitation effect [24].

### Adsorption of surfactants on the magnetite particles

To estimate the adsorption capacity of surfactants onto electrostatically stabilized, 10 nm magnetite particles at 80 °C the adsorption isotherm of an ammonium salt of a perfluoropolyether mono carbonic acid with the molecular weight 929, prepared at Ausimont (Italy), was determined.

The amount of the surfactant was increased step-by-step to determine the formation as well as of the mono- and the bilayer by checking the wetting and dispersing properties of the modified particles towards water and to Galden HT 90, a perfluoropolyether of a average molecular weight of 617 and a viscosity of 0.59 cSt at 40 °C.

Experimentally, a strong wetting and dispersing of the magnetite particles in the Galden carrier was found at a surfactant concentration of 0.5 mMol surfactant/g $Fe_3O_4$. This oleophilic behavior should correspond to the monolayer coverage of the particles.

On the other hand, at higher surfactant concentrations the modified particles became more and more hydrophilic, and could be very well dispersed in the aqueous carrier at a coverage of 1.0 mMol surfactant/g $Fe_3O_4$. This concentration should correspond to a bilayer coverage of the particles.

The adsorption curve in Fig. 7 shows that the surfactant can be nearly totally extracted from the carrier under the conditions of formation of the monolayer possibly caused by the high chemical affinity to the surface. In the opposite case the formation of the second, external layer is caused by hydrophobic interaction. There is an equilibrium with the adsorbed amount and the concentra-

**Fig. 7** Adsorption isotherm of an ammonium salt of the perfluoropolyether-acid S2 (mw = 929) at 80°C onto electrostatically stabilized 10 nm magnetite particles

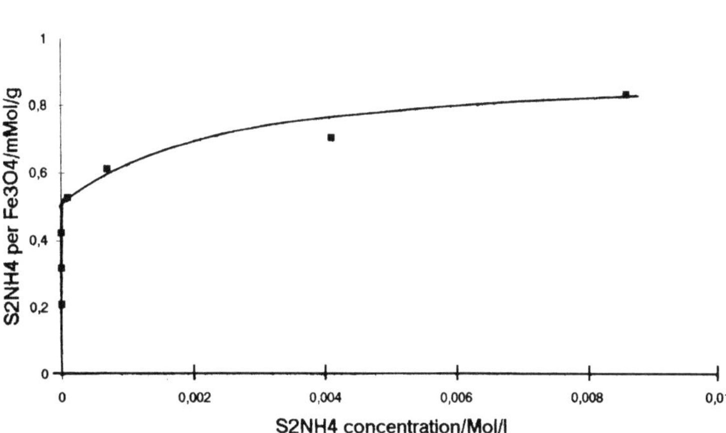


tion of the surfactant in the carrier. The adsorption is not so effective, but the molecules of the second adsorption layer can be extracted by suitable solvents or can be evaporated.

## Conclusions

The colloidal stability of 10 nm particles of magnetite dispersed in aqueous carriers can be calculated by superposition of the individual interparticle interaction energies including the Born repulsion and the induced magnetic attraction.

10 nm magnetite particles can be stabilized both electrostatically and/or by Born repulsion.

The calculations were in agreement with the experimental results. Therefore, stable magnetic fluids with positively charged magnetite particles and with bilayer stabilized magnetite particles could be prepared.

The particles can be used to eliminate harmful molecules or solids from waste water, polluted air or contaminated soil by adsorption, extraction or heterocoagulation:

The adsorption measurements show that at conditions of monolayer formation 0.5 mMol perfluoropolyether-surfactant can be fixed on 1 g magnetite, and that it is possible to nearly totally eliminate the surfactant from the water carrier. During the adsorption process the modified particles coagulate and can be separated by a simple laboratory magnet.

For recycling the modified magnetite particles can be dissolved in hydrochloric acid, the insoluble PFPE-surfactant is separated, and the dissolved iron chloride solution can be used to prepare new magnetite particles.

The adsorption can be carried on to formation of a double layer adsorption. 1.0 mMol of the PFPE-surfac-

tants can be fixed on 1 g magnetite, but the water is not fully free from the surfactant. The second adsorption layer is fixed on the magnetite particles by hydrophobic interaction. It can be extracted by a suitable solvent or evaporated.

A combination of processes b) with a) should give an optimal filtration effect for harmful components.

For extraction of harmful components monolayer or bilayer covered particles can be used: Aqueous magnetic fluids, stabilized by bilayers of hydrocarbons can solve organic molecules, i.e. benzene or chloroform within their layers. During this process the particles flocculate and can be magnetically separated. In principle the separated particles can be recycled after separation of the adsorbed organic molecules, i.e., by distillation.

The heterocoagulation of solids or the covering of droplets on the modified magnetite particles depends in first approximation on the interface tension ratio: the interface tension of the adhered different particles must be smaller than the sum of the interface tensions of the different particles to the carrier. It is possible to vary the individual interface tensions by the kind and amount due to mono or bilayer coverage of the particles. That should make it possible to realize a specific adsorption and specific magnetic separation of the harmful components.

The advantages of this separation technology are that the harmful ingredients together with the magnetic particles can be eliminated from the polluted system by a simple magnetic field. After magnetic separation, the harmful components can be often removed from the magnetite particles, and the magnetite particles can be used again.

We think that the described processes are an alternative for the effective extraction of harmful molecules or particles from polluted systems. Nevertheless, some basic research seems to be still necessary to make this process effective for a technical application.

## References

1. Rosensweig RE (1979) Adv in Electronics and Electron Phys 48:103—199
2. Prospect of Ferrofluidics Corp Nashua, NH, USA
3. Shimoiizaka J, Nakatsuka K, Fujita T, Kounosu A (1980) IEEE Trans Mag, MAG-16:368—371
4. Kaiser R (1974) US patent 3805449
5. Kaiser R (1971) US patent 3796660
6. Mikami TJ (1992) Dispersion Sci Technol 13(2):145—168
7. Silver GL (1983) US patent, Appl Nr 174064
8. Chen W-J, Tarng MR (1991) IEEE Trans Mag 27(6):4642—4644
9. Brozek M, Nowakowski K, Oruba E, Pilch W (1989) Fizykochem Probl Mineralurgii 21:181—189; CA 112 (10) 8254w
10. Gonel DI, Liberti PA (1992) US patent 5108933
11. Verwey EJ, Overbek JThG (1948) "Theory of the Stability of Hydrophobic Colloids" Elsevier Publ Comp Amsterdam
12. Mc Cartney-Levine (1969) J Coll Interf Sci 30:345
13. Hamaker HC (1937) Physica, Utrecht 4:1058
14. Kneller E (1966) In: "Handbuch der Physik" 18(2), Springer, Berlin, p 475
15. Vold MJ (1961) J Colloid Sci 16:1
16. Feke, Prabka, Mann (1984) J Phys Chem 88:5735
17. Strenge K (1993) software program "Interact, Interads", private communication
18. Scholtens PC (1978) In: Berkowsky B "Thermomechanics of Magnetic Fluids" p 1, Hemisphere Publ Co, Washington DC
19. Khalafalla SE, Reimers GW, Roll SA (1984) US patent 4208294
20. Khalafalla SE, Reimers GW (1972) US patent 3764540
21. Massart R (1981) IEEE Trans Mag MAG-17 2:1247—1248
22. Buske N, Götze T (1981) DD patent 235791
23. Günther D (1993) DE patents, Appl Nr P 4325 386.5 and P 4327 826.4
24. Günther D, Buske N (1994) DE patent, Appl Nr P 4407864.1

# Author Index

# Subject Index